International
REVIEW OF
Neurobiology
Volume 44

Neuronal Processing of Optic Flow

International
REVIEW OF
Neurobiology

Volume 44

SERIES EDITORS

RONALD J. BRADLEY
Department of Psychiatry, School of Medicine
Louisiana State University Medical Center
Shreveport, Louisiana, USA

R. ADRON HARRIS
Department of Pharmacology, University of Colorado
Health Sciences Center, Denver, Colorado, USA

PETER JENNER
Biomedical Sciences Division, King's College, London, UK

EDITORIAL BOARD

Neuronal Processing of Optic Flow

EDITED BY

MARKUS LAPPE
Ruhr-Universitat Bochum
Allgemeine Zoologie und Neurobiologie
Bochum, Germany

ACADEMIC PRESS

A Harcourt Science and Technology Company

San Diego San Francisco New York Boston London Sydney Tokyo

This book is printed on acid-free paper. ♾

Academic Press
A Harcourt Science and Technology Company
525 B Street, Suite 1900, San Diego, California 92101-4495, USA
http://www.apnet.com

Academic Press
24-28 Oval Road, London NW1 7DX, UK
http://www.hbuk.co.uk/ap/

International Standard Book Number: 0-12-366844-1

Printed and bound in the United Kingdom
Transferred to Digital Print 2011

CONTENTS

PART I
PERCEPTION

Human Ego-Motion Perception

A. V. van den Berg

PART II
EYE MOVEMENTS

Optic Flow and Eye Movements

Markus Lappe and Klaus-Peter Hoffmann

The Role of MST Neurons during Ocular Tracking in 3D Space

KENJI KAWANO, YUKA INOUE, AYA TAKEMURA, YASUSHI KODAKA, AND FREDERICK A. MILES

PART III
ANIMAL BEHAVIOR AND PHYSIOLOGY

Visual Navigation in Flying Insects

MANDYAM V. SRINIVASAN AND SHAO-WU ZHANG

Neuronal Matched Filters for Optic Flow Processing in Flying Insects

HOLGER G. KRAPP

A Common Frame of Reference for the Analysis of Optic Flow and Vestibular Information

BARRIE J. FROST AND DOUGLAS R.W. WYLIE

Optic Flow and the Visual Guidance of Locomotion in the Cat

HELEN SHERK AND GARTH A. FOWLER

PART IV
CORTICAL MECHANISMS

Stages of Self-Motion Processing in Primate Posterior Parietal Cortex

FRANK BREMMER, JEAN-RENÉ DUHAMEL, SULIANN BEN HAMED, AND WERNER GRAF

Optic Flow Analysis for Self-Movement Perception

CHARLES J. DUFFY

Neural Mechanisms for Self-Motion Perception in Area MST

RICHARD A. ANDERSEN, KRISHNA V. SHENOY, JAMES A. CROWELL, AND DAVID C. BRADLEY

Computational Mechanisms for Optic Flow Analysis in Primate Cortex

Markus Lappe

Human Cortical Areas Underlying the Perception of Optic Flow: Brain Imaging Studies

Mark W. Greenlee

What Neurological Patients Tell Us about the Use of Optic Flow

Lucia M. Vaina and Simon K. Rushton

CONTRIBUTORS

Numbers in parentheses indicate the pages on which the authors' contributions begin.

Richard A. Andersen (219), Division of Biology, California Institute of Technology, Pasadena, California 91125

David C. Bradley (219), Division of Biology, California Institute of Technology, Pasadena, California 91125

Frank Bremmer (173), Department of Zoology and Neurobiology, Ruhr University Bochum, Bochum D-44780, Germany

James A. Crowell (219), Division of Biology, California Institute of Technology, Pasadena, California 91125

Charles J. Duffy (199), Department of Neurology, Neurobiology and Anatomy, Ophthalmology, and Brain and Cognitive Science, the Center for Visual Science, University of Rochester, Rochester, New York 14642

Jean-René Duhamel (173), LPPA College de France, Paris, France; and Institute Science Cognitives, C.N.R.S., Bron, France

Garth A. Fowler (141), Department of Biological Structure, University of Washington, Seattle, Washington 98195

Barrie James Frost (121), Department of Psychology, Queen's University, Kingston, Ontario, K7L 3N6 Canada

Werner Graf (173), LPPA College de France, Paris, France

Mark W. Greenlee (269), Department of Neurology, University Freiburg, Freiburg 79106, Germany

Suliann Ben Hamed (173), LPPA College de France, Paris, France; and Institute Science Cognitives, C.N.R.S, Bron, France

Klaus-Peter Hoffmann (29), Department of Zoology and Neurobiology, Ruhr University Bochum, Bochum D-44780, Germany

Yuka Inoue (49), Neuroscience Section, Electrotechnical Laboratory, Tsukuba-shi, Ibaraki 305-8568, Japan

Kenji Kawano (49), Neuroscience Section, Electrotechnical Laboratory, Tsukuba-shi, Ibaraki 305-8568, Japan

Yasushi Kodako (49), Neuroscience Section, Electrotechnical Laboratory, Tsukuba-shi, Ibaraki 305-8568, Japan

Holger G. Krapp (93), Lehrstuhl für Neurobiologie, Universität Bielefeld, Germany

Markus Lappe (29, 235), Department of Zoology and Neurobiology, Ruhr University Bochum, Bochum D-44780, Germany

Frederick A. Miles (49), Laboratory of Sensorimotor Research, National Eye Institute, National Institutes of Health, Bethesda, Maryland 20892

Simon K. Rushton (293), Cambridge Basic Research, Nissan Research and Development, Inc., Cambridge, Massachusetts 02142; and Department of Clinical Psychology, Astley Ainslie Hospital, Grange Loan, Edinburgh EH9 2HL, Scotland, United Kingdom

Krishna V. Shenoy (219), Division of Biology, California Institute of Technology, Pasadena, California 91125

Helen Sherk (141), Department of Biological Structure, University of Washington, Seattle, Washington 98195

Mandyam V. Srinivasan (67), Center for Visual Sciences, Research School of Biological Sciences, Australian National University, Canberra A.C.T. 2601, Australia

Aya Takemura (49), Neuroscience Section, Electrotechnical Laboratory, Tsukuba-shi, Ibaraki 305-8568, Japan

Lucia M. Vaina (293), Brain and Vision Research Laboratory, Department of Biomedical Engineering and Neurology, Boston University, and Department of Neurology, Harvard Medical School, Brigham and Womens' Hospital and Massachusetts General Hospital, Boston, Massachusetts 02215

A. V. van den Berg (3), Helmholtz School of Autonomous Systems Research, Department of Physiology, Faculty of Medicine, Erasmus University, Rotterdam, the Netherlands

Douglas Richard Wong Wylie (121), Department of Psychology, University of Alberta, Edmonton, Alberta T6G 2E9, Canada

Shao-Wu Zhang (67), Center for Visual Sciences, Research School of Biological Sciences, Australian National University, Canberra A.C.T. 2601, Australia

FOREWORD

The term "optic flow" was coined by James Gibson and came into common use after the publication of his *The Perception of the Visual World* in 1950. However, the study of optic flow is much more ancient: Probably the first treatise on the topic was Euclid's *Optics*. *Optics* is indeed a remarkable text. During the Western Renaissance the work (unfortunately for science!) came to be regarded as a faulty treatise on linear perspective (later projective geometry). Indeed, from the linear perspective of Brunelleschi and Alberti, *Optics* makes little sense. In fact, *Optics* must be read as a treatise in "natural perspective," that is, the study of visual angles rather than of distances over the canvas. Even more exciting, Euclid treats not only static but also dynamic situations. That is, many of the theorems apply to changes in the apparent configuration when the observer moves with respect to the scene. Several observations by Gibson can readily be found in Euclid. Of course, Euclid's contribution was of a purely theoretical nature because empirical sciences didn't really exist in 300 B.C. (No doubt Euclid conceived of many theorems when thinking over empirical observations, though.) The first empirical evidence that optic flow might be instrumental in guiding animal behavior dates from the late 19th century. Especially well known is the remark by Helmholtz, who noticed that, when you find yourself in a forest, you cannot readily discern which branch is in front of which except when you move: Then the scene becomes as if it were transparent and depth relations are every bit as clear "as when you view a stereoscopic rendering of the scene." Because Helmholtz was raised in an era when the stereoscope was a cherished item in any bourgeois family, this was a very strong statement indeed!

There has been relatively little work on optic flow in the first half of this century, partly due to practical obstacles. Today, it is easy enough to present stimuli on a computer screen, but in the old days (I can still remember them), producing dynamical stimuli of any complexity was a major undertaking. I became interested in optic flow myself in the early seventies after reading Gibson's 1950 book. Being a physicist by training, I was convinced that Gibson had very interesting ideas (indeed, the man was clearly a genius of some sort although he evidently didn't have the faintest notion of mathematics) but had no idea of how things ought to be done. So I started theoretical developments together with my wife, Andrea van Doorn. My prior experience with flows (drawing weather maps) helped a great deal. It proved easy enough to work out the first-order theory. Although interesting, this was only a preliminary success because much of the really interesting structure is in the second order or—when you are also interested in the relation to shading (as everyone should be)—even the third order. What turned out to be the main problem was not the science: It was the problem of "selling solutions." Soon we had generated more solutions than the field had questions! Although we had formulated answers to many open problems, nobody was especially interested and we met with a cold reception. In fact, only years later (the eighties, in some respects the early nineties even, after we left the field) was our work noticed at all and mainly because some people (slowly) began to reinvent the wheel. Things turned out better for us in the nineties (when we had already forgotten about optic flow and were pursuing quite different problems), although in a most unexpected way. We were interested in vision, especially human vision (we tended to think of animal studies as "models" of the real thing!), but the first people who started to notice our work were working in (the then new) field of machine vision/robotics. This field has developed Gibson's original ideas into something Gibson wouldn't recognize himself. The progress has been monumental. Today, you can walk through an urban environment with a video camera, feed the tape in a program, and obtain a model of the scene that will allow you to experience arbitrary walk-throughs in virtual reality. Autonomous vehicles can navigate on the basis of optic flow much in the way Gibson originally envisaged.

Especially interesting developments due to the labor of people in machine vision are coordinate-free affine and projective methods. These go far beyond what Gibson could imagine. Perhaps disappointingly, such developments have yet to be taken up by the biological vision community. What has become common are the computer graphics methods, though. These are partly an offspin from the machine vision developments, partly due to the enormous revolution of hardware. When we started off, the only tolerable stimuli were produced (at great cost) for trainers for the U.S. Air Force (especially those for high-speed low-altitude fighters and assault helicopters). Today, any PC game easily beats what we marveled at in the early eighties. This truly enables innovative work in the cognitive neurosciences that would have been completely out of Gibson's reach.

As I have already remarked, the progress made in machine vision has hardly filtered through to the animal and human vision community (yet). In fact, there the gap has widened to such an extent that it may well take a decade or more (I'm optimistic by nature) to catch up. A wealth of readily testable models of various visuomotor functions have been fully worked out and explored, both formally and in computer simulations and real life (machine) demonstrations. Of course, it is no minor undertaking to test (and adapt) such models in psychophysical and neurophysiological experiments. But with all that is needed in place, the prospects have never been as bright.

The present book came as a surprise to me. It is indeed a pleasure to notice how neuroscience is now (finally) in the process of catching up. The book's timely appearance will make it a lasting milestone, and it will be most interesting to compare the present material to what will have become known say 10 years hence. The book presents an excellent overview of all aspects of optic flow processing in animals (ranging from insects to primates) and humans. The contributors are all well-respected experts in their fields, these fields covering the neurosciences broadly, from psychophysics and neurology to neurophysiology of the relevant brain centers. The important interrelations with the vestibular system are also covered. I recommend the book to specialists, but especially to newcomers who want to gain a quick overview of the relevant research questions in this field.

Jan J. Koenderink
Helmholtz Institute
Utrecht, The Netherlands

PREFACE

Goal-directed spatial behavior relies to a large extent on vision. Vision provides essential information about the environment and the consequences of our actions. When we move forward through the world, the pictures on our retinae are also in motion. J. J. Gibson termed this self-induced visual motion the *optic flow* and sparked a long line of research on its use for visual navigation. Optic flow provides visual input to monitor self-motion, navigate and guide future movements, and avoid obstacles along the path. Within the past 10 years, neurophysiology has begun to take the question of the neuronal mechanisms of optic flow processing seriously. Physiological evidence for the use of optic flow has been accumulated from a wide variety of animals ranging from flies and birds to higher mammals and primates. This book provides a thorough review of this investigation and the results obtained, and relates it to parallel developments in psychophysics and computational modeling.

A substantial body of knowledge about how humans analyze optic flow has been accumulated in psychophysical studies. It is well accepted that humans can in principle use optic flow to determine their direction of heading and that the visual system combines optic flow analysis with a multitude of other sensory signals to infer self-motion. Gibson already noted that a forward translating observer experiences visual motion in the "optic array surrounding the observer" that contains a "singularity," or "singular point," an idealized point in the flow field at which visual motion is zero. For an observer moving in a straight line, the destination of travel is such a point, because all visual motion seems to expand radially from this point. Gibson termed this singularity the "focus of expansion." He suggested that the visual system might directly use the focus of expansion to determine heading by analyzing the global optic flow structure.

Yet, the issue is not that simple, because any natural self-motion might be composed of eye, head, or body movements that have different effects on the retinal image. Although expansional optic flow is useful for guidance of self-motion, it also raises issues of visual stability and the generation of eye movements. Obviously, self-motion tends to disturb the stability of the retinal image as it induces optic flow. Eye movement systems such as the optokinetic and ocular following reflexes have evolved to stabilize the retinal image during self-motion. Recent research has shown that optic flow also elicits involuntary vergence and tracking eye movements. Such eye movements, however, have implications for the analysis of the optic flow field because they superimpose additional visual motion. Depending on the structure of the visual scene and on the type of eye movement, the retinal flow field can differ greatly from the simple radial structure of the optic flow, making a direct search for the focus of expansion impossible.

The problem that the visual system faces with regard to optic flow is thus twofold. On one hand, the visual system needs to maintain stable visual images. On the other hand, self-motion must be estimated and controlled. The contributions in this book circle around both of these aspects of optic flow processing. They do this in such diverse animals as flies, pigeons, and primates.

In primate cerebral cortex, the processing of visual motion is attributed to a series of areas in the so-called dorsal stream pathway, which is believed to be specialized in the analysis of motion and spatial actions. Motion information proceeds from the primary visual cortex (V1) to the middle temporal area (MT), the medial superior temporal area (MST), and several higher areas in the parietal cortex. From electrophysiological studies, much evidence indicates that area MST in the superior temporal sulcus is a key structure for optic flow processing. Neurons in area MST of the macaque respond differentially to the location of the focus of expansion in large-field optic flow stimuli, suggesting that an encoding of the direction of heading in the responses of the MST population is possible. Access to oculomotor signals provides a way in which area MST can analyze optic flow during eye movements. This is especially interesting because area MST is also involved in the control and guidance of eye movements, and it is connected to subcortical centers of gaze stabilization.

Other cortical areas of the macaque brain that respond to optic flow are the middle temporal area, the ventral intraparietal area, the superior temporal polysensory area, and area 7A. Together these areas form a network of information flow that transforms retinal motion information into high-level spatial parameters that are used to direct and control spatial behavior. The existence of several different motion processing areas for different visual motion tasks has also been demonstrated in functional imaging studies of human cortex and neurological studies in human patients. These patients have impaired performance in some visual motion tasks but are normal in others.

But the usefulness of optic flow is not restricted to primates; it also applies to a large number of other animal species. Comparative studies of optic flow processing are interesting not only because of the possible comparison of structures and functions, but also because the requirements of optic flow processing may be quite different in different animals. Airborne animals such as birds and flying insects are faced with different problems in the visual control of self-motion than ground-dwelling animals such as cats and monkeys. Indeed, Gibson's original interest in optic flow was linked to studies on how pilots are able to fly an aircraft. It is very interesting to see how supposedly simple organisms such as flying insects deal with the complexity of optic flow analysis. Behavioral studies have shown that flies and bees use optic flow for the control of flight parameters. Recent electrophysiological recordings demonstrated that individual neurons in the fly's horizontal system do in fact act as decoders for optic flow fields that are evaluated during flight control. Similarly, there is behavioral evidence that birds make use of optic flow, and neuronal optic flow selectivity has been described in the accessory optic system (AOS) of the pigeon.

It is obvious that much experimental progress has been made in the identification of the neuronal basis of optic flow processing. However, because of the complexity of the task, a full understanding can only be achieved by a tight coupling between single unit neurophysiology, behavioral and psychophysical observations, and theoretical considerations. This approach should lead to models that take into account not only the computational requirements but also their neuronal implementation. From the comparison of the predictions of such models to the neuronal data, a unified account of the neuronal processing of optic flow fields in the visual system of animals may be reached.

This book brings together a wealth of experimental data on the processing of optic flow and views of models tightly connected to these data. I thank all the contributors and everybody who helped in this effort. Much of the contributed work was made possible by financial support from grants of the Human Frontier Science Program.

Markus Lappe

PART I

PERCEPTION

HUMAN EGO-MOTION PERCEPTION

A. V. van den Berg

Helmholtz School for Autonomous Systems Research, Department of Physiology, Faculty of Medicine, Erasmus University, Rotterdam, the Netherlands

I. Introduction

A seemingly simple task like walking an empty corridor without hitting the walls becomes very difficult when asked to do so blindfolded. Toddlers who have just learned to walk tip over when the walls of a movable room are set into motion (Stoffregen *et al.*, 1987). Walking on a treadmill that is towed around at speeds different than the treadmill's speed result in changes of the felt walking speed (Rieser *et al.*, 1995). These examples illustrate that the interplay between vision, kinaesthetic, and vestibular information is of major importance to the control of locomotion.

In order to serve locomotion, the visual system needs to represent ego-motion in a format that is useful to act in the environment. Thus, one needs to specify what sort of visual information is relevant to locomotion and if—and how—this visual information is acquired. Because locomotion is a broad description of many different tasks that require different elements of visual information (e.g., walking toward a target, making a turn, and avoiding obstacles), the required visual information is to some extent task-specific. For example, to prevent bumping into an obstacle, it is useful to perceive whether it is on one's future path and how much time is left for corrective action. The distance to the object is

0074-7742/00 $30.00

not relevant except in proportion to the speed of forward motion. Consequently, much attention has been given in the psychophysical literature to the visual perception of heading and judgments of the time to contact. In this review, I will concentrate on the first of these tasks: the perception of heading.

Gibson (1966, 1986) recognized that the visual motion field contains useful information for such tasks. He observed that the pattern of direction lines that connects a vantage point with objects in the environment expands when the vantage point moves forward. Only that direction line that coincides with the direction of forward motion remains stationary. Thus, the moving vantage point receives an expanding motion pattern that radiates outward from the direction of heading. This pattern of motion is called the optic flow, and its center is called the focus of outflow. In Gibson's view, the focus of outflow labels the object or the location in the environment to which one is heading. There is no need for a specification of a reference frame for the measured flow. The array of visual objects serves as the frame with respect to which the heading direction is visually specified. These ideas of Gibson have served as a useful starting point for the analysis of visual perception of heading. One can find an excellent review of older literature in Warren (1995).

II. Retinal Flow and Optic Flow

Even when the observer is moving on a linear track, the flow on the retina will rarely be a purely expanding motion pattern. This holds because the retina is placed on top of a series of mobile supports (e.g., the hips, the torso, the head and the eye), which can all rotate relative to one another. It is useful therefore, to make a clear distinction between retinal and optical flow, the former depending on the translational and the rotational movements of the eye, whereas the latter only involves the translatory component of the eye. Both types of flow fields are typically represented by a collection of angular motion vectors, each attributed to a particular visual direction line (Fig. 1). This representation of the flow field is appropriate for heading analysis (Warren *et al.*, 1991a), but derivatives of the flow field may be more appropriate for other tasks like shape from flow (Koenderink, 1986).

The eye's translation causes angular motion away from the direction of heading with a magnitude that is inversely proportional to the distance. The eye's rotation generates flow that consists of parallel motion across the retina with a magnitude that is independent of the distance. Its direction and magnitude merely depend on the orientation of the

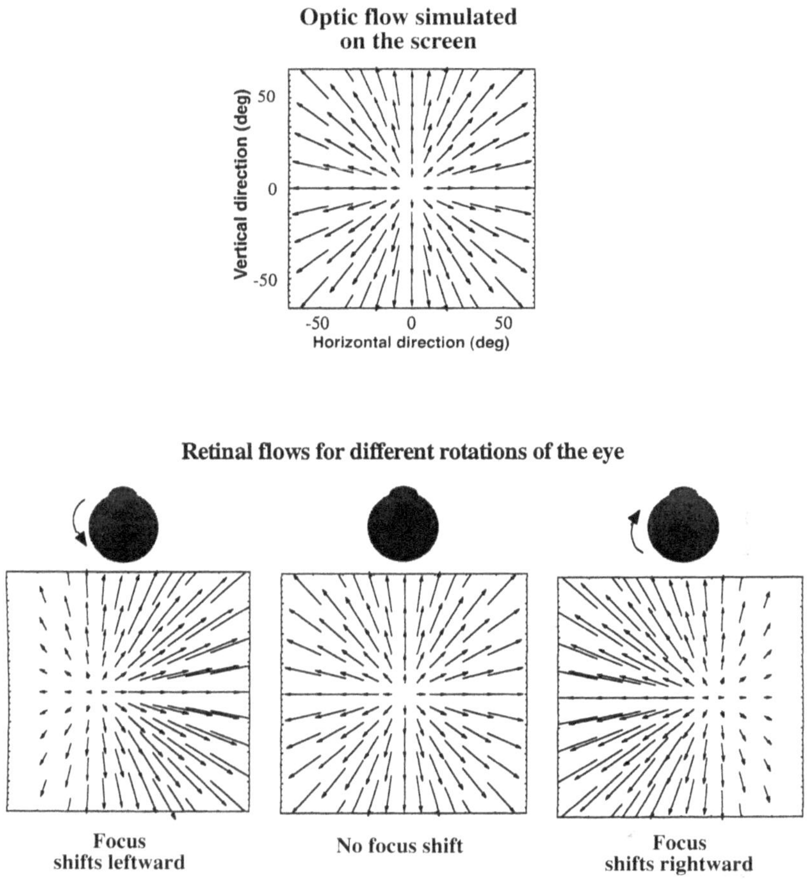

FIG. 1. The retinal motion pattern depends on the pattern of motion on the screen and the eye's rotation. If the motion pattern on the screen simulates the eye's approach of a wall (upper panel), the effect of the eye rotation will be to shift the center of expansion on the retina relative to the center on the screen. One of the moving dots on the screen will be stable on the retina, whereas the dots that correspond to the focus on the screen will be moving relative to the retina. The shift on the retina will be in the same direction as the eye's rotation (left and right panels). Its magnitude depends on the simulated speed of approach, the eye's rotation and the simulated distance to the wall. If the simulated scene is not a wall, there may be no clear focus on the retina, yet there may be an apparent focus that is consistent with a "best fit" of an expanding flow field to the actual retinal flow.

axis of rotation and the rotational velocity. More importantly, it does not even depend on the location of the rotational axis relative to the eye. This gives rise to an ambiguity in the relation between the instantaneous flow field and the eye's motion through the environment. Moreover, the rotations usually change over time in direction and magnitude as does forward motion, leading to nonstationary flow. Yet, current research has mostly dealt with stationary flow patterns (but see Cutting *et al.*, 1992). For the moment, we ignore these difficulties and discuss various studies that have dealt with heading perception from pure expanding retinal motion.

III. Basic Properties of Heading Perception

Studies of ego-motion perception have greatly profited from the advent of affordable fast graphics workstations that can simulate 3D scenes in real time. Typically, one simulates the retinal flow for an eye that moves through scenes without recognizable features (randomly located dots). Such patterns may evoke vivid perception of self-movement, called linear vection. Vection latency and strength depend on the display size, type of flow, direction of simulated motion (Telford and Frost, 1993) and the richness of motion in depth cues (Palmisano, 1996). Linear vection takes several seconds to build up, but the percept of ego-motion direction or heading occurs well within a second (Crowell *et al.*, 1990; Warren and Kurtz, 1992; Crowell and Banks, 1993; Stone and Perrone, 1997), even when the sense of self movement is still relatively weak.

Simple simulations in heading studies involve motion of an eye on a linear track. This turns out to be a relatively simple task if the eye fixates some stationary target on the screen, resulting in pure retinal expansion. Heading can then be discriminated from a reference target in the scene with a just noticeable difference (jnd) angle of 1–2° (Warren *et al.*, 1988), which is thought to be sufficient for avoidance of obstacles during normal locomotion (Cutting *et al.*, 1992; Cutting, 1986). This performance level is little affected by changes in the layout of the simulated scene (Warren *et al.*, 1998; te Pas, 1996), the presentation time (down to 300 ms: Crowell *et al.*, 1990; down to 228 ms: te Pas *et al.*, 1998) or density of the simulated environment (down to 3 visible dots: Warren *et al.*, 1988). Also, the retinal locus of the simulated heading does not affect discrimination performance very much although there is an accuracy gain of the central region over the periphery (Warren and Kurtz, 1992; Crowell and Banks, 1993; te Pas *et al.*, 1998). Azimuthal and elevational components of heading may have different retinal loci of optimal discriminability. Azimuthal precision is slightly larger in the lower hemi-retina than in the upper half (D'Avossa and Kersten, 1996). In contrast to these rather mild effects of retinal location, there is a clear penalty paid when the focus is off-screen. If the flow within a small aperture is nearly parallel (because the focus is very eccentric), finding the focus of the flow vectors is strongly affected by noise of the visual processing (Koenderink and van Doorn, 1987). Indeed, the jnd between two heading directions increases by nearly two orders of magnitude (up to about 30°) when the focus is moved out from the center of a 10° diameter display to 60° eccentricity (Crowell and Banks, 1993).

Thus, consistent with Gibson's hypothesis, the pattern of expanding flow vectors provides the information for heading direction, and the well-known retinal inhomogeneity has a relatively minor effect on the performance.

IV. The Rotation Problem

Of course, the eye often rotates relative to the environment as we habitually turn our eyes and/or head to pursue targets in our environment or because we are moving on a curved trajectory. This adds a rotational component to the expansion flow, which destroys the focus at the direction of heading. For special layouts of the environment, like an extended wall, a new singular point appears in the direction of eye rotation (Fig. 1). Responding to this retinal focus would lead to biases in perceived heading.

Because only the translational component (the expansion) carries information on heading, one has wondered to what extent that component can be retrieved from the retinal flow. Usually, the rotational component is accompanied by nonvisual signals of vestibular or motor (efference copy) origin, which could help to compensate for the effects of the rotation (see discussion that follows). However, the rotation and heading direction could also be retrieved from the retinal flow itself through visual decomposition of the flow. What then are the contributions of visual and extraretinal signals to heading perception? One approach, originated by Warren and Hannon (1988), has been to compare perceived heading during a smooth pursuit eye movement with perceived heading for the same retinal flow simulated in the display and presented to a stationary eye. Extraretinal signals consistent with the rotational flow occur in the former, but not the latter case.

For real eye movement or active head movement, heading errors are invariably small: on the order of 2–4° (Warren and Hannon, 1990; Royden *et al.*, 1992, 1994; van den Berg, 1996; Crowell *et al.*, 1997). This by itself does not mean that extraretinal signals are necessary to perform the decomposition because both extra-retinal and visual information concerning the rotation is present. Theoretically, nearly perfect visual decomposition of the retinal flow is possible (Koenderink and van Doorn, 1987; Heeger and Jepson, 1992). There are, however, limitations. Small fields of view, limited depth range, high rotational velocities, and high noise levels seriously degrade the information in the visual stimulus and may preclude correct heading perception by any measurement system (Koenderink and van Doorn, 1987). In support of visual decomposition, Warren and Hannon (1988, 1990) reported no loss of accuracy when simulated eye rotation was added to retinal expansion, provided the display simulated motion through a scene with depth, like ground planes or clouds of dots. For approach of a wall, however, subjects saw themselves heading toward the retinal focus. Interestingly, when the rotational component was caused by the subject's eye movement, high performance occurred even for approach of a wall, indicating a contribution of an extraretinal eye movement signal. This was further supported by the demonstration, that heading discrimination tolerates more noise for real eye movement than simulated eye movement conditions (van den Berg, 1992). Warren and Hannon (1988, 1990) used conditions that were modeled after natural locomotion: rather slow eye rotations (up to 1.5°/s) were simulated, consistent with fixation of an environmental target at several meters distance during walking. Thus, visual decomposition appears to be sufficiently precise for the rather modest demands set by natural locomotion.

Results for higher simulated eye rotations are more variable. For example, van den Berg (1992) investigated noise tolerance of heading perception using stimuli similar to that of Warren and Hannon. For the lowest noise level and simulated motion through a cloud, errors were about 2.5° for simulated rotation rates up to about 3°/s. Royden *et al.* (1992, 1994) and Banks *et al.* (1996), found much larger bias in the direction of the simulated eye rotation. Bias for cloud stimuli exceeded 15° at 5°/s. Because one expects no systematic errors in heading for perfect decomposition, large and small bias suggest good and faulty visual decomposition, and suggestions of that kind have been made in the literature. Yet, to put observed errors in the right perspective, one needs to compare them to the errors of an observer that does not decompose but rather treats the retinal flow as if it is optical flow. This allows one to compare data in the literature that differ in simulated environment, simulated ego-motion parameters, and response measures. I use a performance measure that can be derived from most published data and directly compared to the performance of the observer that ignores the rotation problem and responds to the retinal focus (retinal focus observer). I excluded from the analysis experiments in which stereoscopic motion was used and studies that simulated motion across a ground plane, because special strategies may be involved (see below).

What kind of errors can we expect for the retinal focus observer? While approaching a frontal plane (at distance d) the retinal focus shifts in the direction of eye rotation by an amount (e), that depends on the speed of ego-translation (T) and rotation rate (R) as:

$$\varepsilon = 0.5 \sin^{-1}(2\ Rd/T)$$

or

$$\varepsilon \approx \frac{d}{T} R$$

if the targets are not too distant.

Thus, the systematic error in heading will grow linearly as the simulated rotation rate increases, with a slope of d/T; alternatively, the error grows as the ratio of R and T increases with a slope d. This analysis holds for a frontal plane. However, it applies more generally, because separating the translational from the rotational component requires increasingly larger precision of retinal motion coding when the R over T ratio increases (Koenderink and van Doorn, 1987). Thus, given the finite precision of visual motion coding, an increase in the number of errors is expected for any environment, as R/T grows (Crowell and Banks, 1996). This was confirmed experimentally for cloud stimuli (Fig. 2b; data from van den Berg, 1997).

I characterized each study by the predicted error-rotation slope (d/T) for the retinal focus observer and compared it to the actual slopes observed in that study. In case of two transparent walls, the distance of the nearest plane was used to compute the predicted focus shift. For a cloud, there is no single distance, and a focus is not clearly defined. Yet, one can estimate a focus by taking the average distance of the visible portion of the cloud.[1] Figure 2 compares the slopes of the heading error versus

[1]For a homogeneous cloud (constant dot density in the simulated environment) one can easily derive that the average distance of a dot to the eye equals 0.75 r_{max}, with r_{max} the far clipping distance of the cloud. For a polar cloud, in which the dots are randomly placed in angular coordinates and ego-centric distance (causing higher density near the eye), the average distance is 0.5 r_{max}. This estimate of the shift of the retinal focus was confirmed with a motion template model (van den Berg and Beintema, 1997). Only those templates that prefer pure expansion on the retina were taken. An array of such templates was used to estimate a focus (even if none is clearly present in the stimulus) by taking the preferred heading direction of the most active template. The focus estimate from the stimulus parameters and the locus of maximal activity in the array of templates differed only marginally. Our estimates are consistent also with a least-squares estimate of the focus used by Royden (1994).

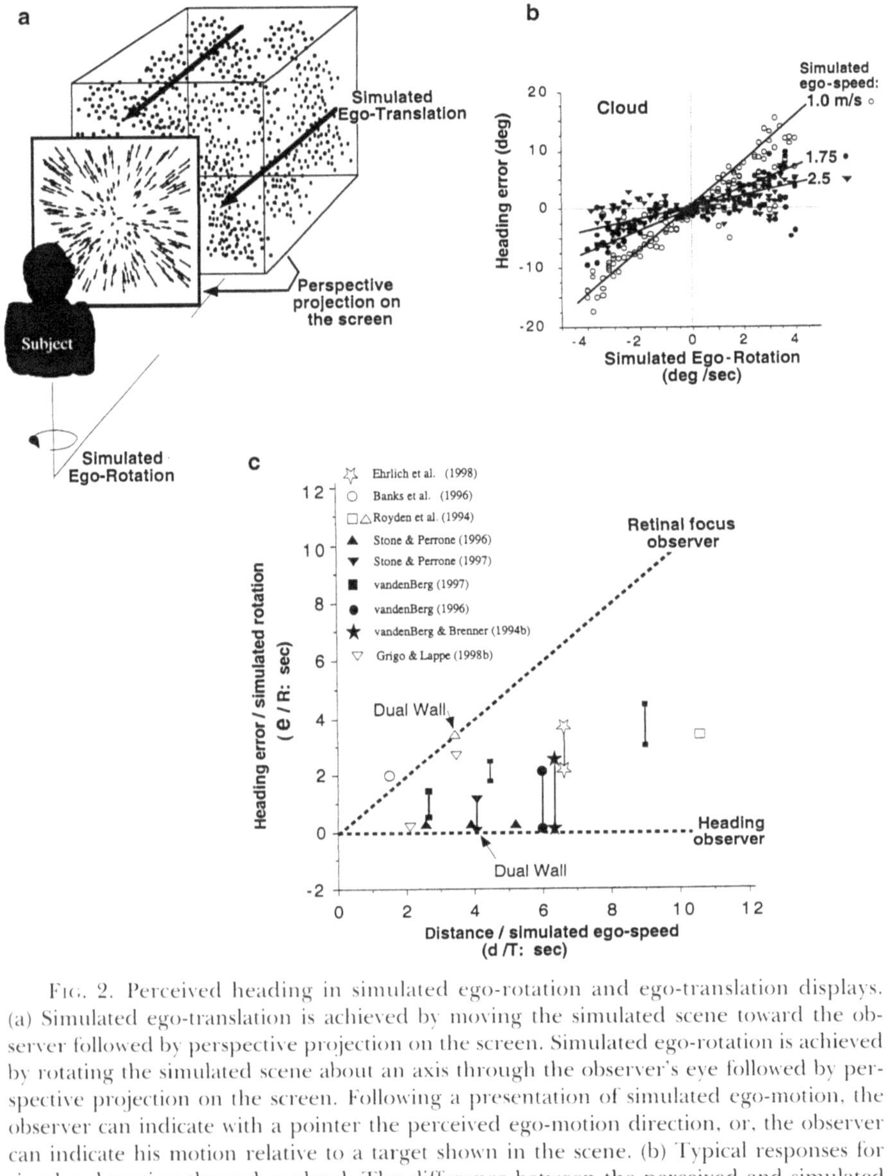

FIG. 2. Perceived heading in simulated ego-rotation and ego-translation displays. (a) Simulated ego-translation is achieved by moving the simulated scene toward the observer followed by perspective projection on the screen. Simulated ego-rotation is achieved by rotating the simulated scene about an axis through the observer's eye followed by perspective projection on the screen. Following a presentation of simulated ego-motion, the observer can indicate with a pointer the perceived ego-motion direction, or, the observer can indicate his motion relative to a target shown in the scene. (b) Typical responses for simulated motion through a cloud. The difference between the perceived and simulated heading direction (heading error) increases linearly as a function of the simulated ego-rotation. When the simulated ego-speed increases (1 → 2.5 m/s), errors decrease. (c) The slope of the relation between heading error and simulated rotation as a function of the predicted slope between heading error and simulated rotation. That prediction depends on the simulated environment and the simulated ego-speed (T), for the observer

(Continued)

the simulated rotation rate with the predicted slope of the retinal focus observer, for several different studies. These data were selected because a range of rotation rates was investigated, allowing for an estimate of the error slope. Clearly the data range from supporting pure visual decomposition ("heading observer": van den Berg, 1996; Stone and Perrone, 1996), to no decomposition ("retinal focus observer": dual wall data of Royden *et al.*, 1994; Banks *et al.*, 1996). The variation in the responses is the more important message of this figure because it points to important factors other than the stimulus parameters that affect the responses for simulated eye rotation.

V. Special Visual Strategies to Solve the Rotation Problem

Fixation of a stationary object of the environment constrains the direction of eye rotation to one optical flow meridian, reducing the search for the heading direction to that meridian. There is no clear evidence that this constraint by itself improves performance. Yet, in combination with other special conditions, it may lead to increased performance. For example, fixating a point in the ground plane with a visible horizon opens the possibility of responding to the intersection of the meridian of the pursuit movement and the horizon. This point corresponds exactly to the direction of heading when one moves parallel to the ground. Indeed, several investigators (van den Berg, 1992; Royden *et al.*, 1994; Banks *et al.*, 1996) found evidence that observers could use such a cue.

FIG. 2. (CONTINUED) that responds to the retinal focus. The characteristic simulated distance in that prediction *d* depends on the type of environment (see text). Because of the simulated forward motion, *d* is not constant during a presentation. I computed *d* from the simulated distances halfway during the presentation. Data for different reports have been combined in this figure. Some points have been labeled (wall, dual wall) to indicate the type of simulated environment. For points that have not been labeled, clouds of dots were used. The legend specifies the reports by name in the reference list. From (Stone and Perrone, 1997) data of experiment 2 were used. From (Ehrlich *et al.*, 1998) data from their Fig. 10 were used. From (Royden *et al.*, 1994), data of experiments 2, 3 and 7 were used. From van den Berg (1996), I used data from Fig. 7, and from Banks *et al.* (1996), I used data of their Figs. 4 and 7. Data from van den Berg and Brenner (1994b) was based on their Fig. 3. For some data sets (Stone and Perrone, 1997; van den Berg, 1996, 1997; van den Berg and Brenner, 1994b; Ehrlich *et al.*, 1998), the range of reported error-rotation rate slopes for different subjects in a single experiment is indicated by a vertical line between symbols. The data of Grigo and Lappe (1998b) refer to single wall stimuli with two different presentation times (0.4 s: lower left; 3.2 s: upper right).

Banks *et al.* (1996) remarked that the intersection is visually defined by the alignment of the local motion vectors along the meridian of the pursuit eye movement. This cue is, however, not necessary because similarly good performance was found for a display with a low signal-to-noise ratio (van den Berg and Brenner, 1994a), which disrupted vector alignment. Presumably, the direction of rotation as specified by the flow at the horizon defines a visual constraint line which is combined with the horizon (van den Berg, 1992; Zwiers *et al.*, 1997).

Fixating a stationary point in the environment not only constrains the possible heading directions to a single meridian but also makes the eccentricity of the heading direction relative to the fixation point (H) dependent on the speed of eye rotation (R) as:

$$H = \sin^{-1}\left(\frac{R}{T/d}\right).$$

Hence, for a constant ego speed (T) and fixation distance (d), the rotation rate becomes predictive of the heading direction. When this correlation was broken by variation of the simulated ego speed (van den Berg, 1997), a steeper increase of the heading error as a function of the rotation rate was found, compared to a previous experiment (van den Berg, 1996) in which a single ego speed was used (see Fig. 2b). This suggests that subjects may use such correlations between rotation rate and heading direction. Nevertheless, heading percepts were fairly accurate (heading errors less than half of the retinal focus observer) when this correlation was broken.

Ground planes provide depth cues that are independent of the retinal flow. As distance increases with height in the scene, and as the flow of distant points is dominated by rotational flow (the translational flow falls of as 1/distance), independent depth cues can improve the visual estimate of ego-rotation by emphasizing the contribution of distant points. A number of observations support such a role for depth. Truncation of the depth in ground planes caused subjects to underestimate the retinal eccentricity of heading by 20–25% (van den Berg and Brenner, 1994a). Also, stereoscopic depth improves noise tolerance of heading perception in a cloud of dots, that lacks the perspective cues of the ground plane. The effect remains when stereoscopic motion-in-depth cues are removed (van den Berg and Brenner, 1994b). However, at low noise levels, there is little or no advantage of static depth cues (van den Berg, 1992; van den Berg and Brenner, 1994b; Ehrlich *et al.*, 1998). The functional coupling between stereo and optic flow processing is also supported by a recent finding that an illusory shift of the focus of expanding motion by uniformly moving dots is larger when the inducing dots are moving behind the expanding dots than vice versa (Grigo and Lappe, 1998a). Interestingly, the perception of object motion during simulated self-motion appears to depend on the object's retinal motion relative to the most distant objects in the scene, rather than its immediate surroundings, suggesting that also in this case compensation for the effect of the eye's rotation is based on the most distant objects (Brenner and van den Berg, 1996).

Points at different depths that are aligned do not remain so when the observer translates unless they are located in the heading direction. Rotations do not affect this locus of zero motion-parallax and if present, it provides a reliable cue to heading (cf. Cutting *et al.*, 1992, 1997).

VI. Circular Heading and Curved Motion Path Percept

Humans are very sensitive to discriminate straight path from curved path motion. Threshold angular velocity of rotation rises linearly with simulated translatory speed, indicating a constant path curvature threshold of 0.0004 m^{-1} (Turano and Wang, 1994). This corresponds to moving on a circle with a 5-km diameter! This is an interesting result because the corresponding threshold-change in simulated heading was only 20 minarc or lower (i.e., at least three times lower than the best heading discrimination thresholds). This suggests the existence of special mechanisms that serve to detect deviations from straight motion (cf. Warren *et al.*, 1991b).

For moderate path-curvatures, observers can discriminate with similar precision as for straight path motion whether an object in the environment is located on their future path or not. For sharper bends, both inside and outside biases may occur, depending on the layout and path curvature (Warren *et al.*, 1991a). As for translational heading, the perception of circular heading depends on successive independent flow fields (Warren *et al.*, 1991b) because no performance gain occurs when dot lifetime is increased over the minimum of two frames, that is required to define a motion field. Interestingly, a curved motion path percept may even occur in the absence of a rotational component in the flow. For pure expansion with different magnitude in nonoverlapping parts of the visual field, curved motion toward the hemifield with slower motion is perceived (Dyre and Andersen, 1997). This points to contributions of flow field asymmetries to the circular ego-motion percept. Unfortunately, most of these studies were done with free eye movements, which complicates their interpretation in relation to the retinal flow.

Stone and Perrone's studies, in which simulated circular motion was used, do bear on the question of visual decomposition of retinal flow (1996, 1997). They asked subjects to indicate the perceived tangent to their circular motion path with a pointer, while fixating a stationary target. To do so, the rotation needs to be disregarded. This turned out to be a difficult task, and some subjects required training with enriched stimuli. Nevertheless, much smaller errors were found than would be expected on the basis of pointing to the retinal focus (Fig. 2c). In fact, errors are among the smallest reported for visual decomposition.

The retinal flow fields for simulated translation + rotation and for circular heading are very similar, although they grow apart over time. It may therefore come as no surprise that observers confuse the two conditions.This is especially the case for faster simulated eye rotation and happens for both simulated pursuit of an environmental target (van den Berg, 1996) and simulated pursuit of a freely moving target (Royden *et al.*, 1994; Banks *et al.*, 1996).This means that even if the flow field can be decomposed, the resulting ego rotation and heading constrain the eye's path through space only partially: the locomotor path belongs to a family of straight and curved paths. Royden (1994) could explain the errors of her subjects by assuming that they perceived motion on a circular path (with the same angular path velocity and tangential velocity as the simulated eye rotation and eye translation) and that they indicated the location some 4 s ahead on that path. She proposed that the extraretinal signal served to distinguish the linear path + eye rotation (extraretinal signal present) from the circular movement (extraretinal signal absent) condition, as shown in Fig. 3a. An alternative explanation (van den Berg, 1996) with links to motion parallax accounts of heading perception (Cutting *et al.*, 1992, 1997) proposes that different types of visual decomposition are done in parallel, consistent with rotation about an axis through the eye, or consistent with rotation about an axis outside the eye. The presence of the extraretinal signal again is supposed to bias responses to the ego-centric decomposition (Fig. 3b).

An important implication is that heading errors observed in simulated rotation + translation displays may be caused by errors in visual decomposition, errors in path extrapolation, or both. Conversely, heading errors in simulated translation + rotation displays cannot be taken as direct evidence for inadequate decomposition. Experimenters have tried to reduce the effects of path extrapolation errors by asking the subjects to make retinal heading judgments as opposed to judgments relative to the environment (Stone and Perrone, 1996, 1997) or to judge their motion relative to the fixation point (van den Berg, 1996). This was successful because low slopes were reported for error versus rotation

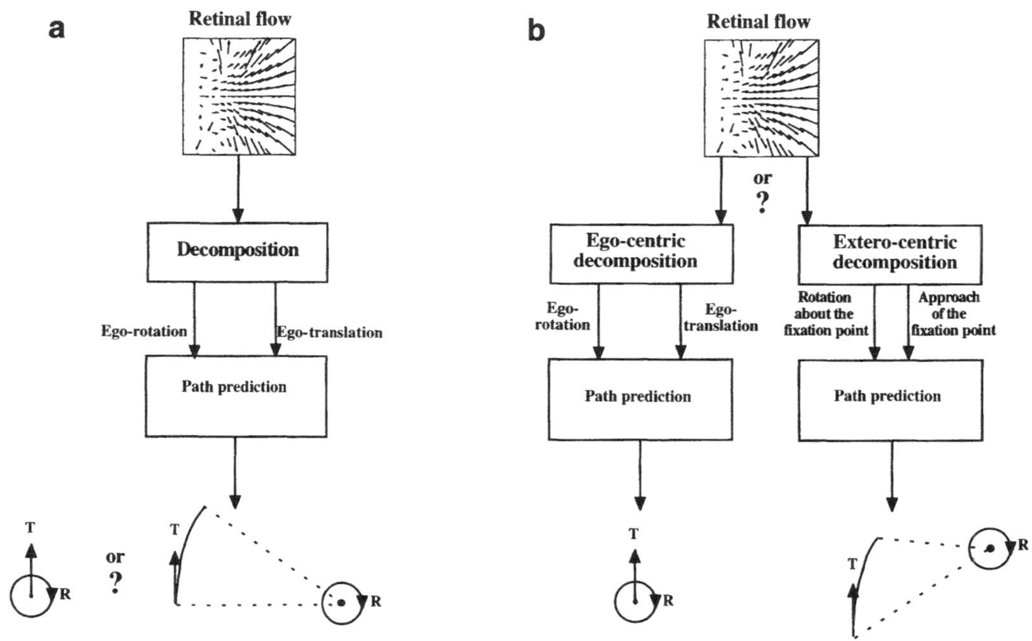

FIG. 3. Path errors and heading perception. Perceived heading is not unambiguously related to the motion parameters that can be estimated from the retinal flow. Whereas the instantaneous retinal flow for scenes with depth specifies unambiguously the rate and direction of ego-rotation and the direction of ego-translation, it does not specify where the axis of rotation is located: in the eye or outside the eye. This ambiguity can be resolved over time, but subjects confuse the two motion conditions for short presentations (1–1.5 s) as used in many heading studies. The subject's choice in these studies may be explained at the level of the path prediction process using an extraretinal signal to distinguish between straight or curved motion paths [a, Royden *et al.* (1994)]. Alternatively, the responses may be explained at the decomposition level [b, van den Berg (1996)], with an extraretinal signal influencing the probability that an ego-centric decomposition (leading to prediction of a straight path) or an extero-centric decomposition (linked to a curved predicted path) takes precedence.

rate compared to the retinal focus observer. This suggests that accurate visual decomposition is possible even for rotation rates much higher than 1.5°/s.

VII. Heading Perception and the Pattern of Flow

The optic flow for ego motion on a straight path or on a curved path is markedly different. As mentioned earlier, human observers can distinguish the two conditions very easily. Both types of movement result in an optic flow with a stationary structure. The pattern is radial in case of motion on a straight path. It consists of a pattern of hyperbolae when the observer moves along a circular path. In either case, the pattern of directions of the flow is constrained by the self motion type. This does not hold for the pattern of local speeds because the latter also depends strongly on the distances in the scene. This raises the question whether the magnitudes and the directions of the local flow vectors are equally important to the discrimination of different types and directions of self motion.

For simulated ego motion on a linear track, Warren *et al.* (1991a) observed that randomly changing the direction of local flow vectors abolishes the heading percept completely. The discrimination threshold for heading direction was unaffected, though, by random changes in speed. Apparently, the pattern of vector directions and not the pattern of vector magnitudes carries the information for heading in retinal expansion displays. In a similar vein, Kim and Turvey (1998) investigated which components of the optic flow were most important to determine one's direction of movement on a curved path across the ground. Again, randomly changing the speeds by which the points moved along the hyperbolical trajectories on the display did not affect discrimination performance. This held for simulated motion on circular and elliptical paths. Thresholds for discrimination of left- or rightward passage of an environmental target were less than 2° (Kim and Turvey, 1998). However, when the flow field was perturbed by addition of dots that moved along circular paths with centers of rotation different from the ego movement, discrimination performance was at chance level. Thus, discrimination of heading direction for circular and linear motion alike is primarily based on the pattern of motion directions in the flow and not on the speeds. A cautionary note should be made: the responses were analyzed in terms of the optic flow in these studies. Thus one assumed that the rotational flow on the retina as caused by eye movements during the experiment was accounted for, possibly through an extraretinal signal.

There is general agreement that for low rotational speeds (typically less than 1.5°/s), heading can be perceived accurately without need for an extraretinal signal. Is the pattern of motion directions on the retina also more important than the pattern of speeds on the retina in this case? If a simulated eye rotation is added to the simulated translation of the eye, both the direction and the magnitude of each local flow vector is affected. This suggests that when the visual system attempts to retrieve heading from the pattern of retinal flow, it should take into account both the magnitude and the direction of the local flow vectors to decompose the retinal flow into its translatory and rotational components. This was investigated in four subjects (van den Berg, unpublished results) for simulated rotation and translation (3 m/s) across a ground plane (1–40 m; 64 white dots). Subjects fixated a red point in the ground plane to the side of their simulated path. They were asked to discriminate the heading direction as left- or rightward with respect to an environmental target that was presented at the end of the 1.67-s motion sequence. To ensure that decomposition was feasible, the average eye rotation rate was 2°/s or less. Again, it was tested whether speed and direction noise had dissimilar effects on the perception of heading in the presence of simulated eye rotation. To investigate the relative importance of speed and direction, each local motion vector was perturbed by a noise velocity in proportion to its magnitude:

$$\mathrm{SNR} = \frac{|\text{Unperturbed local flow}|}{|\text{Noise}|}.$$

Thus, the magnitude of noise velocity depended on the location in the flow field. Direction noise occurred when the noise vector was perpendicular to the local flow. Speed noise was made by addition of a noise term aligned with the local flow vector. Figure 4 shows the discrimination threshold, averaged for the four subjects as a function of the SNR. Clearly, when the perturbation increased (decreasing SNR), the discrimination threshold rose. This decline in performance was the same, however, for speed (○) and direction noise (△). There was a significant main effect of SNR on the threshold (ANOVA: $F(4,30) = 11.1$; $p < 0.0001$), but neither the type of noise nor its interaction with SNR was significant ($p > .5$). Thus, to perceive heading in the presence of eye rotation on the basis of the retinal flow, both the speed and the direction are necessary.

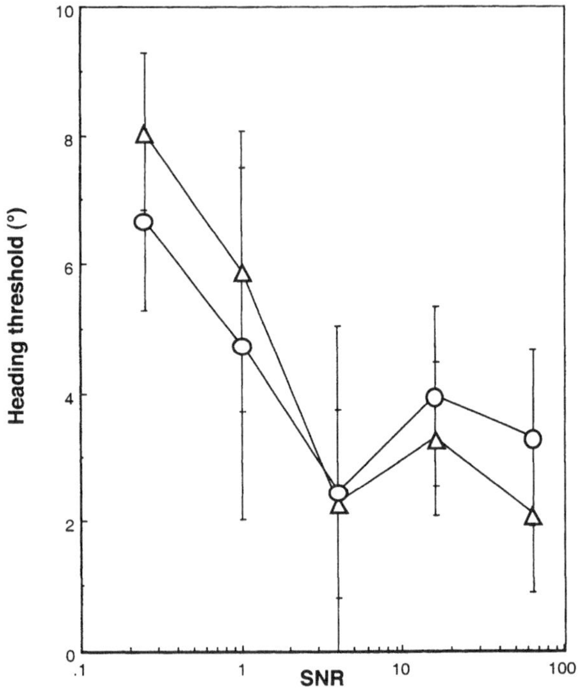

FIG. 4. Heading discrimination thresholds as a function of the signal-to-noise ratio (SNR). Thresholds were based on the last 6 of 12 turn points of a staircase procedure, that searched for the 75% correct threshold. Data points indicate the average threshold across four subjects. Error bars indicate the across-subject SD of the threshold. Thresholds were collected for all conditions (5 SNR levels × 2 noise types) simultaneously, through randomly interleaved presentation of the trials of the different conditions.

VIII. Temporal Properties of Heading Perception

Psychophysical experiments have established that heading discrimination in pure expansion displays deteriorates when presentation time is reduced below about 300 ms (Crowell *et al.*, 1990; te Pas *et al.*, 1998). This is about three times longer than the time required to discriminate expansion from contraction flow (De Bruyn and Orban, 1993). Yet, through the minimal presentation time of the flow one may underestimate the processing time for the heading direction. Because no masking stimulus followed the flow stimulus in the studies by Crowell and te Pas, visual processing may have continued after the stimulus had ended. To side step this issue Hooge *et al.* (1999) used a different paradigm to estimate the processing time for the heading direction. They asked subjects to make a saccade towards the perceived heading direction during a 1.5-s presentation of expanding flow. Saccadic latency varied over a range of several hundreds of milliseconds. Error at the end of the saccade declined when the saccade's latency was longer. Because a saccade cannot be modified some 70 ms before the saccade starts (Hooge *et al.*, 1996), its direction and amplitude must have been based on the visual processing prior to that instant. The decline of the error for larger latencies thus reflects the longer processing time that was available for programming the saccade. The error saturated at about 500 ms after stimulus onset. Hence, Hooge *et al.* (1999) estimated processing time for the heading direction at about 430 ms. This long processing time is of the same order as the integration time for the perception of coherent motion in displays with random perturbations of dot motions (Watamaniuk and Sekuler, 1992).

When perceived heading lags the actual heading direction by several hundred milliseconds, one may expect considerable errors in perceived heading for a changing heading direction over time. Indeed, when the heading direction is stepped across the retina, the perceived heading direction at the end of the sequence of steps is biased in the direction opposite to the steps. The error is proportional to the stepping rate (van den Berg, 1999). Corresponding processing times range from 300 to 600 ms for different subjects. Interestingly, when the same steps of the heading direction were presented to a moving eye, the errors were countered when the direction of the eye movement and the direction of the steps were opposite. Thus, the errors due to the processing time appear to be compensated normally by an extraretinal signal.

Few studies have investigated the processing time for heading direction in simulated rotation and translation displays. Heading discrimination becomes less accurate when the lifetime of dots that carry the flow is reduced below eight frames or 150 ms (van den Berg, 1992). This is more than the minimum dot lifetime reported for pure expansion displays (two frames; Warren *et al.*, 1991a). However, because the total presentation time consisted of many dot lifetimes, these studies are not informative of the minimal presentation time that is required to find the heading direction. Recently, Grigo and Lappe (1998b) investigated the perception of heading for simulated approach of a large (90° × 90°) fronto-parallel plane. When the simulation contained rotation and translation (presented to a stationary eye) the errors decreased in the majority of subjects, when the presentation time was reduced from 3.2 to 0.4 s (cf. Fig. 2c). This unexpected result points to important questions regarding how the interaction between visual and extraretinal estimates of eye rotation build up over time. Taken together, it appears that finding the heading direction from the retinal flow is a relatively slow visual process that can take several hundred milliseconds.

IX. Heading Perception and Moving Objects

Heading perception from optic flow must rely on the assumption that the visual motion is caused solely by the observer's movement. The flow of other objects that are moving in the environment cannot be a reliable indicator of ego-movement. A large, single moving object does not affect the perceived direction of heading unless it covers the observer's focus of outflow (Warren and Saunders, 1995; Royden and Hildreth, 1996; Vishton and Cutting, 1995). In that case, biases are small (about 1°) and may be similarly or oppositely directed to the lateral motion of the object, depending on the object's path and the subject's response strategy (Royden and Hildreth, 1996).

Even when the environment is filled with a mixture of noisily moving and stationary objects, little effect is found on heading accuracy, especially when the erratic motion is much faster than the coherent flow. For example, when the number of noisy points in a ground plane equals the number of coherently moving points, heading discrimination angles are 2–4° for linear motion (van den Berg, 1992) and 1–2° for circular motion (Kim and Turvey, 1998). Yet, when the distractor dots move coherently in one direction, the perceived center of expanding dots on the screen shifts in the motion direction of the distractor dots (Duffy and Wurtz, 1993). These errors vary in magnitude between subjects, ranging from 0.3 to 1.2° shift of the focus per degree per second motion of the distractor dots. Increasing speed of the expanding dots or decreasing speed of the uniformly moving dots reduces the shift of the focus in most subjects. One explanation is that the subjects respond to the pattern of relative motion between the two sets of dots. This pattern has a focus that is shifted in the direction of the distractor dots (Meese *et al.*, 1995). Another explanation attributes the uniform dot motion to an ego-rotation, that is compensated, which causes an illusory shift of the focus. Separating the two sets of dots in different depth planes stereoscopically modulates the illusion in a way that is consistent with stereoscopic effects on induced motion (Grigo and Lappe, 1998a), supporting the relative motion interpretation.

Taken together, the effects of independently moving objects on heading perception are quite small, suggesting that such objects are removed largely from the heading computations.

X. The Reciprocal Relation between Optic Flow and Ego-Motion

The perceptual studies discussed so far simulate highly simplified forms of ego-motion. What kind of flow patterns are received in more natural situations? During walking, the head does not move on a linear track but follows a waving course, with horizontal and vertical undulations that differ in frequency. Horizontal frequency is about half that of the vertical undulations (Grossman *et al.*, 1988; Pozzo *et al.*, 1990; Crane and Deamer, 1997). This reflects the body motion during the step-cycle with two moments of lift-off during one period of lateral sway. Major frequency of vertical head motion is about 2 Hz during walking, with head displacements of several centimeters, peak-to-peak head rotations of about 7°, and peak angular velocities of 30°/s and over (Pozzo *et al.*, 1990; Crane and Deamer, 1997; Das *et al.*, 1995). Because head rotation and head displacements tend to be antiphase, the net vertical displacement of the orbita is reduced and retinal slip of a fixated distant target is only about 4°/s during running (Crane and Demer, 1997). Thus, walking movements introduce "bounce" and "sway" components in the retinal flow. Such stride-related oscillations of the direction of heading do not affect perceptual judgments (Cutting *et al.*, 1992). Yet, when simulated, such oscillations do evoke postural responses during walking (Warren *et al.*, 1996; Bardy *et al.*, 1996). Induced sway is usually less than half of the driver amplitude, and some walkers hardly respond. Expansion-contraction flow, motion parallax, and uniform motion induce sway (Bardy *et al.*, 1996) in decreasing order of magnitude. Nonuniform depth distributions, as occur when large surfaces dominate the view (corridors, ground planes), lead to predictable anisotropies of induced sway because motion parallax is less effective to specify the motion parallel to the surface (Bardy *et al.*, 1996). Flow patterns affect walking speed too. Walking velocity is reduced in response to optic flow of faster walking (Pailhous *et al.*, 1990; Konzak, 1994), primarily through a reduced stride length with stride frequency remaining relatively unaffected (Prokop *et al.*, 1997). There may be a division of labor with high frequency changes in the translatory component of flow used to control balance and very-low-frequency components affecting the perceived direction of a walk.

When walking on a curved path, the head is consistently deviated to the inside of the circle (Grasso *et al.*, 1996). This is qualitatively consistent with directing the center of the oculomotor range to the focus in the flow which is shifted inside relative to the tangent to the circular path. Similarly, when steering a car through a bend, the driver's gaze is directed to the tangent point on the inside of the road, which is argued to be informative of the bend's curvature (Land and Horwood, 1994). Only a narrow horizontal strip of visible road, some 2 s ahead of the current location, is sufficient to steer accurately through a winding road. Many abrupt steering corrections are necessary, though, unless also a near strip of road is visible, suggesting that the nearer parts of the road are informative of the lateral position (Land and Lee, 1995). Thus, different parts of the flow field serve different control functions.

In a recent study, the response to simulated ego-rotation and translation was investigated with a steering task. Rushton *et al.* (1998a) found that steering toward a stationary target did not show a performance gain for depth cues, in contrast to perceived heading (van den Berg and Brenner, 1994a,b) and strength of linear vection (Palmisano, 1996). These results indicate that perceived heading may not always be the single source of information for steering actions. Even walking toward a target may use different information additional to perceived heading from the retinal flow: walkers whose visual field is displaced by a prism walk in a curved path toward a target (Rushton *et al.*, 1998b). This suggests that subjects adjusted their course in response to the perceived angle between the target and the body's sagittal plane. A strategy based on keeping the focus of outflow aligned with the target was not used, as it would predict a straight walk to the target. These results suggest that perceived heading from the retinal flow may be no more than one piece of visual information used to guide our actions in the environment.

Acknowledgment

This work was supported by the Netherlands Organisation for Scientific Research (grant SLW805.33.171) and Human Frontier (grant RG 34 / 96B).

References

D'Avossa, G., and Kersten, D. (1996). Evidence in human subjects for independent coding of azimuth and elevation for direction of heading from optic flow. *Vision Res.* **36,** 2915–2924.

Banks, M. S., Ehrlich, S. M., Backus, B. T., and Crowell, J. A. (1996). Estimating heading during real and simulated eye movements. *Vision Res.* **36,** 431–443.

Bardy, B. G., Warren, W. H, Jr., and Kay, B. A. (1996). Motion parallax is used to control postural sway during walking. *Exp. Brain Res.* **111,** 271–282.

van den Berg, A. V. (1992). Robustness of perception of heading from optic flow. *Vision Res.* **32,** 1285–1296.

van den Berg A. V. (1996). Judgements of heading. *Vision Res.* **36,** 2337–2350.

van den Berg, A. V. (1997). Perception of heading or perception of ego-rotation? *Invest. Ophthalm. Vis. Sci. Abstr.* **37,** 380.

van den Berg, A. V. (1999). Predicting the present direction of heading. Vision Res., in press.

van den Berg, A. V., Brenner, E. (1994a). Humans combine the optic flow with static depth cues for robust perception of heading. *Vision Res.* **34,** 2153–2167.

van den Berg, A. V., Brenner, E. (1994b). Why two eyes are better than one for judgements of heading. *Nature* **371,** 700–702.

van den Berg, A. V. , Beintema, J. A. (1997). Motion templates with eye velocity gain fields for transformation of retinal to head centric flow. *Neuroreport* **8,** 835–840.

De Bruyn, B., and Orban, G. A. (1993). Segregation of spatially superimposed optic flow components. *J. Exp. Psychol. Human Percept. Perform* **19,** 1014–1127.

Brenner, E., and van den Berg, A. V. (1996). The special role of distant structures in perceived object velocity. *Vision Res.* **36,** 3805–3814.

Crane, B. T., and Demer, J. L. (1997). Human gaze stabilization during natural activities: Translation, rotation, magnification, and target distance effect. *J. Neurophysiol.* **78,** 2129–2144.

Crowell, J. A., and Banks, M. S. (1996). Ideal observer for heading judgments. *Vision Res.* **36,** 471–490.

Crowell, J. A., Banks, M. S., Shenoy, K. V., and Andersen, R. A. (1997). Self-motion path perception during head and body rotation. *Invest. Ophthalm. Vis. Sci. Abstr.* **38,** 2224.

Crowell, J. A., Banks, M. S., Swenson, K. H., and Sekuler, A. B. (1990). Optic flow and heading judgements. *Invest. Ophthalm. Vis. Sci. Abstr.* **31,** 2564.

Crowell, J. A., and Banks, M. S. (1993). Perceiving heading with different retinal regions and types of optic flow. *Percept. Psychophys.* **53**(3), 325–337.

Cutting, J. E. (1986). "Perception with an Eye to Motion." MIT Press, Cambridge, MA.

Cutting, J. E., Springer, K., Braren, P. A., and Johnson, S. H. (1992). Wayfinding on foot from information in retinal, not optical, flow. *J. Exp. Psychol. Gen.* **121,** 41–72.

Cutting, J. E., Vishton, P. M., Flückiger, M., Baumberger, B., and Gerndt, J. D. (1997). Heading and path information from retinal flow in naturalistic environments. *Percept. Psychophys.* **59,** 426–441.

Das, V. E., Zivotofsky, A. Z., DiScenna, A. O., and Leigh, R. J. (1995). Head perturbations during walking while viewing a head-fixed target. *Aviat. Space Environ. Med.* **66,** 728–732.

Duffy, C. J., and Wurtz, R. H. (1993). An illusory transformation of optic flow fields. *Vision Res.* **33,** 1481–1490.

Dyre, B. P., and Andersen, G. J. (1997). Image velocity magnitudes and perception of heading. *J. Exp. Psychol. Human Percept. Perform.* **23,** 546–565

Ehrlich, S. M., Beck, D. M., Crowell, J. A., and Banks, M. S. (1998). Depth information and perceived self-motion during simulated gaze rotations. *Vision Res.* **38,** 3129–3145.

Gibson, J. J. (1966). "The Senses Considered as Perceptual Systems." Houghton Mifflin, Boston.

Gibson, J. J. (1986). "The Ecological Approach to Visual Perception." Houghton Mifflin, Boston.

Grasso, R., Glasauer, S., Takei, Y., and Berthoz, A. (1996). The predictive brain: Anticipatory control of head direction for the steering of locomotion. *Neuroreport* **7,** 1170–1174.

Grigo, A., and Lappe, M. (1998a). Interaction of stereo vision and optic flow processing revealed by an illusory stimulus. *Vision Res.* **38,** 281–290.

Grigo, A., Lappe, M. (1998b). An analysis of heading towards a wall. *In:* "Vision and Action" (L. R. Harris and M. Jenkin, Eds.), pp. 215–230. Cambridge University Press, Cambridge, UK.

Grossman, G. E., Leigh, R. J., Abel, L. A., Lanska, D. J., and Thurston, S. E. (1988). Frequency and velocity of rotational head perturbations during locomotion. *Exp. Brain Res.* **70,** 470–476.

Heeger, D. J., and Jepson, A. (1992). Subspace methods for recovering rigid motion I: Algorithm and implementation. *Int. J. Computer Vision* **7,** 95–117.

Hooge, I. Th. C., Beintema J. A. and van den Berg A. V. (1999). Visual search of heading direction. *Exp. Brain Res.,* in press.

Hooge, I. Th. C., Boessenkool, J. J. and Erkelens, C. J. (1996). Stimulus analysis times measured from saccadic responses. *In:* "Studies in Ecological Psychology," A. M. L. Kappers, C. J. Overbeeke, G. J. F. Smets, and P. J. Stappers, Eds., Proceedings of the Fourth European Workshop on Ecological Perception, pp. 37–40.

Kim N., and Turvey, M. T. (1998). Visually perceiving heading on circular and elliptical paths. *J. Exp. Psychol. Human Percept. Perform.* **24,** 1690–1704.

Koenderink, J. J. (1986). Optic flow. *Vision Res.* **26,** 161–179.

Koenderink, J. J., and van Doorn, A. J. (1987). Facts on optic flow. *Biol. Cybern.* **56,** 247–254.

Konzak, J. (1994). Effects of optic flow on the kinematics of human gait: A comparison of young and older adults. *J. Mot. Behav.* **26,** 225–236.

Land, M. F., and Horwood, J. (1994). Where we look when we steer. *Nature* **369,** 742–744.

Land, M. F., and Lee, D. N. (1995). Which parts of the road guide steering? *Nature* **377,** 339–340.

Meese, T. S., Smith, V., and Harris, M. G. (1995). Speed gradients and the perception of surface slant: Analysis is two-dimensional not one-dimensional. *Vision Res.* **3,** 2879–2888.

te Pas, S. F. (1996). Perception of structure in optical flow fields. PhD thesis, University of Utrecht.

te Pas, S. F., Kappers, A. M. L., and Koenderink, J. J. (1998). Locating the singular point in first-order optical flow fields. *J. Exp. Psychol. Human Percept. Perform.* **24,** 1415–1430.

Pailhous, J., Ferrandez, A. M., Flückiger, M., and Baumberger, B. (1990). Unintentional modulations of human gait by optical flow. *Behav. Brain Res.* **38,** 275–281.

Palmisano, S. (1996). Perceiving self-motion in depth: the role of stereoscopic motion and changing-size cues. *Percept. Psychophys.* **58,** 1168–1176.

Pozzo, T., Berthoz, A., and Lefort, L. (1990). Head stabilization during various locomotor tasks in humans. I. Normal subjects. *Exp. Brain Res.* **82,** 97–106.

Prokop, T., Schubert, M., and Berger, W. (1997). Visual influence on human locomotion. Modulation to changes in optic flow. *Exp. Brain Res.* **114,** 63–70.

Rieser, J. J., Pick, H. L. Jr, Ashmead, D. H., and Garing, A. E. (1995). Calibration of human locomotion and models of perceptual-motor organization. *J. Exp. Psychol. Human Percept. Perform.* **21,** 480–497.

Royden, C. S., Banks, M. S., and Crowell, J. A. (1992). The perception of heading during eye movements. *Nature* **360,** 583–585.

Royden, C. S., Crowell, J. A., and Banks, M. S. (1994). Estimating heading during eye movements. *Vision Res.* **34,** 3197–3214.

Royden, C. S. (1994). Analysis of misperceived observer motion during simulated eye rotations. *Vision Res.* **34,** 3215–3222.

Royden, C. S., and Hildreth, E. C. (1996). Human heading judgments in the presence of moving objects. *Percept. Psychophys.* **58,** 836–856.

Rushton, S. K., Harris, J. M., and Wann, J. P. (1998a). Active control of heading and the importance of 3D structure, 2D structure and rotation rates. *Invest. Ophthalm. Vis. Sci. Abstr.* **37,** 379.

Rushton, S. K , Harris J. M., Lloyd, M. R., and Wann, J. P. (1998b). Guidance of locomotion on foot uses perceived target location rather than optic flow. *Current Biol.* **8,** 1191–1194.

Stoffregen, T. A., Schmuckler, M. A., and Gibson, E. J. (1987). Use of central and peripheral optical flow in stance and locomotion in young walkers. *Perception* **16,** 113–119.

Stone, L. S., and Perrone, J. A. (1997). Human heading estimation during visually simulated curvilinear motion. *Vision Res.* **37,** 573–590.

Stone, L. S., and Perrone, J. A. (1996). Translation and rotation trade off in human visual heading estimation. *Invest. Ophthalm. Vis. Sci. Abstr.* **37,** 2359.

Telford, L., and Frost, B. J. (1993). Factors affecting the onset and magnitude of linear vection. *Percept. Psychophys.* **53,** 682–692.

Turano, K., and Wang, X. (1994). Visual discrimination between a curved and straight path of self motion: effects of forward speed. *Vision Res.* **34,** 107–114.

Vishton, P. M., and Cutting, J. E. (1995). Wayfinding, displacements, and mental maps: Velocity fields are not typically used to determine one's aimpoint. *J. Exp. Psychol. Human Percept. Perform.* **21,** 978–995.

Warren, W. H. Jr., (1995). Self-motion: Visual perception and visual control. *In:* "Perception of Space and Motion" (W. Epstein and S. Roger, Eds.), pp. 263-325. Academic Press, San Diego.

Warren, W. H. Jr, Morris, M. W., and Kalish, M. (1988). Perception of translational heading from optical flow. *J. Exp. Psychol. Human Percept. Perform.* **14,** 646–660.

Warren, W. H. Jr, and Hannon D. J. (1988). Direction of self-motion is perceived from optical flow. *Nature* **336,** 162–163.

Warren, W. H. Jr, Blackwell, A. W., Kurtz, K. J., Hatsopoulos, N. G., and Kalish, M. L. (1991a). On the sufficiency of the velocity field for perception of heading. *Biol. Cybern.* **65**(5), 311–320.

Warren, W. H. Jr, Mestre, D. R., Blackwell, A. W., and Morris, M. W. (1991b). Perception of circular heading from optical flow. *J. Exp. Psychol. Human Percept. Perform.* **17,** 28–43.

Warren, W. H., and Kurtz, K. J. (1992). The role of central and peripheral vision in perceiving the direction of self-motion. *Percept. Psychophys.* **51,** 443–454.

Warren, W. H. Jr, and Saunders, J. A. (1995). Perceiving heading in the presence of moving objects. *Perception* **24,** 315–331.

Warren, W. H. Jr, and Hannon, D. J. (1990). Eye movements and optical flow. *J. Opt. Soc. Am.* **A7,** 160–169.

Warren, W. H., Kay, B. A., and Yilmaz, E. H. (1996). Visual control of posture during walking: Functional specificity. *J. Exp. Psychol. Human Percept. Perform.* **22**(4), 818–838.

Watamaniuk, S. N. J., and Sekuler, R. (1992). Temporal and spatial integration in dynamic random-dot stimuli. *Vision Res.* **32,** 2341–2347.

Zwiers, M., Brenner, E., and van den Berg, A. V. (1997). Direction of pursuit and heading perception. *Invest. Ophthalm. Vis. Sci. Abstr.* **38,** 860.

PART II

EYE MOVEMENTS

OPTIC FLOW AND EYE MOVEMENTS

Markus Lappe and Klaus-Peter Hoffmann

Department of Zoology and Neurobiology, Ruhr University Bochum, Bochum, Germany

I. Introduction

Eye movements are an integral part of many visually guided behaviors. We typically shift our gaze to a new object of interest twice every second. These gaze shifts are used to obtain essential visual information through foveal vision. During self-motion, eye movements have a further important function for visual perception. Because self-motion induces image motion on the retina, eye movements are needed to counteract the induced visual motion and stabilize the image of the object that is fixated. Eye movements during self-motion have important consequences for the processing of optic flow. On the one hand, they may help optic flow analysis in a task-dependent manner. On the other hand, they introduce complications for optic flow analysis because they add further retinal image motion. In the following, we will first look at the distribution of gaze during self-motion. Then we will review work on gaze stabilization, mainly during linear (forward) motion. Finally, we will describe the consequences of eye movements on the retinal flow pattern during self-motion.

0074-7742/00 $30.00

II. Gaze during Self-Motion

Reliable and accurate recording of gaze direction during self-motion is a difficult technical problem. First of all, many eye movement recording systems cannot be easily taken along with a moving subject. Second, the gaze movements of freely moving subjects are composed of movements of the eye in the head, movements of the head on the trunk, and movements of the trunk with respect to the feet. It is quite challenging—but not impossible, see Solomon and Cohen (1992a, 1992b)—to measure all these components simultaneously. A way to circumvent this problem is to measure the position of the eye in the head along with a headcentric view of the visual scene by a camera fixed to the head (Land, 1992; Patla and Vickers, 1997).

Presumably because of the technical problems involved only a few studies have examined eye movements during active self-motion. Naturally, most interest in the allocation of gaze during self-motion and the percentage of time spent on different parts of the visual field has come from applied psychological research on driving behavior in automobilists (Shinar, 1978). The next section presents some basic results of this research. After that, studies of gaze measurements during walking are reviewed.

A. Gaze during Driving

Basic results early on showed that gaze during open road driving is typically directed straight ahead, or to the far scenery on the side, to other vehicles, or (very infrequently) to the near parts of the road (Mourant *et al.*, 1969). The percentage of time spent in these gaze directions increases in this order. But it also depends on the scene and on the task or objective of the driver. More gaze shifts to eccentric positions are made when the driver is asked, for instance, to attend to all the road signs, memorize the travel area, and the like (Mourant *et al.*, 1969; Hughes and Cole, 1988; Luoma, 1988). Frequent and large gaze shifts occur when crossing an intersection (Land, 1992).

During straight driving, gaze stays mostly close to the focus of expansion or the heading of the car (Mourant *et al.*, 1969; Land and Lee, 1994), presumably because it is important to monitor constantly the way ahead, particularly at the high travel speed in a car. A further characteristic and consistent relationship between gaze direction and driving behavior has been described for the negotiation of curves (Land and

Lee, 1994). While approaching and driving through a curve, gaze is directed toward a specific point at the inner edge of the road. This point has been termed the tangent point because it is the point where the tangent to the edge of the road reverses direction. It is also the innermost point of the road edge seen from the driver. The tangent point is a characteristic point of the visual projection of the curve in the driver's display, not a fixed point on the curve in space. As such, the tangent point moves on the edge of the road as the driver continues to pass the curve. During driving in a curve, gaze is directed toward the tangent point on average 80% of the time. Land and Lee propose that this gaze strategy eases the task of steering because the motion and position of the tangent point provides visual information to estimate the curvature. Thus the fixation of the tangent point could be a special visual strategy for the requirements of driving.

B. Gaze during Walking

Locomotion on foot is composed of entirely different visuomotor characteristics and requirements than driving a car. The important param-eter that needs to be controlled is the placement of the foot in the step cycle. Hollands *et al.* (1995, 1996) and Patla and Vickers (1997) reported that gaze in walking human subjects was mostly directed toward future landing positions of the feet. Hollands *et al.* (1995, 1996) measured eye movements in human observers who had to traverse a course of stepping stones. The subjects were required to place their feet on particular predetermined target positions. Gaze alternated between left and right targets in correlation with the step cycle. Patla and Vickers (1997) recorded gaze direction in a task where subjects had to step over small obstacles along their locomotor path. Most of the time, gaze was directed to a location on the ground several steps ahead of the current position. Only 20% of the time did subjects fixate an obstacle immediately before stepping over it. They concluded that information about the stepping positions is obtained well ahead of time.

Wagner *et al.* (1981) investigated the gaze behavior of walking humans in an outdoor environment. Rather than measure gaze positions with an instrument, they simply asked their subjects to report what they looked at as soon as a certain auditory signal was sounded. They took 58 measurements from each of 16 subjects. The results indicated that most often gaze was directed to objects close to the observer. The maximum of the distribution of gaze points lay between 1.5 and 3 m from the observer. From an analysis of this distribution, one might conclude that only a small proportion (< 10%) was near the focus of expansion. The majority of gaze directions deviated quite substantially from the focus of expansion (median deviation about 20°). Wagner *et al.* also classified the types of objects at which gaze was directed. In almost half of the cases, subjects looked at moving objects such as other people, vehicles, or animals. Thirteen percent of gazes were directed to the ground in front of the subject.

Solomon and Cohen (1992a,b) studied eye movements of walking monkeys. They used a setup in which the monkey ran on a circular platform, being tethered to a radial arm that could move about a centered pole. They simultaneously recorded eye-in-head, head-on-body, and body-in-space positions. The direction of gaze in space could be recovered from these measurements. The two monkeys in these experiments usually fixated a point in the environment and kept their gaze directed toward this point for a period of several hundred milliseconds. Then they shifted gaze to a new target.

From these studies, one may conclude three basic findings. First, normal self-motion is accompanied by a large number of eye movements. This in not surprising since eye movements are an integral part of many behaviors and are needed to obtain necessary visual information to guide these behaviors. Second, the distribution of gaze depends on the task that is required from the observer. Third, and related to the second point, the pattern of gaze movements is different for driving a car and for walking. In the first case, there is a consistent relation between driving parameters and gaze direction. Gaze is kept near the focus of expansion for straight driving and near the tangent point of the curve during turns. In contrast, during walking, gaze is typically not directed at the focus of expansion but rather more variably at objects in the near vicinity along the path of travel.

III. Ocular Reflexes during Self-Motion

Section II concerned the distribution of gaze and of fast, saccadic gaze shifts during self-motion. A further concern are slow eye movements that occur between gaze shifts. During self-motion, the visual image of the world on the retinae of the eyes is also in motion. This retinal image motion creates a problem for stable vision. In order to perceive the environment accurately, it is desirable to have a clear and stationary visual image. Several types of compensatory eye movement reflexes exist that attempt to counteract the self-motion-induced visual motion and to keep the retinal image stable (Carpenter, 1988). These gaze stabilization eye movements use vestibular, proprioceptive, or visual signals to achieve this task.

Rotations and translations of the head in space induce corresponding signals in the semicircular canals and otholiths of the vestibular organs. These signals are directly used to move the eyes opposite to the movement of the head. These eye movements are called the rotational and translational vestibuloocular reflexes. The cervicoocular reflex uses signals from the neck muscles to determine head movement and the corresponding compensatory eye movement.

The optokinetic and ocular following reflexes in contrast use the retinal image motion directly. In this case, the eye movement follows the motion on the retina in order to minimize retinal image slip and generate a stable image. The optokinetic reflex (OKR) acts as a feedback loop system which adapts eye velocity to the velocity of the retinal image motion. The ocular following reflex (OFR) describes the initial (60–150 ms) ocular reaction to the onset of motion of a visible scene. In this case, the eyes follow the visual motion of the scene in an open loop manner. A recent review of the properties of these reflexes in relation to self-motion can be found in Miles (1998).

The requirements for gaze stabilization are very different for rotational and translational self-movements. Because the rotation is the simpler part, we will first look at reflectory eye movements induced by self-rotation and then proceed to eye movements induced by self-translation and the associated expansional optic flow.

A. Ocular Compensation for Rotational Movements

For rotations of the head or body, the entire visual scene moves with a single angular velocity. The rotational vestibuloocular reflex (rVOR) compensates for rotations of the head by rotating the eyes opposite to the head rotation. The speed of the eyes in the rVOR closely matches the speed of the head movement such that very good image stabilization is achieved. This is particularly true for fast head movements (e.g., head oscillations in the 2- to 8-Hz range). For slower head movements, ocular compensation increasingly relies on the optokinetic reflex. The optokinetic reflex tries to null retinal image motion by adjusting the eye speed of the visual motion. It works best for low visual speeds. A combination of the two reflexes, which is the normal situation during active movement, results in almost complete image stabilization during head rotations.

B. Vestibuloocular Compensation for Translational Movements

Translations of the head in space also induce vestibularly driven compensatory eye movements. This is called the translational vestibuloocular reflex (tVOR). For lateral or up-and-down head shifts, the eyes are again rotated against the head movement. Unlike in the case of head rotations, however, the required speed of the eye movement cannot be determined from the head movement alone. Accurate image stabilization in this case requires to take into account the geometry of the visual scene. If one considers, for instance, lateral head movements in front of a large object, the induced visual speed of the object depends on its distance from the eye. If the object is close to the eye, the same head movement would induce a much larger visual motion than if the object is farther away. Hence to achieve accurate image stabilization, the compensatory eye speed must be different, depending on the viewing distance.

This situation has been investigated by Schwartz et al. (1989, 1991). They recorded eye movements of rhesus monkeys placed on a sled that moved laterally in the dark. Immediately before the movement, the animals were required to fixate a small spot of light that could be placed at various distances from the animal. This fixation target was extinguished before the sled movement started and merely served to enforce a defined state of vergence at the beginning of the movement. Nevertheless, the speed of the induced vestibuloocular eye movements changed with the viewing distance such that compensation for head movement was always near the optimum. A similar scaling of eye speed with viewing distance also occurs for the ocular following reflex (Busettini *et al.*, 1991). Both findings have been confirmed for humans (Busettini *et al.*, 1994).

The requirements for gaze stabilization become even more complicated when forward movement is considered instead of lateral or up-and-down movement. During forward motion, it is physically impossible to stabilize the entire retinal image. Forward motion induces an expanding pattern of optic flow in the eyes (Fig. 1). Points in different parts of the visual field move in different directions. Hence it is only possible to stabilize part of the visual image. This should be the part at which gaze is directed. For motion along a nasooccipital axis, the tVOR of squirrel monkeys indeed depends on the viewing direction. Eye movement is rightward when gaze is directed to the right and leftwards when gaze is directed to the left (Paige and Tomko, 1991). The speed of the tVOR eye movements in this situation varies with viewing distance and with gaze eccentricity. This variation is consistent with the increasing speed of the optic flow at eccentric locations (Paige and Tomko, 1991).

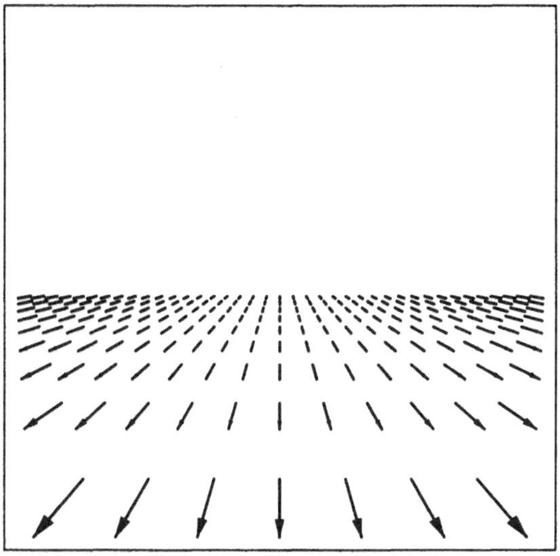

FIG. 1. Optic flow field for linear forward movement over a flat ground plane.

IV. Optic Flow Induced Eye Movements

The aforementioned studies suggest that the translational vestibuloocular reflex is well adapted to the requirements of gaze stabilization during linear motion. We have recently demonstrated the existence also of optokinetic responses to radial optic flow fields, which are associated with linear forward translation (Lappe *et al.*, 1998, 1999; Niemann *et al.*, 1999).

We recorded spontaneous optokinetic eye movements of humans and macaque monkeys that were watching a large-field radial optic flow stimulus presented on a large projection screen in front of them. The stimulus simulated movement across a ground plane (Fig.1). The typical oculomotor response in this situation is shown in Fig. 2. It consists of regularly alternating slow tracking phases and saccades, or quick phases, at a frequency of about 2 Hz. In the following discussion, we will first describe the properties of the slow phases and then those of the saccades.

A. Optokinetic Tracking Movements

During the visual scanning of a radial optic flow stimulus, the visual motion pattern arriving on the retina depends on the direction of gaze.

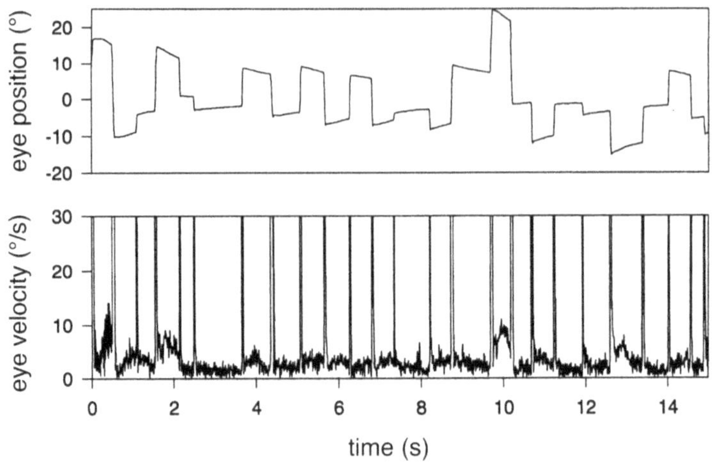

FIG. 2. Horizontal eye position and eye velocity recorded from a monkey that watched a radial optic flow stimulus. The stimulus in this case consisted of a contraction corresponding to backward movement. A regular pattern of alternating tracking phases and saccades can be seen. The eye movement in the tracking phases follows the stimulus motion in gaze direction. Eye movement direction depends on gaze direction. All tracking phases move toward the center.

For instance, if one looks directly at the focus of expansion, the visual motion pattern is symmetric, and there will be no motion in the direction of gaze. If one instead looks in a different direction, retinal slip on the fovea will occur, the direction and speed of which will depend on the gaze direction. Therefore, the eye movement behavior needs to depend on the direction of gaze, too.

Eye movements in the slow phases follow the direction of motion that is present at the fovea and parafovea. The slow phases stabilize the retinal image in a small parafoveal region only. Figure 3a shows a vector field plot of the optokinetic tracking phases of a monkey viewing a radial flow pattern. Each line depicts the direction and speed of a single slow phase eye movement that occurred while the animal looked at a specific location in the flow pattern. Figure 3b shows for comparison the optic flow stimulus (i.e., the visual motion vectors that occurred at these positions in the stimulus). One can see a nice correspondence of eye movement direction and local motion direction in most cases. This close correspondence was confirmed in several quantitative measurements regarding the deviation of the eye movement direction from the local motion direction, all of which indicated a very low deviation (Lappe *et al.*, 1998).

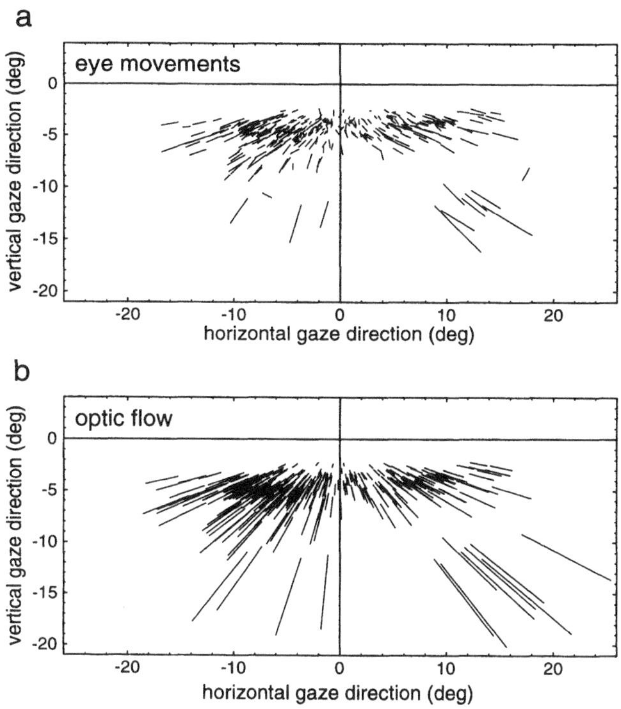

FIG. 3. (a) Vector field illustration of eye movements of a monkey that watched an expanding optic flow stimulus. This stimulus simulated movement over a ground plane consisting of a flat layer of random dots. (b) The visual motion that occurred in the stimulus. Each line in panel a indicates starting position, mean direction, and mean speed of a single optokinetic tracking movement. Each line in panel b represents the local speed and direction in the stimulus at a given eye position from panel a. One can observe that the direction of eye movement is in very good agreement with the local motion in the direction of gaze. The differing vector lengths demonstrate that eye speed is often lower than the corresponding stimulus speed.

However, it is also apparent from Fig. 3 that the speed of the eye movement is often considerably lower than the corresponding local stimulus speed. We defined the gain of the eye movement as the ratio between the eye speed in the direction of the local flow on the fovea and the speed of the foveal motion, averaged across the entire slow phase eye movement. On average, the gain reached a value of 0.5 in both humans and monkeys (Lappe *et al.*, 1998; Niemann *et al.*, 1999). Thus, eye speed was only about half as fast as the speed of the local, foveal image motion.

This discrepancy is resolved, however, if one considers not only the foveal motion but also the motion from within the parafoveal region.

The optokinetic system is known to integrate visual motion signals from throughout the visual field with a special emphasis on the fovea and a parafoveal area of up to 5 or 10° eccentricity (Hoffmann *et al.*, 1992). Thus the visual input that drives these eye movements most likely consists of the spatial average of motion signals from the fovea and parafovea. For a ground plane flow field, this averaged motion signal has a substantially smaller speed than the foveal motion (Lappe *et al.*, 1998). Therefore, the low gain with respect to the foveal motion might be explained from an integration process in the optokinetic system. A much higher gain (close to unity) can be observed, however, when subjects are instructed to actively perform a smooth pursuit movement to follow a single element of the flow field (Niemann *et al.*, 1999).

To summarize, the direction of involuntary optokinetic tracking movements elicited by radial optic flow stimulation well matches the direction of the foveated part of the flow field. The speed of these eye movements is predicted by the averaged speed from the motion in a foveal and parafoveal area.

B. Saccades and Optokinetic Quick Phases

When optokinetic nystagmus is normally evoked by presentation of full-field uniform motion or by a drum rotating around the subject, slow phase tracking movements and saccadic quick phases are very stereotyped. An initial saccade against the direction of the stimulus motion is followed by a slow phase that tracks the stimulus motion in order to stabilize the retinal image. After the eye has moved a certain distance, another saccade against the stimulus motion occurs; it repositions the eye and compensates for positional change during the slow phase. Saccades in this situation serve two functions (Carpenter, 1988). The first is to orient gaze toward the direction from which the stimulus motion originates. The second is to reset eye position after the slow phase tracking movement.

In the case of radial optic flow stimulation, the slow phase tracking movements largely reflect this passive, stereotyped behavior. They are mainly determined by the local stimulus motion. In contrast, the saccades do not share the reflectory nature of the slow phases but rather support an active exploration of the visual scene (Lappe *et al.*, 1999). During forward locomotion, it is necessary to monitor the environment constantly and to identify possible obstacles along the path. Saccades in this situation must serve the ocular scanning of the visual scene instead of merely resetting the eye position.

We calculated that less than 20% of the total distance covered by all saccadic amplitudes in our experiments were required to compensate the positional changes resulting from the tracking phases (Lappe *et al.*, 1999). Hence, most saccadic activity must be attributed to exploration behavior. The distribution of saccades and gaze directions depended on the direction of simulated self-motion (the location of the focus of expansion) and the structure of the visual scene.

Gaze clustered near the horizon and was biased toward the location of the focus of expansion (Lappe *et al.*, 1998, 1999). This bias was stronger in human subjects than in monkeys (Niemann *et al.*, 1999). But in both cases, gaze often deviated by several degrees from the focus location. When we presented a flow field simulating movement through a tunnel instead of a ground plane, the pattern of saccadic directions changed accordingly. While in the ground plane environment, most saccades were directed parallel to the horizon, for the tunnel environment saccade directions were equally distributed in all directions (Lappe *et al.*, 1999).

More recent experiments in human subjects showed that the pattern of saccades and the distribution of gaze depend very much on the task given to the subject. In this study, we used a flow stimulus that simulated movement across a textured ground plane. On this plane, a number of black 2-D shapes that simulated holes in the surface were placed. In the simulation, subjects were driven along a zig-zag course over the surface such that the direction of self-motion changed unpredictably. In successive trials, three different instructions were given to the subjects: (a) passive viewing with no specific task to do, (b) active tracking of the direction of self-motion by pointing gaze toward the focus of expansion, and (c) identifying whether self-motion is toward any of the holes in the surface. This latter condition combines the task of heading detection with the task of obstacle detection.

When the subjects merely viewed the flow stimulus without any specific task, gaze was clustered near the focus of expansion. The same was found when the subjects were explicitly instructed to look into the focus. In contrast, when the subjects were required to identify obstacles along the simulated path of self-motion, saccades were directed to the obstacles or to the ground plane immediately in front of the subject. Virtually no saccade was directly targeted at the focus of expansion. An example scanpath is shown in Fig. 4.

Saccadic parameters are affected by optic flow. Saccadic latencies to the onset of independent object motion are higher during optic flow stimulation than for a stationary background (Niemann and Hoffmann, 1997). Saccades directed to the focus of expansion typically undershoot

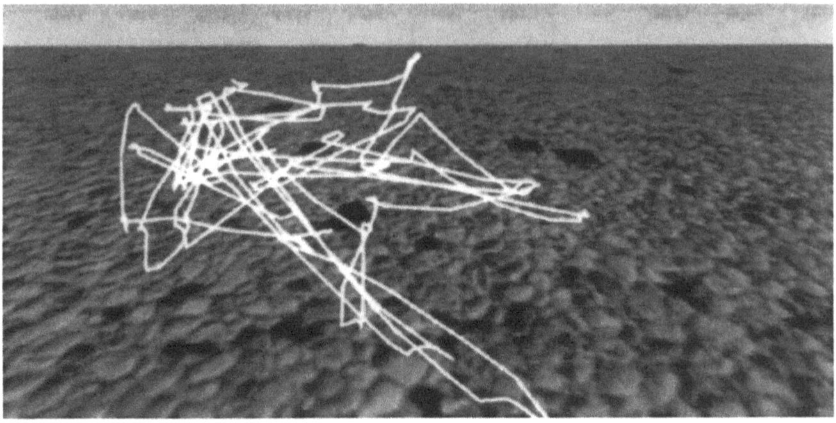

FIG. 4. Scan path of gaze of a subject performing a combined heading detection and obstacle avoidance task. The subject viewed an optic flow stimulus that simulated movement on top of a textured ground plane on which a number of black 2-D shapes were attached. The figure shows a static frame of the stimulus. In the simulation, subjects were driven along a zig-zag course over the surface such that the direction of self-motion changed unpredictably. The task of the subject was to monitor constantly whether self-motion was toward any of the black elements. The white line gives the gaze directions of the subjects over the course of the trial. One can see that gaze was mostly directed to the part of the plane immediately in front of the subject. The focus of expansion was never looked at.

the distance by as much as 40% (Hooge *et al.*, in press). These saccades are much less accurate than those toward a target in front of a stationary background. A sequence of several saccades is required to orient gaze directly into the focus of expansion.

The visual scanning of the optic flow field by saccadic eye movements also introduces complications for the gaze stabilization eye movements between two saccades. With each saccade, the direction and speed of the visual motion on the fovea changes. The dependence of stimulus motion on gaze direction demands a rapid adjustment of eye velocity after each saccade. Due to the latency of signals in the visual system, such an adjustment cannot be done instantly. Appropriate parameters for the eye movement after the saccade can be determined only after a delay of several tens of milliseconds. Within the first 50–100 ms after a saccade, the direction of the eye movement is inappropriate for accurate gaze stabilization (Lappe *et al.*, 1998). Deviation of up to 180° between the local motion direction and the eye movement directions were observed in individual cases. the mismatch seems to occur because of a tendency for the eye to (a) keep the direction of motion that was used before the saccade and (b) direct the eye movement after the saccade in an opposite direction of the saccade itself. Both factors are reminiscent of the oculomotor behavior during regular optokinetic nystagmus evoked by large-field unidirectional motion. In regular optokinetic nystagmus, slow phases are always in the same direction and always opposite in direction to the quick phases.

C. Voluntary Tracking

Passive viewing of radial optic flow fields elicits reflectory optokinetic tracking movements with a low gain of about 0.5. A much higher gain is observed when human subjects are asked to pick a single point of the flow field and actively track it with the eyes (Niemann *et al.*, 1999). In this case, the targets can be pursued almost perfectly. This is remarkable for two reasons. First, the motion of each dot within a flow field is accelerating over time. The eye movements nicely match this acceleration. Second, each dot in a flow field is surrounded by many other dots which move with different speeds and directions and might be considered a source of noise for the pursuit system. Nevertheless, the motion of the chosen point is tracked accurately. These results show that the gain of the tracking eye movements is under voluntary control. The higher gain for voluntary pursuit compared to reflectory optokinetic responses could reflect the restriction of the stabilization to a smaller, more foveal area instead of a parafoveal integration.

D. Vergence Responses

Radial optic flow is normally associated with forward movement. In this case, the distance between the objects in the environment and the eyes of the observer become smaller over time. Hence accurate stabilization of gaze onto an environmental object requires not only the tracking of the motion of that object by version eye movements but also by vergence eye movements to keep both eyes aligned on the object. Interestingly, radial optic flow stimuli elicit such vergence eye movements even in the absence of a distance change. Busettini *et al.* (1997) used a brief, two-frame expansion stimulus to elicit transient open-loop oculomotor responses. Such an expansion step resulted in short-latency vergence eye movements. Vergence changes began approximately 80 ms after the motion step and peaked 30–50 ms later. These findings demonstrate that gaze stabilization reflexes are truly adapted to motion in a three-dimensional environment.

V. Implications of Eye Movements for Optic Flow Processing

The eye movements that occur during self-motion and that are induced by the optic flow in turn influence the structure of the optic flow that reaches the retina. Any eye movement induces motion of the retinal image. Thus, on the retina, movements of the eye superimpose onto movements in the optic flow. The retinal motion pattern during forward movement hence becomes a combination of radial optic flow with retinal slip induced by eye movement.

A consequence of this is that the motion pattern of the retina might look very different from the simple expansion that one normally associates with optic flow. In particular, eye movements usually destroy or transpose the focus of expansion on the retina (e.g., Regan and Beverley, 1982; Warren and Hannon, 1990). It is therefore appropriate to distinguish retinal flow from optic flow clearly and define retinal flow as the actual motion pattern seen on the retina during combined self-motion and eye movement (Warren and Hannon, 1990). Retinal flow is the starting point for any process of flow field analysis in the visual system.

Figure 5 illustrates how eye movements modify the structure of the retinal flow even when self-motion remains constant (following Lappe and Rauschecker, 1994, 1995). Several examples essentially depict the same observer translation but with different types of eye movement. Figures 5a and b depict the general scenario. The observer moves across a ground plane. In this and all following plots, the direction of observer movement is identified by a cross. During the movement his gaze (indicated by a circle) could be pointed to different locations in space. The direction of gaze defines the center of the coordinate system in which the retinal flow is represented.

Figure 5b shows a view of the optic flow field in a body-centered coordinate system. This is the flow which would be seen by a body fixed camera pointed along the direction of travel. All motion is directed away from the focus of expansion which coincides with the heading of the observer. The projection of this flow field onto the retina of the observer, the retinal flow, depends on the direction of gaze and on the motion of the eye. The examples in Figs. 5c–f correspond to four different combinations of gaze and eye movement. The points at which gaze is directed in these four situations are indicated by circles in Fig. 5b and labeled in correspondence with the associated figures. Three of the points (c, d, f) are located at the horizon. One point is located on the ground close to the observer (e).

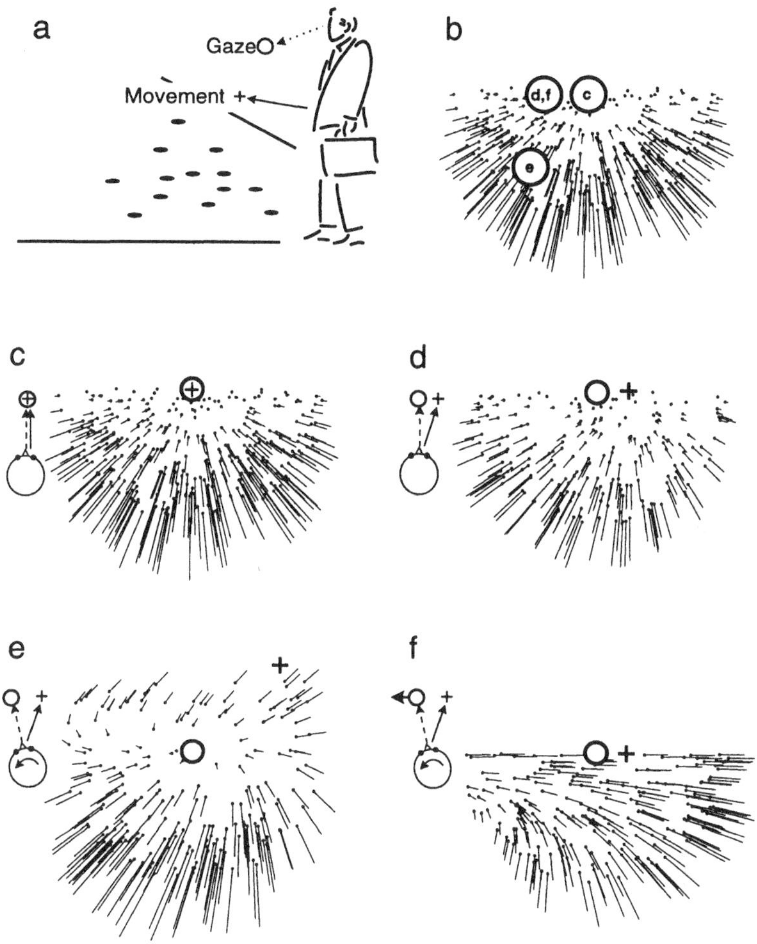

FIG. 5. The influence of gaze direction and eye movements on the structure of the retinal flow field. See text for a detailed explanation.

Figures 5c and d show the results of a gaze shift on the retinal projection of the optic flow. They mainly consist of an offset or shift of the full visual image. Figure 5c shows the retinal flow when the direction of gaze and the direction of movement coincide (i.e., when the observer looks straight ahead into the direction of movement). In this case, the focus of expansion is centered on the retina. In Fig. 5d, the observer now looks off to the side from the direction of movement. Gazing at some fixed point on the horizon allows him to keep his eyes stationary (i.e., no eye movements occur). Again, the focus of expansion is visible and indicates heading, but now it is displaced from the center of the visual field.

Figure 5e shows a situation in which the observer's gaze is directed at some element of the ground plane located in front of him and to the left. There are two consequences of this change in gaze direction. The first is an opposite displacement of the retinal image. The horizon has moved up in the visual field. The second, more serious consequence is that the point at which the gaze is directed is now in motion. This is unlike the situation in Figs. 5c and d, where gaze was directed toward the horizon, which is motionless in the optic flow. The visual motion in gaze direction now enforces a rotation of the eye in order to track the foveal image motion and stabilize gaze on this point. Direction and speed of this eye movement are related to the observer's movement. Since direction is determined by the direction of the flow on the fovea, it is always away from the focus of expansion. Eye speed, however, might be less well defined, depending on the gain of the eye movement (see Sections IV.A and IV.C).

The eye movement induces full-field retinal image motion in the opposite direction. This retinal image motion is combined with the radial motion pattern of the optic flow. The resulting retinal flow field somewhat resembles a distorted spiraling motion around the fovea. The focus of expansion is lost as an indicator of heading. The retinal flow field instead contains a new singular point which is related to the stabilizing eye movement. Perfect stabilization gaze (unity gain) would result in a singular point located exactly on the fovea (circle in Fig. 5e). For an optokinetic tracking movement with lower gain, the singular point would lie about midway between the heading point and the fovea.

In the case of gaze stabilization, the eye movement is linked to the motion and the scene. This is different when the observer looks at a target that undergoes independent motion, such as a moving vehicle or another moving person. In Fig. 5f the observer is assumed to track an object that moves leftward along the horizon. In this case, the retinal flow again has a different structure. This leftward pursuit induces rightward retinal image motion. The combination with the radial optic flow results in a motion pattern that resembles a curved movement. No focus of expansion is visible.

These examples show that the visual signal available to a moving observer can change fundamentally during eye movements, although self-motion remains unchanged. Therefore, mechanisms of optic flow processing that want to recover self-motion must deal with the presence of such involuntary eye movements. For a presentation and discussion of these mechanisms, the reader is referred to van den Berg (this volume), Andersen *et al.* (this volume), and Lappe (this volume).

The close interaction between gaze stabilization eye movements and optic flow is reflected in an overlap of the neural pathways for optic flow processing and eye movement control. For both purposes, the medial superior temporal (MST) area in macaque cortex plays an important role. Different aspects of the involvement of area MST in optic flow processing are discussed in several chapters in this volume (Andersen *et al.*, this volume; Bremmer *et al.*, this volume; Duffy, this volume; Lappe, this volume). But area MST is also an important structure for the generation and control of various types of slow eye movements (recent review in Ilg, 1997). The contribution of area MST to the generation of reflectory gaze stabilization in the ocular following paradigm is presented in detail in Kawano *et al.* (this volume). In this paradigm, the responses of MST neurons closely parallel the generation of ocular following eye movements, their dependence on the vergence state of the eyes, and the generation of short-latency vergence responses to radial optic flow. Area MST also contributes to the optokinetic reflex. The main pathway of the optokinetic system is through the pretectal nucleus of the optic tract (NOT) (Hoffmann, 1988) and the nuclei of the accessory optic system (AOS) (Mustari and Fuchs, 1989). Besides direct retinal afferents, this pathway receives specific cortical input from the middle temporal area (MT) and area MST (Hoffmann *et al.*, 1992; Ilg and Hoffmann, 1993). Currently unpublished experimental results suggest that neurons in this pathway also respond to radial optic flow stimuli.

VI. Conclusion

Eye movements are common during self-motion. Saccadic gaze shifts are used to scan the environment and to obtain important visual information for the control of self-motion. The scanning behavior depends on the requirements of the motion task. During car driving, the location of the focus of expansion is important for straight driving, and the tangent point along the road edge, for driving in a curve. During walking or stepping over obstacles, gaze is directed at the ground in front of the observer.

Between gaze shifts, reflectory eye movements stabilize gaze on the fixated target and reduce retinal image motion in the center of the visual field. Different types of gaze stabilization eye movements are driven by vestibular signals and by visual motion in the optic flow. These eye movements in turn induce additional retinal image motion and thus influence the structure of the retinal flow field. On the retina, the uniform visual motion originating from the eye movement is superimposed on the radial motion pattern of the optic flow. The combination leads to complicated nonradial flow fields which require complex mechanisms for their analysis.

Acknowledgment

This work was supported by the Human Frontier Science Program and grants from the Deutsche Forschungsgemeinschaft.

References

Busettini, C., Masson, G. S., and Miles, F. A. (1997). Radial optic flow induces vergence eye movements with ultra-short latencies. *Nature* **390,** 512–515.

Busettini, C., Miles, F. A., and Schwarz, U. (1991). Ocular responses to translation and their dependence on viewing distance. II. Motion of the scene. *J. Neurophysiol.* **66,** 865–878.

Busettini, C., Miles, F., Schwartz, U., and Carl, J. (1994). Human ocular responses to translation of the observer and of the scene: Dependence on viewing distance. *Exp. Brain Res.* **100,** 484–494.

Carpenter, R. H. S. (1988). "Movement of the Eyes," 2nd ed. Pion Ltd., London.

Hoffmann, K.-P. (1988). Responses of single neurons in the pretectum of monkeys to visual stimuli in three dimensional space. *Ann. NY Acad. Sci.* **545,** 180–186.

Hoffmann, K.-P., Distler, C., and Ilg, U. (1992). Callosal and superior temporal sulcus contributions to receptive field properties in the macaque monkey's nucleus of the optic tract and dorsal terminal nucleus of the accessory optic tract. *J. Comp. Neurol.* **321,** 150–162.

Hollands, M., and Marple-Horvat, D. (1996). Visually guided stepping under conditions of step cycle-related denial of visual information. *Exp. Brain Res.* **109,** 343–356.

Hollands, M., Marple-Horvat, D., Henkes, S., and Rowan, A. K. (1995). Human eye movements during visually guided stepping. *J. Mot. Behav.* **27,** 155–163.

Hooge, I. T. C., Beintema, J. A. and van den Berg, A. V. (1999). Visual search of the heading direction. *Exp. Brain Res.*, in press.

Hughes, P. K., and Cole, B. L. (1988). The effect of attentional demand on eye movement behaviour when driving. *In:* "Vision in Vehicles II." (A. G. Gale, Ed.), Elsevier, New York.

Ilg, U. J. (1997). Slow eye movements. *Prog. Brain Res.* **53,** 293–329.

Ilg, U. J., and Hoffmann, K.-P. (1993). Functional grouping of the cortico-pretectal projection. *J. Neurophysiol.* **70**(2), 867–869.

Land, M. F. (1992). Predictable eye-head coordination during driving. *Nature* **359,** 318–320.

Land, M. F., and Lee, D. N. (1994). Where we look when we steer. *Nature* **369,** 742–744.

Lappe, M., Pekel, M., and Hoffmann, K.-P. (1998). Optokinetic eye movements elicited by radial optic flow in the macaque monkey. *J. Neurophysiol.* **79,** 1461–1480.

Lappe, M., Pekel, M., and Hoffmann, K.-P. (1999). Properties of saccades during optokinetic responses to radial optic flow in monkeys. *In:* "Current Oculomotor Research: Physiological and Psychological Aspects" (W. Becker, H. Deubel, and T. Mergner, Eds.), pp. 45–52. Plenum, New York.

Lappe, M., and Rauschecker, J. P. (1994). Heading detection from optic flow. *Nature* **369,** 712–713.

Lappe, M., and Rauschecker, J. P. (1995). Motion anisotropies and heading detection. *Biol. Cybern.* **72,** 261–277.

Luoma, J. (1988). Drivers' eye fixations and perceptions. *In:* A. G. Gale (Ed.), "Vision in Vehicles II." Elsevier, New York.

Miles, F. A. (1998). The neural processing of 3-d visual information: Evidence from eye movements. *Eur. J. Neurosci.* **10,** 811–822.

Mourant, R. R., Rockwell, T. H., and Rackoff, N. J. (1969). Driver's eye movements and visual workload. *Highway Res. Rec.* **292,** 1–10.

Mustari, M. J., and Fuchs, A. F. (1989). Response properties of single units in the lateral terminal nucleus of the accessory optic system in the behaving primate. *J. Neurophysiol.* **61,** 1207–1220.

Niemann, T., and Hoffmann, K.-P. (1997). Motion processing for saccadic eye movements during the visually induced sensation of ego-motion in humans. *Vision Res.* **37,** 3163–3170.

Niemann, T., Lappe, M., Büscher, A., and Hoffmann, K.-P. (1999). Ocular responses to radial optic flow and single accelerated targets in humans. *Vision Res.* **39,** 1359–1371.

Paige, G. D., and Tomko, D. L. (1991). Eye movement responses to linear head motion in the squirrel monkey. II. visual-vestibular interactions and kinematic considerations. *J. Neurophysiol.* **65,** 1184–1196.

Patla, A. E., and Vickers, J. N. (1997). Where and when do we look as we approach and step over an obstacle in the travel path? *NeuroReport* **8,** 3661–3665.

Regan, D., and Beverley, K. I. (1982). How do we avoid confounding the direction we are looking and the direction we are moving? *Science* **215,** 194–196.

Schwarz, U., Busettini, C., and Miles, F. A. (1989). Ocular responses to linear motion are inversely proportional to viewing distance. *Science* **245,** 1394–1396.

Schwarz, U., and Miles, F. A. (1991). Ocular responses to translation and their dependence on viewing distance. I. Motion of the observer. *J. Neurophysiol.* **66,** 851–864.

Shinar, B. (1978). "Psychology on the Road." Wiley, New York.

Solomon, D., and Cohen, B. (1992a). Stabilization of gaze during circular locomotion in darkness: II. Contribution of velocity storage to compensatory head and eye nystagmus in the running monkey. *J. Neurophysiol.* **67**(5), 1158–1170.

Solomon, D., and Cohen, B. (1992b). Stabilization of gaze during circular locomotion in light: I. Compensatory head and eye nystagmus in the running monkey. *J. Neurophysiol.* **67**(5), 1146–1157.

Wagner, M., Baird, J. C. and Barbaresi, W. (1981). The locus of environmental attention. *J. Environ. Psychol.* **1,** 195–206.

Warren, Jr., W. H., and Hannon, D. J. (1990). Eye movements and optical flow. *J. Opt. Soc. Am. A* **7**(1), 160–169.

THE ROLE OF MST NEURONS DURING OCULAR TRACKING IN 3D SPACE

Kenji Kawano, Yuka Inoue, Aya Takemura, Yasushi Kodaka, and Frederick A. Miles

Neuroscience Section, Electrotechnical Laboratory, Tsukubashi, Ibaraki, Japan; and Laboratory of Sensorimotor Research, National Eye Institute, National Institutes of Health, Bethesda, Maryland

Whenever we move around in the environment, the visual system is confronted with characteristic patterns of visual motion, termed optic flow. The optic flow contains important information about self-motion and the 3D structure of the environment as discussed in other chapters in this book. On the other hand, visual acuity is severely affected if the images of interest on the retina move at more than a few degrees per second. A major function of eye movements is to avoid such retinal motion and thereby improve vision. The observer's movements activate the vestibular organs and are then compensated by vestibuloocular reflexes. However, the vestibuloocular reflexes are not always perfect, and the residual disturbances of gaze are compensated by the visual tracking system(s). Until recently, visual stabilization of the eyes was regarded only in terms of providing backup to the canal-ocular vestibular reflexes, which deal solely with rotational disturbances of the observer. This is reflected in the stimulus traditionally used to study these visual mechanisms: The subject is seated inside a cylinder with vertically striped walls that are rotated around the subject at constant speed, often for periods of a minute or more. Because of the cylinder's bulk, it is usual to first bring the cylinder up to speed in darkness and to then suddenly expose the subject to the motion of the stripes by turning on a light. The ocu-

0074-7742/00 $30.00

lar responses associated with this stimulus consist of so-called slow phases, which are the tracking responses proper, interrupted at intervals by so-called quick phases, which are resetting saccades, and this pattern is termed optokinetic nystagmus (OKN). The slow tracking phases in turn have two distinct components: an initial rapid rise in eye speed—the early component, OKNe—and a subsequent more gradual increase—the delayed component, OKNd (Cohen *et al.*, 1977). Miles and co-workers (see Miles, 1998, for recent review) have emphasized the need to consider translational disturbances also, and it is now clear that there is a translational vestibuloocular reflex (TVOR) in addition to the rotational vestibuloocular reflex (RVOR). In fact, these workers have suggested that OKNe and OKNd reflect quite different functional systems that operate as visual backups to the TVOR and RVOR, respectively. Most recently, these workers have found that there are several distinct visual tracking systems dealing with translational disturbances and all operate in machinelike fashion to generate eye movements with ultra-short latencies. One of these systems—termed ocular following—seems designed to deal best with the visual stabilization problems confronting the observer who looks off to one side (Fig. 1A). Two other systems—termed radial-flow vergence and disparity vergence—deal with the problems of the observer who looks in the direction of heading (Fig. 1B). Our laboratory has been investigating the role of the medial superior temporal

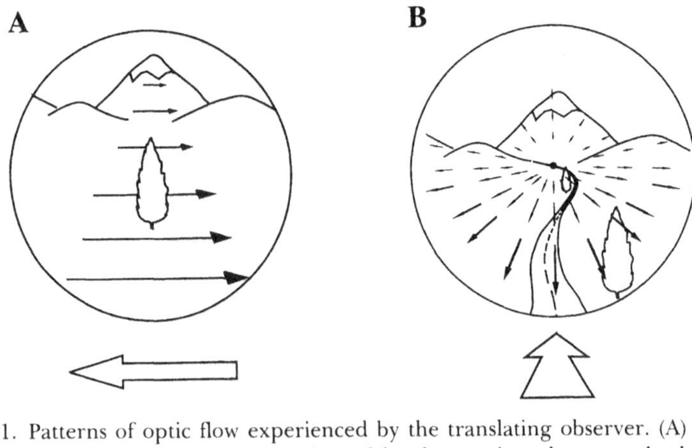

FIG. 1. Patterns of optic flow experienced by the translating observer. (A) A cartoon showing the pattern of optic flow experienced by the moving observer who looks off to one side but does not make compensatory eye movements and so sees images move in inverse proportion to their viewing distance (from Miles *et al.*, 1992). (B) A cartoon showing the radial pattern of optic flow experienced by the observer who looks in the direction of heading (from Busettini *et al.*, 1997).

area of cortex (MST) in the etiology of these visual tracking responses (Kawano *et al.*, 1994). Our electrophysiological experiments suggest that, despite their short latency, ocular following responses, and possibly all three visual tracking mechanisms, are mediated by MST. This chapter will review our latest findings.

I. Neuronal Activity in MST during Short-Latency Ocular Following

A. Ocular Following Responses

To study neuronal responses in MST during ocular following (OKNe), we projected a random-dot pattern onto a translucent tangent screen in front of a monkey and moved it at a constant speed (test ramp) (Kawano *et al.*, 1994). Figure 2 shows sample responses of the activity of

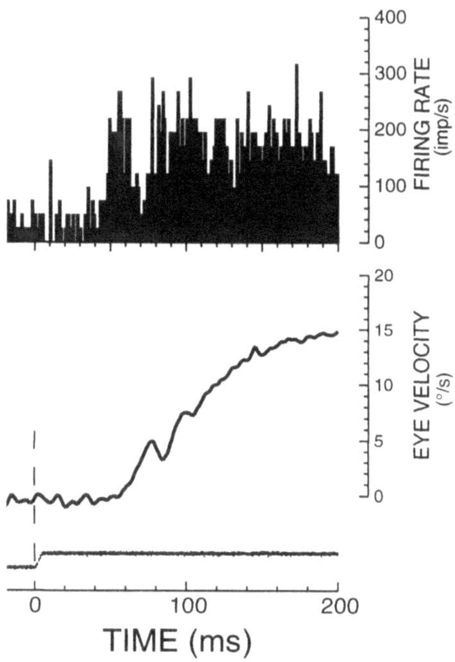

FIG. 2. Response of an MST neuron to large-field image motion that elicited ocular following. Traces, all aligned on stimulus onset, from top to bottom: peristimulus histogram (1 ms binwidth), averaged eye velocity and stimulus velocity profiles. Stimuli: 40°/s right-upward.

an MST neuron and the ocular following responses during 40°/s right-upward test ramps. It is evident that the firing rate of the neuron increased abruptly ~40 ms after the onset of stimulus motion, and the eyes began moving ~10 ms later. This neuron showed a strong preference for motion of the scene at high speed (80°/s). Most of the other MST neurons studied showed similar response properties, which are exactly the properties expected for neurons mediating the earliest ocular following responses: vigorous activation by movements of large patterns at latencies that precede eye movements by 10 ms or more, showing strong directional preferences together with a preference for high speeds (Kawano *et al.*, 1994). Taken together with our finding that early ocular following is attenuated by chemical lesions in the MST area (Shidara *et al.*, 1991), these data suggest that neurons in MST are involved in the genesis of ocular following.

Additional single-unit recordings in the brain stem and the cerebellum suggest that the visual information abstracted in the MST area concerning the moving visual scene is delivered via the dorsolateral pons to the ventral paraflocculus which then computes the motor information needed to drive the eyes (Kawano *et al.*, 1996).

B. Effect of Disparity

When the observer undergoes translational disturbance through the environment (e.g., when looking out from a moving train), the image motion on the retina depends on the 3D structure of the scene. The task confronting the visual stabilization mechanisms here is to select the motion of particular elements in the scene and to ignore all the competing motion elsewhere. Recently, Busettini *et al.* (1996a) have shown that the earliest ocular following is sensitive to binocular disparity in humans. We have sought to determine if the direction-selective neurons in MST that have been implicated in the generation of ocular following show a similar dependence on binocular disparity.

The dependence of ocular following on horizontal disparity steps was studied in two monkeys, and its associated unit discharges in MST were studied in one of them (Takemura *et al.*, 1998). A dichoptic viewing arrangement was used to allow the images seen by the two eyes to be positioned and moved independently (Fig. 3A). Mirror galvanometers controlled the horizontal disparity of the two images. Horizontal disparity steps (crossed and uncrossed, ranging in amplitude from 0.5 to 6.0°) were applied during a centering saccade (Fig. 3B). Fifty milliseconds after the saccade, both patterns were moved together at a constant rate for 150 ms (conjugate ramp).

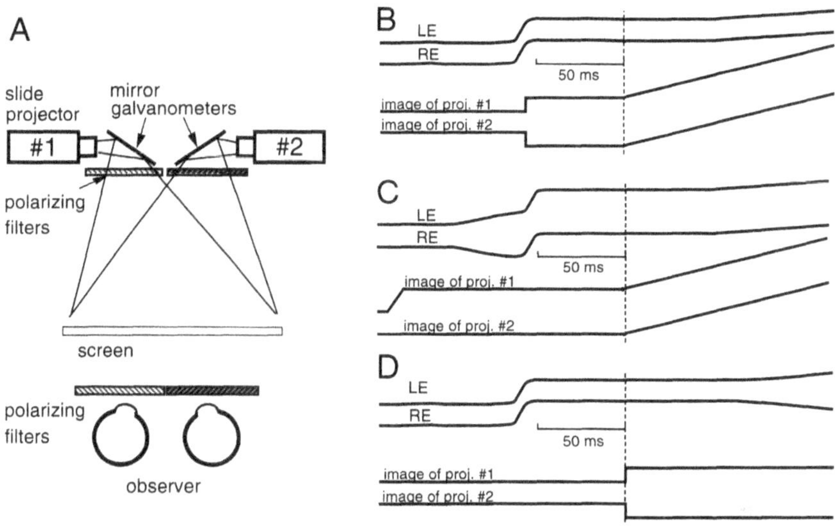

FIG. 3. Schematic drawing of the experimental setup. (A) Diagram of the optical arrangements. Two identical visual images were produced with two identical projectors (#1, #2) with orthogonal polarizing filters. The animal viewed the scene through matching polarizing filters so that the left eye saw only the image produced by projector #1 and the right eye saw only the image produced by projector #2. (B–D) The experimental paradigms to study ocular tracking and their associated neuronal activity. In each panel, traces, all aligned on stimulus onset (vertical dotted line), from top to bottom: left eye position, right eye position, position of the image produced by projector #1, position of the image produced by projector #2. (B) Paradigm to study dependence of ocular following on horizontal disparity. (C) Paradigm to study dependence of ocular following on ocular vergence. (D) Paradigm to study disparity vergence.

The initial ocular following responses showed clear dependence on the disparity imposed during the preceding centering saccade (Fig. 4A). Based on the change in eye position over the period 60–93 ms (measured from stimulus onset), the disparity tuning curves for two monkeys peaked at small crossed disparities, one showing a trough at uncrossed disparities (Fig. 4B). To study the response properties of MST neurons, the images with disparities were moved together in the preferred direction and at the preferred speed for each neuron. Figure 5A shows the superimposed mean discharge rate profiles of an MST neuron with ocular-following-related activity when the binocular disparity was $+0.4^\circ$, $+4^\circ$ (crossed disparity), and -4° (uncrossed disparity): the activity of the neuron showed increased modulation when the disparity was small ($+0.4^\circ$) and smaller modulation when the disparity was large ($+4^\circ$, -4°). The early neuronal responses had a disparity tuning curve (Fig. 5B, continuous line) similar to that for the initial ocular following responses (Fig. 5B, dotted line), peaking with a small crossed disparity ($+0.4^\circ$). Most of the neurons were

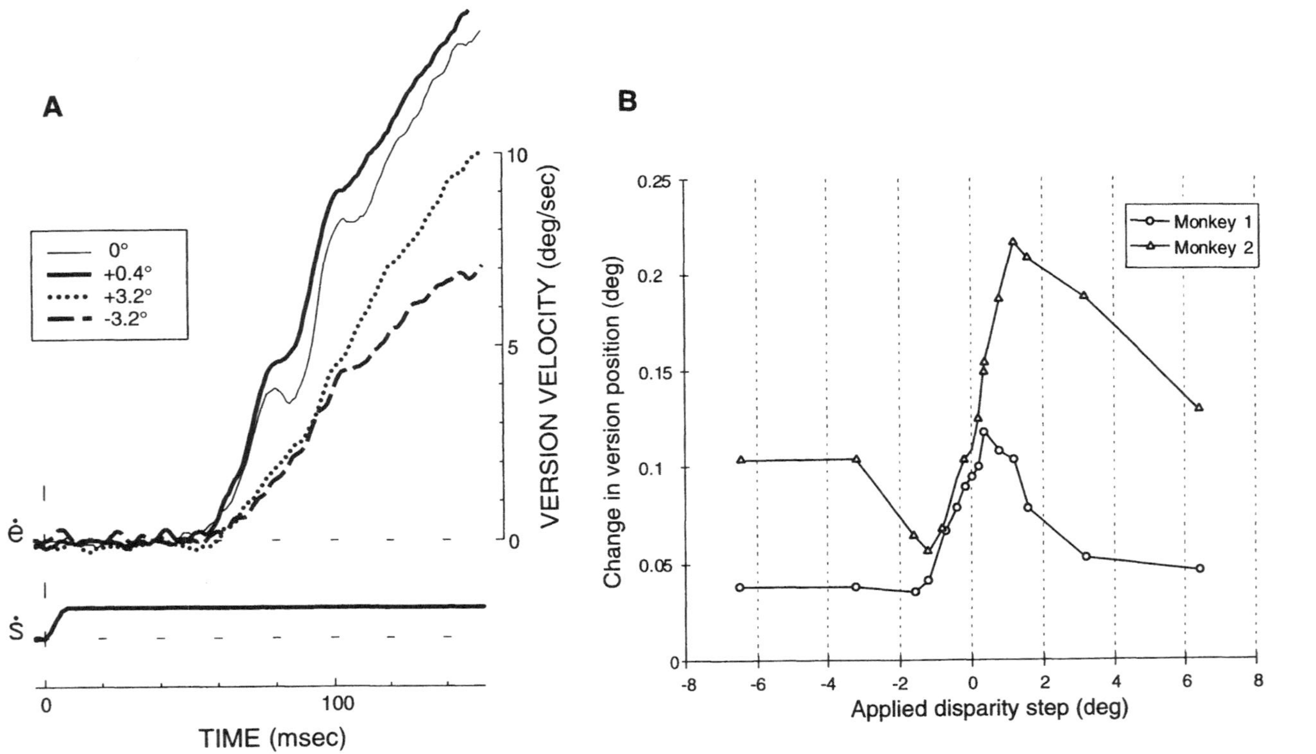

FIG. 4. Dependence of ocular following responses on the horizontal disparity of the tracked images. (A) Superimposed version velocity profiles in response to the conjugate ramps (rightward 60°/s) with various amplitude disparity steps; +3.2° (dotted line), +0.4° (thick continuous line), 0° (thin continuous line), and −3.2° (dashed line). (B) Disparity tuning curves of the initial version responses to the conjugate ramps (downward 60°/s) for two monkeys.

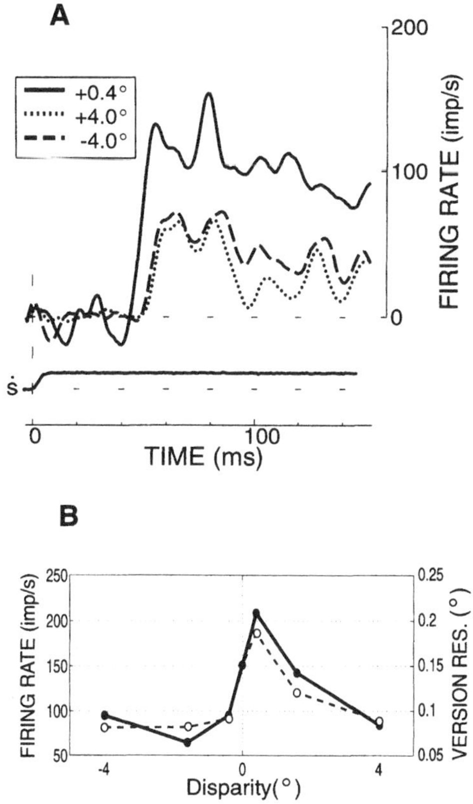

FIG. 5. Dependence of neuronal activity in MST on the horizontal disparity of the tracked images. (A) The superimposed mean discharge rate profiles of an MST neuron in response to the conjugate ramps (downward 60°/s) with various disparities; +4° (dotted line), +0.4° (continuous line), and −4° (dashed line). (B) Comparison between disparity tuning curves for neuronal responses (continuous line with closed circles) and initial version responses (dashed line with open circles).

as sensitive to the disparity as the ocular following was, and more than half of them (~60%) had disparity tuning curves resembling those of ocular following, suggesting that most of the modulation of ocular following with disparity is already evident at the level of MST.

C. Effect of Vergence Angle

It is important to remember that the primary mechanisms compensating for the observer's head motion and thereby helping to stabilize gaze are vestibular and that the visual mechanisms, such as ocular following, compensate only for the residual disturbance of gaze. The rotational vestibuloocular reflex compensates for angular accelerations of the head, which are sensed through the semicircular canals, and the translational vestibuloocular reflex compensates for linear accelerations of the head, which are sensed through the otolith organs. The gain of the RVOR is known to be near unity, whereas the output of the TVOR is inversely proportional to viewing distance (as required by the optical geometry) and far from perfect, tending to overcompensate at far and undercompensate at near viewing distances (Schwarz and Miles, 1991). As mentioned earlier, it has been suggested that ocular following functions as a backup to this imperfect TVOR. Support for this idea comes from the finding that ocular following shares the TVOR's dependence on the inverse of the viewing distance (Busettini *et al.*, 1991), a property that has been attributed to shared anatomical pathways, commensurate with shared function.

We have investigated the dependence of ocular following and its associated neuronal activity in MST on a major cue to viewing distance, ocular vergence (Inoue *et al.*, 1998a). Again, a dichoptic viewing arrangement was used (Fig. 3A), and at the beginning of each trial the image seen by one eye was slowly moved (horizontally) to a new position to induce the monkey to adopt a new convergence angle (Fig. 3C). Then, 50 ms after a centering saccade, both patterns were moved together at constant velocity to elicit ocular following. Using visual stimuli moving in the preferred direction and at the preferred speed for each neuron, it was apparent that the discharge modulations of many MST neurons showed clear dependence on the convergence angle. Figure 6 shows the superimposed mean discharge rate profiles of one such MST neuron when the desired convergence was 0, 2 and 4 m^{-1}: the activity of the neuron showed increased modulation with increased convergence. About half of the MST neurons were like the neuron in Fig. 6 and responded more vigorously when the animal was converged ("near viewing" neurons), whereas ~10% responded more vigorously when the animal was not converged ("far viewing" neurons). The remaining cells showed no significant modulation with convergence state. The mean percentage modulation with vergence $\{[(\text{max} - \text{min})/\text{min}] \times 100\%\}$ was 44% for the "near viewing" neurons and 52% for the "far viewing" neurons. The result indicates that changes in the vergence state, which for example occur when viewing objects at different distances, alter the level of activation of many MST neurons during ocular following. However, none of the neurons were as sensitive to vergence as was the ocular following (~180%), suggesting that most of the modulation of ocular following with vergence occurs downstream of MST.

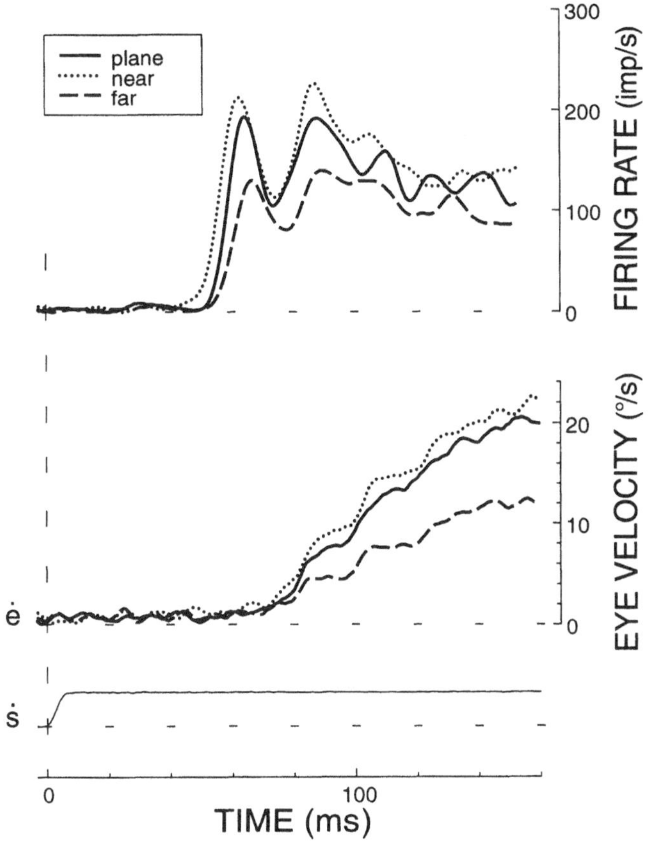

FIG. 6. Dependence of neuronal activity in MST on vergence. Traces, from top to bottom: the superimposed mean discharge rate profiles, averaged eye velocity profiles. Eyes converged on binocular images at 4 m^{-1} (dotted lines), 2 m^{-1} (continuous lines), and 0 m^{-1} (dashed lines). Stimuli in all cases: 40°/s upward.

II. Neuronal Activity in MST during Short-Latency Vergence

An observer moving forward through the environment experiences a radial expansion of the retinal image resulting in a centrifugal pattern of optic flow (Fig. 1B), and an observer moving backward through the environment experiences a radial contraction of the retinal image with a centripetal pattern of optic flow. If gaze is eccentric with respect to the focus of expansion/contraction, then the slow eye movements follow the direction of local stimulus motion in the fovea and parafovea (Lappe *et al.*, 1998), suggesting that the spatial average of motion signals from the central retina elicits ocular following. Since the direction and speed of the visual motion in the central retina change with each saccade to a new part of the optic flow field, the direction of the resulting ocular response constantly changes, preventing any buildup in the neural integrator responsible for the OKNd. Thus when the moving observer looks off to one side, ocular following (or OKNe) is used to track that part of the optic flow field occupying the central retina.

A. Radial Flow Vergence

When the moving observer looks in the direction of heading, the radial pattern of optic flow is centered on the fovea. Since this situation is associated with a change in viewing distance, at some point the observer must change the vergence angle of his/her eyes to keep the object of most interest imaged on both foveae. Recent experiments have shown that radial optic flow elicits vergence eye movements with ultra-short latencies in both human (~80 ms, Busettini *et al.*, 1997) and monkey (~60 ms, Inoue *et al.*, 1998b). Sample vergence velocity profiles in a monkey are shown in Fig. 7. For this experiment, a normal binocular viewing arrangement was used instead of the dichoptic one. A random-dot pattern was projected on the screen in front of the animal. Fifty milliseconds after a centering saccade, this first pattern was replaced by a new one that showed the same image viewed from a slightly different distance (two-frame movie with a looming step). Centrifugal flow, which signals a forward approach and hence a decrease in the viewing distance, elicited rightward movement of the left eye and leftward movement of the right eye, resulting in convergence (Fig. 7A). On the other hand, centripetal flow, which signals a backward motion, resulted in divergence (Fig. 7B). The results agreed with the findings in human subjects of Busettini *et al.* (1997). As shown in velocity profiles in Fig. 7, vergence responses were always transient, generally lasting <60 ms. The vergence responses were very sensitive to the magnitude of the looming stimulus and generally largest with a step-change in viewing distance of 2% (Inoue *et al.*, 1998b).

Interestingly, it has been reported that there are neurons in the MST that are selectively sensitive to radial optic flow patterns (Duffy and Wurtz, 1991; Saito *et al.*, 1986). However, the role of these neurons in generating the short-latency vergence eye movements remains to be investigated.

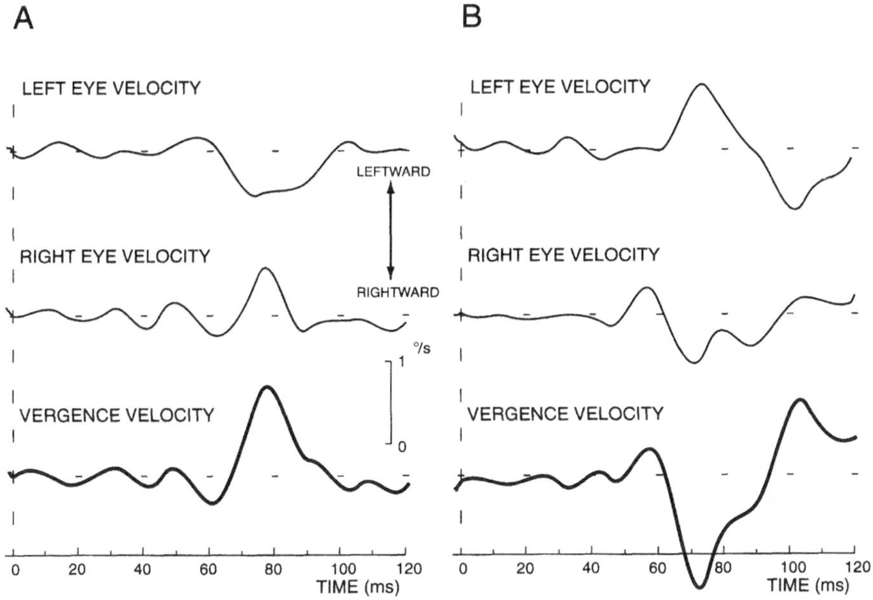

FIG. 7. Vergence responses to looming steps. Traces, from top to bottom: averaged left eye velocity, averaged right eye velocity, averaged vergence velocity. Vergence angle increased in response to centrifugal steps (A), whereas vergence angle decreased in response to centripetal steps (B). Looming stimuli simulated a sudden change in viewing distance of 2%.

B. Disparity Vergence

The radial pattern of optic flow is only one of several cues which indicate the forward motion of the observer. Another very important cue, which also generates vergence eye movements, is horizontal binocular disparity. Recently, it has been shown that sudden changes in the horizontal disparity of a large textured scene result in vergence responses with ultrashort latencies of ~60 ms in monkeys and ~85 ms in humans (Busettini *et al.,* 1994, 1996b).

We recorded neuronal activity in MST during horizontal disparity steps that evoked these vergence responses (Takemura *et al.,* 1997). Again, a dichoptic viewing arrangement (Fig. 3A) was used. Horizontal disparity steps (crossed and uncrossed, ranging in amplitude from 0.5 to 6.0°) were applied 50 ms after a centering saccade (Fig. 3D). Very-short-latency vergence responses (60–70 ms) were consistently observed, confirming the result of Busettini *et al.* (1996b). We found neurons that

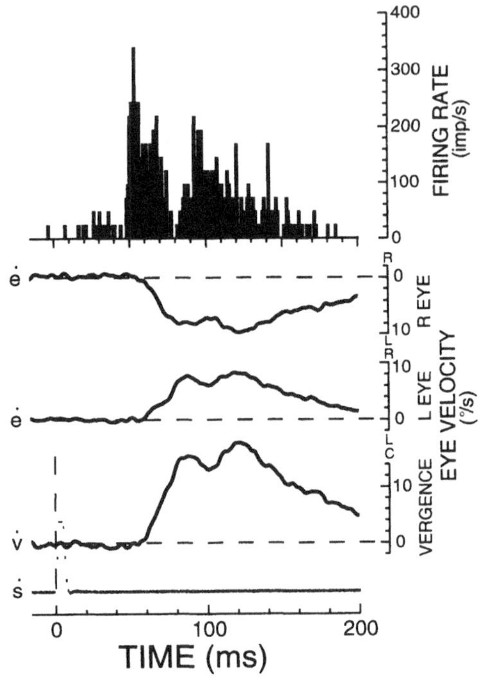

FIG. 8. Responses of an MST neuron and vergence eye movements to disparity steps applied to a large-field image. Traces, all aligned on stimulus onset, from top to bottom: peristimulus histogram (1-ms binwidth), averaged right eye velocity, averaged left eye velocity, and vergence eye velocity profiles (L, leftward; R, rightward; C, convergence). Stimuli were horizontal 3° crossed disparity steps.

are activated 50–70 ms after the onset of the horizontal disparity steps. Figure 8 shows a sample of responses of activity of an MST neuron and the short-latency vergence responses to 3° cross-disparity steps. It is evident that the firing rate of the neuron increased abruptly ~50 ms after the onset of stimulus motion and that both eyes began converging ~10 ms later. Approximately 20% of the neurons studied modulated their discharges in relation to the disparity steps. Disparity tuning curves based on the initial change in the unit discharge rate were obtained for these MST neurons. At one extreme, some units (~30%) could be grouped into classes similar to "disparity-selective" neurons described previously in the V1, V2, and MT areas of the monkey (Maunsell and van Essen, 1983; Poggio *et al.*, 1988). At the other extreme were neurons (~20%) that discharged more closely in relation to the motor behavior (vergence) than to the sensory stimuli (disparity), and so were classified as "vergence-related." The remaining neurons (~50%) showed intermediate discharge behavior and were classified as "intermediate" type. However, it seems that the neurons represent a continuum rather than distinct classes, perhaps linked to successive stages of processing, beginning with the disparity-selective types and ending with the vergence-related types. The disparity tuning curves of the "vergence-related" neurons closely resembled those of the short-latency vergence responses described by Busettini *et al.* (1996b)—with a characteristic S-shape and nonzero asymptote—and some of these neurons discharged early enough to have some role in producing the very earliest of the vergence responses.

III. Role of MST Neurons during Ocular Tracking in 3D Space

Despite the rapid, reflexive nature of the visual tracking responses, we think that they are mediated by MST neurons sensitive to optic flow stimuli. That the ocular-following-related discharges in MST showed a weaker dependence on vergence than did ocular-following responses suggests that further modulation with extraretinal information (vergence) occurs downstream of MST. These results suggest that the MST area decodes optic flow and extracts information to stabilize gaze, and the downstream structures modify and transform its output to the motor command needed to drive the eyes, including short-term and long-term adaptive gain control (Miles and Kawano, 1986).

IV. Tracking Objects Moving in 3D Space

Visual tracking neurons, which discharge when the animal pursues with its eyes an object moving through its visual field, were originally described by Mountcastle and Hyvärinen's group in the mid 1970s as a group of neurons recorded in the posterior parietal cortex of monkeys (Hyvärinen and Poranen, 1975; Mountcastle et al., 1975). However, further intensive study of the visual tracking neurons in the posterior parietal cortex by Sakata *et al.* (1983) showed that visual tracking neurons were mainly encountered in the posterior part of area 7a, especially in the bank of the superior temporal sulcus. This area almost overlaps the area where Komatsu and Wurtz (1988) recorded pursuit neurons (i.e., the MST area). Since further studies on pursuit-related neurons in the

MST always required animals to pursue a small spot projected and moved on a tangential screen, detailed information is available only about the response properties of MST neurons during pursuit of a spot moving in the front-parallel plane (Erickson and Dow, 1989; Ferrera and Lisberger, 1997; Kawano *et al.*, 1994; Komatsu and Wurtz, 1988; Thier and Erickson, 1992). Thus although Sakata *et al.* (1983) reported that some of the visual tracking neurons were activated while tracking targets approaching or receding in depth, the role of these neurons in tracking objects moving in 3D space is still unclear.

Acknowledgments

We thank Dr. Charles Duffy for his comments on this manuscript. This work was supported by the Human Frontier Science Program, the Japanese Agency of Industrial Science and Technology, and CREST of Japan Science and Technology Corporation.

References

Busettini, C., Masson, G. S., and Miles, F. A. (1996a). A role for stereoscopic depth cues in the rapid visual stabilization of the eyes. *Nature* **380,** 342–345.

Busettini, C., Masson, G. S., and Miles, F. A. (1997). Radial optic flow induces vergence eye movements with ultra-short latencies. *Nature* **390,** 512–518.

Busettini, C., Miles, F. A., and Krauzlis, R. J. (1994). Short-latency disparity vergence responses in humans. *Soc. Neurosci. Abstr.* **20,** 1403.

Busettini, C., Miles, F. A., and Krauzlis, R. J. (1996b). Short-latency disparity vergence responses and their dependence on a prior saccadic eye movement. *J. Neurophysiol.* **75,** 1392–1410.

Busettini, C., Miles, F. A., and Schwarz, U (1991). Ocular responses to translation and their dependence on viewing distance II. Motion of the scene. *J. Neurophysiol.* **66,** 865–878.

Cohen, B., Matsuo, V., and Raphan, T. (1977). Quantitative analysis of the velocity characteristics of optokinetic nystagmus and optokinetic after-nystagmus. *J. Physiol. (Lond.)* **270,** 321–344.

Duffy, C. J., and Wurtz, R. H. (1991). Sensitivity of MST neurons to optic flow stimuli. I. A continuum of response selectivity to large-field stimuli. *J. Neurophysiol.* **65,** 1329–1345.

Erickson, R. G., and Dow, B. M. (1989). Foveal tracking cells in the superior temporal sulcus of the macaque monkey. *Exp. Brain Res.* **78,** 113–131.

Ferrera, V. P., and Lisberger, S. G. (1997). Neuronal responses in visual areas MT and MST during smooth pursuit target selection. *J. Neurophysiol.* **78,** 1433–1446.

Hyvärinen, J., and Poranen, A. (1975). Function of the parietal associative area 7 as revealed from cellular discharges in alert monkeys. *Brain* **97,** 673–692.

Inoue, Y., Takemura, A., Kawano, K., Kitama, T., and Miles, F. A. (1998a). Dependence of short-latency ocular following and associated activity in the medial superior temporal area (MST) on ocular vergence. *Exp. Brain Res.* **121,** 135–144

Inoue, Y., Takemura, A., Suehiro, K., Kodaka, Y., and Kawano, K. (1998b). Short-latency vergence eye movements elicited by looming steps in monkeys. *Neurosci. Res.* **32,** 185–188.

Kawano, K., Shidara, M., Takemura, A., Inoue, Y., Gomi, H., and Kawato, M. (1996). Inverse-dynamics representation of eye movements by cerebellar Purkinje cell activity during short-latency ocular-following responses. *Ann. NY Acad. Sci.* **781,** 314–321.

Kawano, K., Shidara, M., Watanabe, Y., and Yamane, S. (1994). Neural activity in cortical area MST of alert monkey during ocular following responses. *J. Neurophysiol.* **71,** 2305–2324.

Komatsu, H., and Wurtz, R. H. (1988). Relation of cortical areas MT and MST to pursuit eye movements. I. Localization and visual properties of neurons. *J. Neurophysiol.* **60,** 580–603.

Lappe, M., Pekel, M., and Hoffmann, K.-P. (1998). Optokinetic eye movements elicited by radial optic flow in the macaque monkey. *J. Neurophysiol.* **79,** 1461–1480.

Maunsell, J. H. R., and van Essen, D. C. (1983). Functional properties of neurons in middle temporal visual area of the macaque monkey. II. Binocular interactions and sensitivity to binocular disparity. *J. Neurophysiol.* **49,** 1148–1167.

Miles, F. A., Schwarz, U., and Busettini, C. (1992). Decoding of optic flow by the primate optokinetic system. *In:* "The Head-Neck Sensory Motor System" (A. Berthoz, P. P. Vidal and W. Graf, Ed.), pp. 471–478. Oxford University Press, New York.

Miles, F. A. (1998). The neural processing of 3-D visual information: evidence from eye movements. *Eur. J. Neurosci.* **10,** 811–822.

Miles, F. A., and Kawano, K. (1986). Short-latency ocular following responses of monkey. III. Plasticity. *J. Neurophysiol.* **56,** 1381–1396.

Mountcastle, V. B., Lynch, J. C., Georgopoulos, A., Sakata, H., and Acuna, C. (1975). Posterior parietal association cortex of the monkey: Command functions for operations within extrapersonal space. *J. Neurophysiol.* **38,** 871–908.

Poggio, G. F., Gonzalez, F., and Krause, F. (1988). Stereoscopic mechanisms in monkey visual cortex: Binocular correlation and disparity selectivity. *J. Neurosci.* **8,** 4531–4550.

Saito, H., Yukie, M., Tanaka, K., Hikosaka, K., Fukada, Y., and Iwai, E. (1986). Integration of direction signals of image motion in the superior temporal sulcus of the macaque monkey. *J. Neurosci.* **6,** 145–157.

Sakata, H., Shibutani, H., and Kawano, K. (1983). Functional properties of visual tracking neurons in posterior parietal association cortex of the monkey. *J. Neurophysiol.* **49,** 1364–1380.

Schwarz, U., and Miles, F. A. (1991). Ocular responses to translation and their dependence on viewing distance. I. Motion of the observer. *J. Neurophysiol.* **66,** 851–864.

Shidara, M., Kawano, K., and Yamane, S. (1991). Ocular following response deficits with chemical lesions in the medial superior temporal area of the monkey. *Neurosci. Res.* **14,** S69.

Takemura, A., Inoue, Y., Kawano, K., and Miles, F. A. (1997). Short-latency discharges in medial superior temporal area of alert monkeys to sudden changes in the horizontal disparity. *Soc. Neurosci. Abstr.* **23,** 1557.

Takemura, A., Inoue, Y., Kawano, K., and Miles, F. A. (1998). Discharges in monkey cortical area MST related to initial ocular following: Dependence on horizontal disparity. *Soc. Neurosci. Abstr.* **24,** 1145.

Thier, P., and Erickson, R. G. (1992). Responses of visual-tracking neurons from cortical area MST-l to visual, eye and head motion. *Eur. J. Neurosci.* **4,** 539–553.

PART III

ANIMAL BEHAVIOR AND PHYSIOLOGY

VISUAL NAVIGATION IN FLYING INSECTS

Mandyam V. Srinivasan and Shao-Wu Zhang

Center for Visual Science, Research School of Biological Sciences, Australian National University, Canberra, Australia

I. Introduction

A glance at a fly evading a rapidly descending hand or orchestrating a flawless landing on the rim of a teacup would convince even the most skeptical observer that many insects are not only excellent fliers and navigators but also possess visual systems that are fast, reliable, precise, and exquisitely sensitive to optic flow.

Early studies of the analysis of optic flow by insects concentrated on the so-called optomotor response (rev. Reichardt, 1969). An insect, flying tethered inside a striped drum, will tend to turn in the direction in which the drum is rotated. If the drum rotates clockwise, the insect will generate a yaw torque in the clockwise direction and vice versa. This reaction helps the insect maintain a straight course by compensating for undesired deviations: a gust of wind that causes the insect to veer to the left, for example, would create rightward image motion on the eyes and cause the insect to generate a compensatory yaw to the right. Investigation of this so-called optomotor response over several decades has provided valuable information on some of the characteristics of motion perception by the insect visual system (Borst and Egelhaaf, 1989; Buchner, 1984; Reichardt, 1969).

0074-7742/00 $30.00

More recent studies, carried out primarily with freely flying honeybees, have revealed a number of additional contexts in which optic flow information is analyzed to coordinate flight and to obtain a useful percept of the world. It appears, for example, that bees analyze optic flow in a variety of different ways for negotiating narrow gaps, estimating the distances to objects, avoiding obstacles, controlling flight speed, executing smooth landings, and monitoring distance traveled. Here we describe some of these strategies and attempt to elucidate the properties of the underlying motion-sensitive mechanisms.

Unlike vertebrates, insects have immobile eyes with fixed-focus optics. Therefore, they cannot infer the distance of an object from the extent to which the directions of gaze must converge to view the object or by monitoring the refractive power that is required to bring the image of the object into focus on the retina. Furthermore, compared with human eyes, the eyes of insects are positioned much closer together and possess inferior spatial acuity. Therefore, even if an insect possessed the neural apparatus required for binocular stereopsis, such a mechanism would be relatively imprecise and restricted to ranges of a few centimeters (Collett and Harkness, 1982; Horridge, 1987, Rossell, 1983; Srinivasan, 1993). Not surprisingly, insects have evolved alternative visual strategies for guiding locomotion and for "seeing" the world in three dimensions. Many of these strategies rely on using optic flow as the significant cue. Some of these strategies are outlined here, and references to more complete accounts are provided.

II. Peering Insects

Over a hundred years ago, Exner (1891), pondering the eyestalk movements of crabs, speculated that invertebrates might use image motion to estimate object range. However, the first clear evidence to support this conjecture did not arrive until the middle of the present century, when Wallace (1959) made the astute observation that a locust sways its head from side to side before jumping on to a nearby object (Fig. 1a). Wallace hypothesized that this "peering" motion, typically 5–10 mm in amplitude, was a strategy for measuring object range. To test this hypothesis, he presented a locust with two objects subtending the same visual angle. One object was relatively small in size and was placed close to the locust, whereas the other was larger and situated farther away. He found that the locust, after peering, jumped almost invariably to the nearer object. In a further series of elegant experiments, recently confirmed more quantitatively by Sobel (1990), a target was oscillated

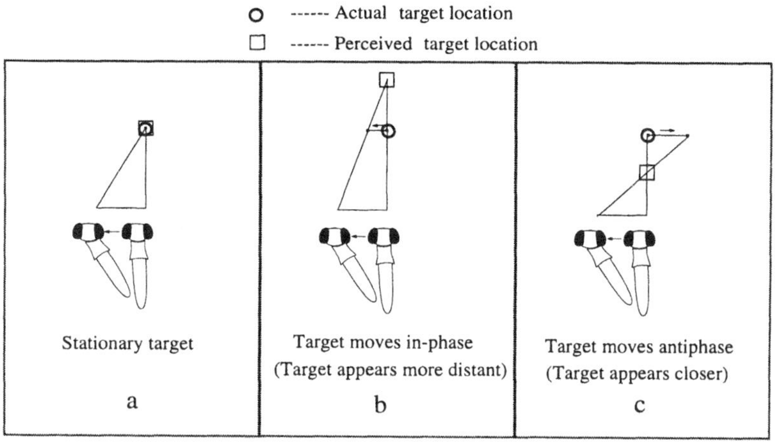

FIG. 1. Experiments investigating how locusts measure the range of a target by peering (i.e., moving the head from side to side). Range is estimated correctly when the target is stationary (a), overestimated when the target is moved in the same direction as the head (b), and underestimated when it is moved in the opposite direction (c). Thus, the range of the target is estimated in terms of the motion of the target's image during the peer. Adapted from *J. Comp. Physiol.*, The locust's use of motion parallax to measure distance, Sobel, E. C., **167,** 579–588, Figs. 3 and 4, 1990, © Springer-Verlag.

from side to side, in synchrony with the insect's peering movements. When the target was oscillated out of phase with the movement of the head, thereby increasing the speed and amplitude of the object's image on the retina, the locust consistently underestimated the range of the target (Fig. 1c); when the target was oscillated in phase with the head, it consistently overestimated the range (Fig. 1b). This showed that the reduced image motion of the target caused the insect to overestimate the target's range, whereas increased motion had the opposite effect. These findings demonstrated convincingly that the peering locust was estimating the range of the target in terms of the speed of the image on the retina. It is now known that certain other insects such as grasshoppers (Eriksson, 1980) and mantids (Horridge, 1986; Kral, 1998; Kral and Poteser, 1997; Poteser *et al.*, 1998) also use peering to measure object range.

III. Flying Insects

Peering, however, is practicable only when an insect is not locomoting. Are flying insects capable of gleaning range information from image motion, and if so, how do they accomplish this? Stable flight in a straight line would seem to be a prerequisite for extracting information on range (Horridge, 1987; Srinivasan, 1993). Research over the past 50 years have uncovered a number of different ways in which insects use image motion to stabilize flight control and to extract useful information about the environment. We shall begin by considering strategies for visual control and stabilization of flight and then proceed to examine the ways in which optic flow is used to glean information about the structure of the environment.

A. Stabilizing Flight

For insects, vision provides an important sensory input for the stabilization of flight. If an insect flying along a straight line is blown to the left by a gust of wind, the image on its frontal retina moves to the right. This causes the flight motor system to generate a counteractive yaw torque, which brings the insect back on course (rev. Reichardt, 1969). Similar control mechanisms act to stabilize pitch and roll (e.g., Srinivasan, 1977). This so-called optomotor response (Reichardt, 1969), has provided an excellent experimental paradigm in which to probe the neural mechanisms underlying motion detection. Largely through studies of the optomotor response in flies, we now know that the direction of image movement is sensed by correlating the intensity variations registered by neighboring ommatidia, or facets, of the compound eye (rev. Reichardt, 1969). Research over the past 30 years has uncovered the existence of a number of motion-sensitive neurons with large visual fields, each responding preferentially to motion in a specific direction (rev. Hausen and Egelhaaf, 1989; Hausen, 1993) or to rotation of the fly about a specific axis (Krapp and Hengstenberg, 1996; Krapp, this volume). These neurons are likely to play an important role in stabilizing flight and providing the fly with a visually "kinaesthetic" sense. Their properties have been reviewed extensively (e.g., Egelhaaf and Borst, 1993; Hausen and Egelhaaf, 1989; Hausen, 1993; Krapp, this volume), and we shall not repeat this here.

B. Hovering

Hoverflies and certain species of bee display an impressive ability to hold a rigid position in midair, compensating almost perfectly for wind gusts and other disturbances. Kelber and Zeil (1997) recently investigated hovering in a species of stingless bee, *Tetragonisca angustula.* Guard bees of this species hover stably in watch near the entrance to their nest, protecting it from intruders. To investigate the visual stabilizing mechanisms, Kelber and Zeil got the bees used to the presence of a spiral pattern mounted on the vertical face of the hive, surrounding the entrance. When the spiral was briefly rotated to simulate expansion, the hovering guard bees darted away from the focus of apparent expansion; when the spiral was rotated to simulate contraction, they moved toward the focus of contraction. These responses were always directed toward or away from the nest entrance, irrespective of the bee's orientation, and therefore irrespective of the region of the eye that experienced the experimentally imposed pattern of image motion. Clearly, then, these creatures were interpreting expansion and contraction of the image as unintended movements toward or away from the nest entrance, and compensating for them.

C. Negotiating Narrow Gaps

When a bee flies through a hole in a window, it tends to fly through its center, balancing the distances to the left and right boundaries of the opening. How does it gauge and balance the distances to the two rims?

One possibility is that it does not measure distances at all but simply balances the speeds of image motion on the two eyes as it flies through the opening. To investigate this possibility, Kirchner and Srinivasan (1989) trained bees to enter an apparatus that offered a reward of sugar solution at the end of a tunnel. Each side wall carried a pattern consisting of a vertical black-and-white grating (Fig. 2). The grating on one wall could be moved horizontally at any desired speed, either toward the reward or away from it. After the bees had received several rewards with the gratings stationary, they were filmed from above as they flew along the tunnel. When both gratings were stationary, the bees tended to fly along the midline of the tunnel (i.e., equidistant from the two walls, Fig. 2a). But when one of the gratings was moved at a constant speed in the direction of the bees' flight—thereby reducing the speed of retinal image motion on that eye relative to the other eye—the bees' trajectories shifted toward the side of the moving grating (Fig. 2b). When the grating moved in a direction opposite to that of the bees' flight—thereby increasing the speed of retinal image motion on that eye relative to the other—the bees' trajectories shifted away from the side of the moving grating (Fig. 2c). These findings demonstrate that when the walls were stationary, the bees maintained equidistance by balancing the speeds of

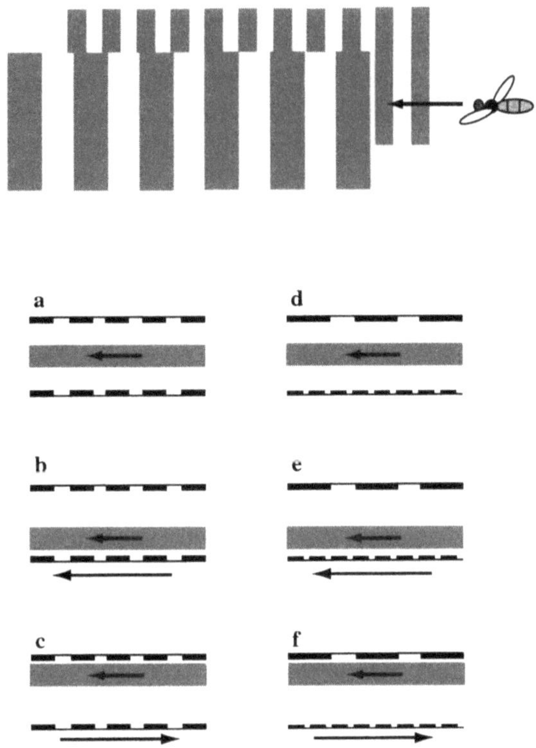

FIG. 2. Experiment investigating how bees fly through the middle of a tunnel (the centering response). Bees are trained to fly through a tunnel (40 cm long, 12 cm wide, and 20 cm high) to collect a reward placed at the far end. The flanking walls of the tunnel are lined with vertical black-and-white gratings of period 5 cm. The flight trajectories of bees, as recorded a by video camera positioned above the tunnel (a–f). In each panel, the shaded area represents the mean and standard deviation of the positions of the flight trajectories, analyzed from recordings of several hundred flights. The dark bars represent the black stripes of the patterns on the walls. The small arrow indicates the direction of bee flight, and the large arrow indicates the direction of pattern movement, if any. When the patterns on the walls are stationary, bees tend to fly close to the midline of the tunnel (a,d). When the pattern on one of the walls is in motion, however, bees tend to fly closer to that wall if the pattern moves in the same direction as the bee (b,e) and farther away from that wall if the pattern moves in the opposite direction (c,f). These results indicate that bees balance the distances to the walls of the tunnel by balancing the speeds of image motion that are experienced by the two eyes, and that they are able to measure image speed rather independently of the spatial structure of the image. Modified from Srinivasan *et al.* (1991), Range perception through apparent image speed in freely flying honeybees. *Visual Neuroscience* **6,** 519–535, Cambrige University Press.

the retinal images in the two eyes. A lower image speed on one eye evidently caused the bee to move closer to the wall seen by that eye. A higher image speed, on the other hand, had the opposite effect.

Were the bees really measuring and balancing image speeds on the two sides as they flew along the tunnel, or were they simply balancing the contrast frequencies produced by the succession of dark and light bars of the gratings? This question was investigated by analyzing the flight trajectories of bees when the two walls carried gratings of different spatial periods. When the gratings were stationary, the trajectories were always equidistant from the two walls, even when the spatial frequencies of the gratings on the two sides—and therefore the contrast frequencies experienced by the two eyes—differed by a factor of as much as four (Fig. 2d). When one of the gratings was in motion, the trajectories shifted toward or away from the moving grating (as described earlier) according to whether the grating moved with or against the direction of the bees' flight (Figs. 2e,f). These results indicate that the bees were indeed balancing the speeds of the retinal images on the two eyes and not the contrast frequencies. These findings are true irrespective of whether the gratings possess squarewave intensity profiles (with abrupt changes of intensity) or sinusoidal profiles (with gradual intensity changes) and irrespective of whether the contrasts of the gratings on the two sides are equal or considerably different (Srinivasan *et al.*, 1991). Further experiments have revealed that—knowing the velocities of the bee and the pattern—it is even possible to predict the position of a bee's flight trajectory along the width of the tunnel, on the assumption that the bee balances the apparent angular velocities on either side of the tunnel (Srinivasan *et al.* 1991). These findings suggest that the bee's visual system is capable of computing the apparent angular speed of a grating independently of its contrast and spatial-frequency content.

More recent studies (Srinivasan *et al.*, 1993; Srinivasan and Zhang, 1997) have investigated this "centering" response further by comparing its properties with those of the well-known optomotor response in an experimental setup which allows the two responses to be compared in one and the same individual under the same conditions. The results indicate that even though the optomotor response is mediated by a movement-detecting system that is direction-sensitive, the centering response is driven by a movement-detecting system that is direction-*insensitive*. Thus, for eliciting a centering response, the image need not necessarily move backward on the eye; an image moving vertically upward, downward, or forward at the same speed has the same effect. In particular, rapid movement of the pattern on one of the walls in the same direction as the flying bee has the same effect as slow movement in the opposite direction (Srinivasan *et al.*, 1991). The results of these studies also reveal that the centering response is sensitive to higher temporal frequencies than is the optomotor response. While the optomotor response exhibits a peak in the vicinity of 25–50 Hz and drops to zero at 100 Hz, the strength of the centering response is approximately constant over the range of 25–100 Hz and is substantial at 100 Hz (Srinivasan *et al.*, 1993; Srinivasan and Zhang, 1997). The centering response may be related to the so-called movement avoidance response, a tendency shown by bees to avoid flying toward rapidly moving objects (Srinivasan and Lehrer, 1984; Srinivasan and Zhang, 1997). The movement avoidance response, like the centering response, is sensitive to a broad range of temporal frequencies. At low temporal frequencies (1–20 Hz), the movement avoidance response depends primarily on image speed; at relatively high frequencies (50–200 Hz), the response depends primarily on temporal frequency (see Srinivasan and Lehrer, 1984, Fig. 5; Srinivasan and Zhang, 1997). It is possible that the centering response that bees exhibit while flying through a tunnel is the result of equal and oppositely directed movement avoidance responses generated by the optic flow experienced by the two eyes (Srinivasan and Zhang, 1997). These studies have also shown that the movement-detecting system that underlies the centering response computes motion within receptive fields whose diameter is no larger than ca. 15° (Srinivasan *et al.*, 1993; Srinivasan and Zhang, 1997).

To summarize, the centering response differs from the optomotor response in three respects. First, the centering response is sensitive primarily to the *angular speed* of the stimulus, regardless of its spatial structure. The optomotor response, on the other hand, is sensitive primarily to the *temporal frequency* of the stimulus; therefore, it confounds the angular velocity of a striped pattern with its spatial period. Secondly, the centering response is *nondirectional*, whereas the optomotor response is *directionally selective.* (It is worth noting, however, that nondirectional motion detection can be achieved by summing the outputs of directionally selective motion detectors with opposite preferred directions.) Thirdly, the centering response is sensitive to higher temporal frequencies than is the optomotor response. Thus, the motion-detecting processes underlying the centering response exhibit properties that are substantially different from those that mediate the optomotor response (Srinivasan *et al.*, 1993; Srinivasan and Zhang, 1997). Models of movement-detecting mechanisms underlying the centering response are described in Srinivasan *et al.* (1999).

Given that the role of the centering response is to ensure that the insect flies through the middle of a gap irrespective of the texture of the side walls, it is easy to see why the response is mediated by a movement-detecting system which measures the angular speed of the image independently of its spatial structure. The movement-detecting system that subserves the optomotor response, on the other hand, does not need to measure image speed in a robust way; it merely needs to signal the *direction* of image motion reliably so that a corrective yaw of the appropriate polarity may be generated.

Why is the centering mechanism sensitive only to the *speed* of the image and not to direction in which the image moves? We can think of two reasons. First, in neural terms, it may be much simpler to build a nondirectional speed detector than to build a detector that computes speed as well as direction of motion. In straight-ahead flight, the direction of image motion along each viewing direction is predetermined (Gibson, 1979; Wehner, 1981, Fig. 5, p. 325) and therefore does not need to be computed. It is the local *speed* that conveys information on range. The insect visual system may thus be adopting a "shortcut" which takes advantage of the fact that the optic flow experienced in straight-ahead flight is constrained in special ways. Second, a nondirectional speed detector offers a distinct advantage over a detector that measures speed along a given axis: the latter can produce large spurious responses when the orientation of an edge is nearly parallel to the detector's axis. For example, a detector configured to measure speed along the horizontal axis will register large horizontal velocities if it is stimulated by a near-horizontal edge moving in the vertical direction. This "obliquity" problem can be avoided by using either a two-dimensional velocity detector, or a nondirectional speed detector—of which the latter offers a simpler, more elegant solution (Srinivasan, 1992).

D. Controlling Flight Speed

Do insects control the speed of their flight, and, if so, how? Work by David (1982) and by Srinivasan *et al.* (1996) suggests that flight speed is controlled by monitoring the velocity of the image of the environment.

David (1982) observed fruit flies flying upstream in a wind tunnel, attracted by an odor of fermenting banana. The walls of the cylindrical wind tunnel were decorated with a helical black-and-white striped pattern so that rotation of the cylinder about its axis produced apparent movement of the pattern toward the front or the back. With this setup, the rotational speed of the cylinder (and hence the speed of the backward motion of the pattern) could be adjusted such that the fly was stationary (i.e., did not move along the axis of the tunnel). The apparent backward speed of the pattern then revealed the ground speed that the fly was "choosing" to maintain, as well as the angular velocity of the image of the pattern on the flies' eyes. In this setup, fruit flies tended to hold the angular velocity of the image constant. Increasing or decreasing the speed of the pattern caused the fly to move backward or forward (respectively) along the tunnel at a rate such that the angular velocity of the image on the eye was always "clamped" at a fixed value. The flies also compensated for headwind in the tunnel, increasing or decreasing their thrust so as to maintain the same apparent ground speed (as indicated by the angular velocity of image motion on the eye). Experiments in which the angular period of the stripes was varied revealed that the flies were measuring (and holding constant) the angular velocity of the image on the eye, irrespective of the spatial structure of the image.

Bees appear to use a similar strategy to regulate flight speed (Srinivasan *et al.*, 1996). When a bee flies through a tapered tunnel, it decreases its flight speed as the tunnel narrows so as to keep the angular velocity of the image of the walls, as seen by the eye, constant at about 320°/s (Fig. 3). This suggests that flight speed is controlled by monitoring and regulating the angular velocity of the image of the environment on the eye. (That is, if the width of the tunnel is doubled, the bee flies twice as fast.) On the other hand, a bee flying through a tunnel of uniform width does not change her speed when the spatial period of the stripes lining the walls is abruptly changed (Fig. 4). This indicates that flight speed is regulated by a visual motion-detecting mechanism which measures the angular velocity of the image largely independently on its spatial structure. In this respect, the speed-regulating system is similar to the centering system. However, it is not yet known whether the regulation of flight speed in bees is mediated by a directionally selective movement-detecting mechanism, or a nondirectional one. Visual stimulation of tethered flies with forward- or backward-moving gratings in the two eyes indicates that flight thrust (which is related to flight speed) is controlled by directionally selective movement detectors (Götz, 1989; Götz and Wandel, 1984; Götz and Wehrhahn, 1984).

An obvious advantage of controlling flight speed by regulating image speed is that the insect would automatically slow down to a safer speed when negotiating a narrow passage. A by-product of this mode of operation is that the speed of flight is significantly higher when the optic-flow cues provided by the environment are weak or absent. Thus, bees flying in a tunnel tend to fly considerably faster when the walls are lined with axial stripes, rather than with cross stripes or a random Julesz texture (Srinivasan *et al.*, 1997, and unpublished observations).

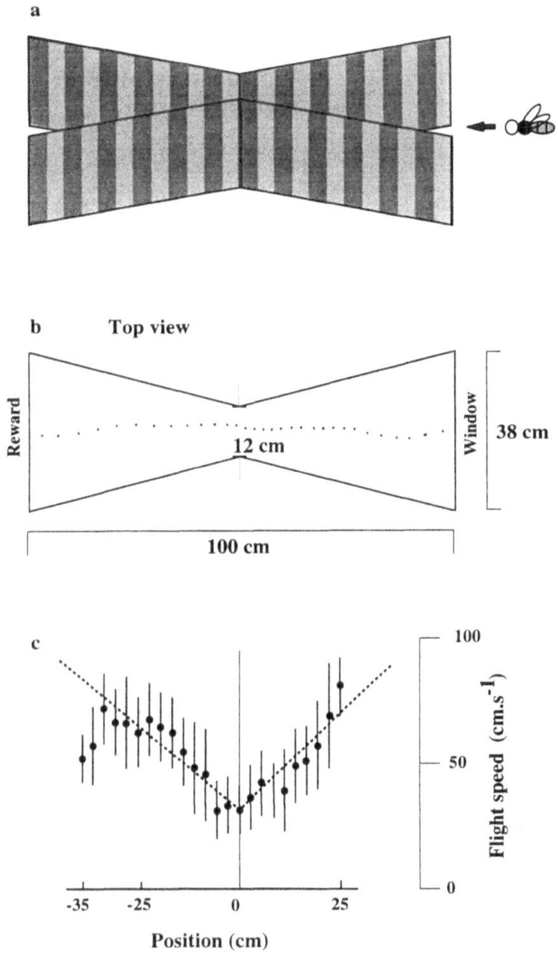

FIG. 3. Experiment investigating visual control of flight speed. (a) Bees are trained to fly through a tapered tunnel to collect a reward placed at the far end. The walls of the tunnel are lined with vertical black-and-white gratings of period 6 cm. (b) A typical flight trajectory, as filmed from above by a video camera, where the bee's position and orientation are shown every 50 ms. (c) Mean and standard deviations of flight speeds measured at various positions along the tunnel (data from 18 flights). The dashed line shows the theoretically expected flight speed profile if the bees were to hold the angular velocity of the images of the walls constant at 320°/s as they fly through the tunnel. The data indicate that bees control flight speed by holding constant the angular velocity of the image of the environment. Adapted from Srinivasan *et al.* (1996), Company of Biologists Ltd.

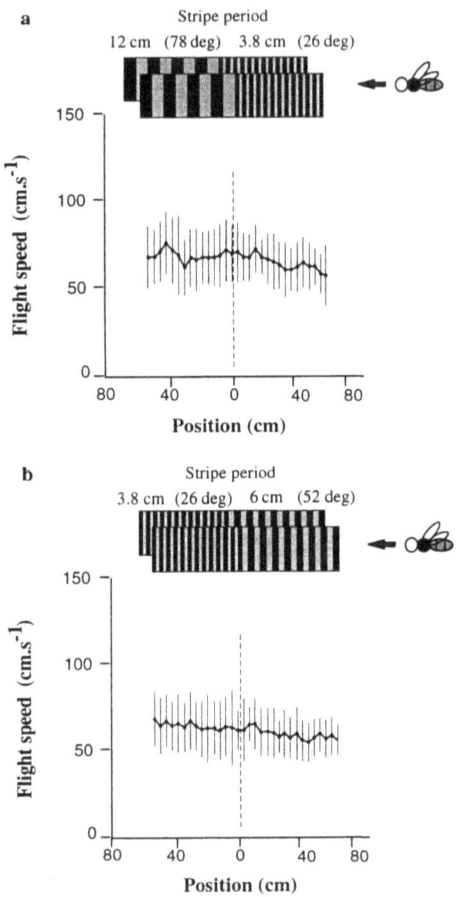

FIG. 4. Two experiments (a and b) examining control of flight speed in tunnels of constant width, each lined with black-and-white gratings whose spatial period changes abruptly in the middle. In each panel, the upper illustration shows the tunnel and the patterns, and the lower illustration shows measurements of flight speed at various positions along the tunnel (0 cm represents the position at which the spatial period changes). Bees flying through such tunnels maintain a nearly constant flight speed regardless of whether the period increases (a) or decreases (b). This suggests that the speed of flight is controlled by a movement-detecting system which measures and holds constant the speeds of the images of the walls accurately regardless of their spatial structure. Data represent mean and standard deviation of 18 flights in (a) and 21 flights in (b). Adapted from Srinivasan *et al.* (1996), Company of Biologists Ltd.

E. Estimating Distance Flown

It is well known that honeybees can navigate accurately and repeatedly to a food source. It is also established that bees communicate to their nestmates the distance and direction in which to fly to reach it, through the "waggle dance" (von Frisch, 1993). But the cues by which bees gauge the distance flown to the goal have been a subject of controversy.

A few years ago, Esch and Burns (1995, 1996) investigated distance measurement by enticing honeybees to find food at a feeder placed 70 m away from a hive in an open field and recording the distance as signaled by the bees when they danced to recruit other nestmates in the hive. When the feeder was 70 m away, the bees signaled 70 m—the correct distance. But when the feeder was raised above the ground by attaching it to a helium balloon, the bees signaled a progressively *shorter* distance as the height of the balloon was increased. This was despite the fact that the balloon was now *farther* away from the hive! Esch and Burns explained this finding by proposing that the bees were gauging distance flown in terms of the motion of the image on the ground below, rather than, for example, through the energy consumed to reach the feeder. The higher the balloon, the lower was the total amount of image motion that the bee experienced en route to the feeder.

This hypothesis was examined by Srinivasan *et al.* (1996, 1997) who investigated the cues by which bees estimate and learn distances flown under controlled laboratory conditions. Bees were trained to enter a 3.2 m long tunnel and collect a reward of sugar solution at a feeder placed in the tunnel at a fixed distance from the entrance. The walls and floor of the tunnel were lined with black-and-white gratings perpendicular to the tunnel axis (Fig. 5a). During training, the position and orientation of the tunnel were changed frequently to prevent the bees from using any external landmarks to gauge their position relative to the tunnel entrance. The bees were then tested by recording their searching behavior in an identical, fresh tunnel which carried no reward and was devoid of any scent cues. In the tests, these bees showed a clear ability to search for the reward at the correct distance, as indicated by the search distribution labeled by the squares in Fig. 5b.

How were the bees gauging the distance they had flown in the tunnel? Tests were carried out to examine the participation of a variety of potential cues, including energy consumption, time of flight, airspeed integration, and inertial navigation (Srinivasan *et al.*, 1997). It turned out that the bees were estimating distance flown by integrating, over

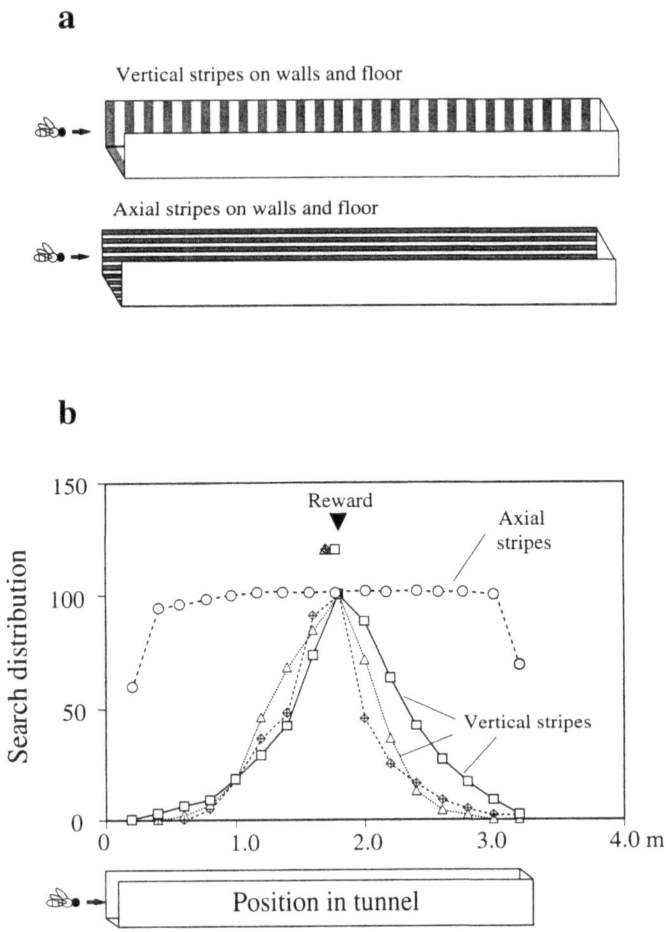

FIG. 5. (a) Experiment investigating how honeybees gauge distance flown to a food source. Bees are trained to find a food reward placed at a distance of 1.7 m from the entrance of a 3.2 m long tunnel of width 22 cm and height 20 cm. The tunnel is lined with vertical black-and-white gratings of period 4 cm. (b) When the trained bees are tested in a fresh tunnel with the reward absent, they search at the former location of the feeder, as shown by the bell-shaped search distributions. This is true irrespective of whether the period of the grating is 4 cm (as in the training, square symbols), 8 cm (triangles), or 2 cm (diamonds). The inverted triangle shows the former position of the reward, and the symbols below it depict the mean values of the search distributions in each case. Bees lose their ability to estimate the distance of the feeder when image-motion cues are removed by lining the tunnel with axial (rather than vertical) stripes (circles). These experiments and others (Srinivasan *et al.*, 1997) demonstrate that (i) distance flown is estimated visually by integrating over time the image velocity that is experienced during the flight, and (ii) the honeybee's odometer measures image velocity independently of image structure. Adapted from Srinivasan *et al.* (1997), Company of Biologists Ltd.

time, the motion of the images of the walls on the eyes as they flew down the tunnel. The crucial experiment was one in which bees were trained and tested in conditions where image motion was eliminated or reduced by using axially oriented stripes on the walls and floor of the tunnel. The bees then showed no ability to gauge distance traveled: in the tests, they searched uniformly over the entire length of the tunnel, showing no tendency to stop or turn at the former location of the reward (see the search distribution identified by the circles, Fig. 5b). Trained bees tended to search for the feeder at the same position in the tunnel, even if the period of the gratings lining the walls and floor was varied in the tests (search distributions identified by triangles and diamonds, Fig. 5b). This indicates that the odometric system reads image velocity accurately over a fourfold variation in the spatial period of the grating.

These results, considered together with those of Esch and Burns (1995, 1996), indicate that the bee's "odometer" is driven by the image motion that is generated in the eyes during translatory flight. Evidently, bees use optic flow information not only to stabilize flight and regulate its speed but also to infer how far they have flown.

What are the consequences of measuring distance traveled by integrating optic flow? One consequence would be that errors in the measurement and integration of image speed accumulate with distance so that larger distances would be estimated with greater error. To test this prediction, Srinivasan *et al.* (1997) examined the accuracy with which bees were able to localize a feeder when it was placed at various distances along a tunnel. The results (Fig. 6) show that the width of the search distribution indeed increases progressively with the distance of the feeder from the tunnel entrance. Thus, the error in estimating distance increases with distance flown, as would be expected of an integrative mechanism.

An integrative mechanism for measuring distance traveled would be feasible only if the cumulative errors are somehow prevented from exceeding tolerable levels. One strategy, which could be employed when traversing familiar routes, would be to recommence the integration of image motion whenever a prominent, known landmark is passed. Do bees adopt such a tactic?

To investigate this, Srinivasan *et al.* (1997) examined the bees' performance when they were again trained to fly to a feeder placed at a large distance into a tunnel (Fig. 6), but now had to pass a prominent landmark (a baffle consisting of a pair of overlapping partitions) occurring en route to the feeder. If these bees reset their odometer at the landmark, they should display a smaller error because they would then only need to measure the distance between the landmark and the feeder.

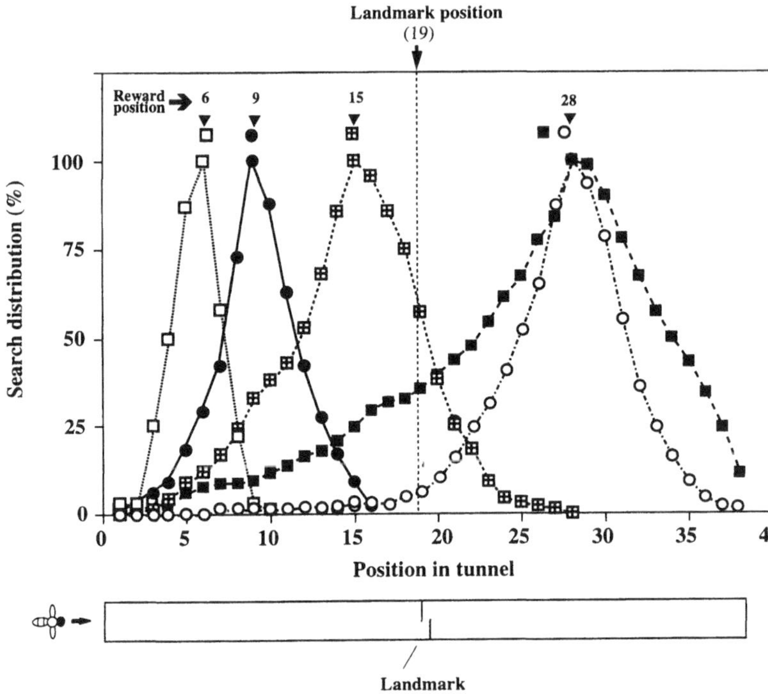

FIG. 6. Experiments investigating how odometric error varies with distance traveled. Bees are trained in a 7.2 m long tunnel to find a feeder placed at a certain distance from the entrance. The tunnel is 22 cm wide and 20 cm high, and lined with a random Julesz texture of pixel size 1 cm × 1 cm. The trained bees are then tested in a fresh tunnel, with the feeder removed. When the feeder is placed close to the entrance in the training (e.g., at position 6), the search distribution in the tests is relatively narrow (open squares). The distribution becomes progressively wider as the training distance is increased (position 9: filled circles; position 15: checkered squares: position 28: filled squares), and is widest when the feeder is at position 28. This cumulative error in the odometer is consistent with that expected of an integrative process of distance measurement. But when bees are trained to position 28 with a prominent landmark (a baffle consisting of a pair of overlapping partitions) introduced at position 19, the search distribution (open circles) is significantly narrower. This indicates that the trained bees are resetting their odometer at the landmark and searching for the feeder at a position 9 units beyond it. The width of the search distribution in this case is comparable to that obtained when the feeder is placed at a distance of 9 units from the tunnel entrance (filled circles). Adapted from Srinivasan *et al.* (1997), Company of Biologists, Ltd.

This is precisely what occurred: the search distribution was then significantly narrower (open circles, Fig. 6). Furthermore, when the trained bees were confronted with a test in which the landmark was positioned closer to the tunnel entrance, the bees' mean search position shifted toward the entrance by almost exactly the same distance (Srinivasan *et al.*,

1997). These results confirm that bees recommence computation of distance when they pass a prominent landmark, and that such landmarks are used to enhance the accuracy of the odometer.

F. Executing Smooth Landings

How does a bee execute a smooth touchdown on a surface? An approach that is perpendicular to the surface would generate strong looming (image expansion) cues which could, in principle, be used to decelerate flight at the appropriate moment. Indeed, work by Wagner (1982) and Borst and Bahde (1988) has shown that deceleration and extension of the legs in preparation for landing are triggered by movement-detecting mechanisms that sense *expansion* of the image (but not contraction).

Looming cues are weak, however, when a bee performs a grazing landing on a surface. By "grazing landings" we mean landings whose trajectories are inclined to the surface at an angle that is considerably less than 45°. In such landings, the motion of image of the surface would be dominated by a strong translatory component in the front-to-back direction in the ventral visual field of the eye.

To investigate how bees execute grazing landings. Srinivasan *et al.* (1996) trained bees to collect a reward of sugar water on a textured, horizontal surface. The reward was then removed and the landings that the bees made on the surface in search of the food were video-filmed in three dimensions.

Analysis of the landing trajectories revealed that the forward speed of the bee decreases steadily as the bee approaches the surface (Fig. 7). In fact, the speed of flight is approximately proportional to the height above the surface, indicating that the bee is holding the angular velocity of the image of the surface approximately constant as the surface is approached. This may be a simple way of controlling flight speed during landing, ensuring that its value is close to zero at touchdown. The advantage of such a strategy is that the control is achieved by a very simple process, and without explicit knowledge of the distance from the surface.

G. Distinguishing Objects at Different Distances

The experiments described earlier show that bees stabilize flight, negotiate narrow passages, and orchestrate smooth landings by using what seem to be a series of simple, low-level visual reflexes. But they do not tell us whether flying bees "see" the world in three dimensions in the

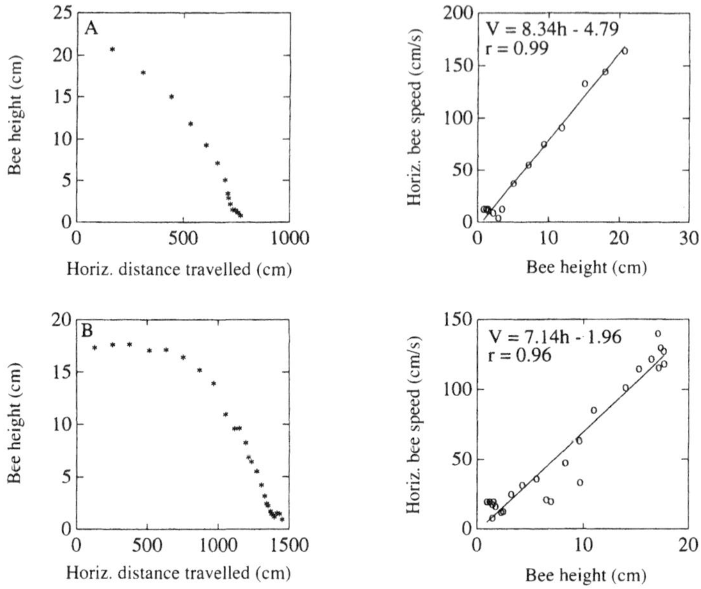

FIG. 7. Experiment investigating how bees make a grazing landing on a horizontal surface. Analyses of landing trajectories are shown for two bees (A) and (B). In each case, the left-hand panel shows the relationship between height (*h*) and horizontal distance traveled, whereas the right-hand panel shows the relationship between horizontal flight speed (*V*) and height (*h*). The landing bee holds the angular velocity of the image of the ground constant at 480°/s in (A) and at 410°/s in (B), as calculated from the slopes of the linear regression lines. Also shown are the values of the correlation coefficient (*r*). Holding the angular velocity of the image of the ground constant during the approach automatically ensures that the landing speed is zero at touchdown. From Srinivasan *et al.* (1996), Company of Biologists Ltd.

way we do. Do bees perceive the world as being composed of objects and surfaces at various ranges? Although this is a difficult question—one that a philosopher might even declare unanswerable—one can at least ask whether bees can be trained to distinguish between objects at different distances. Lehrer *et al.* (1988) trained bees to fly over an artificial "meadow" and distinguish between artificial "flowers" at various heights. The training was carried out by associating a reward with a flower at a particular height. The sizes and positions of the flowers were varied randomly and frequently during the training. This ensured that the bees were trained to associate only the height of the flower (or, more accurately, the distance from the eye) and not its position, or angular subtense, with the reward. Using this approach—details of which are described in Srinivasan *et al.* (1989)—it was possible to train bees to choose either the highest flower, the lowest flower, or even one at intermediate height. Clearly, then, the bees were able to distinguish flowers at different heights. Under the experimental conditions, the only cue that a bee could have used to gauge the height of each flower would be the speed of the flower's image as she flew over it: the taller the flower, the faster the motion of its image. To test if the bees were really using optic flow information to infer flower height, Lehrer and Srinivasan (1992) repeated the same experiment with the flowers presented inside a drum whose inside was lined with vertical stripes (Fig. 8a). Bees trained in this apparatus lost the ability to discriminate height when the drum was rotated. Evidently, the optomotor response evoked by the rotating drum caused the bees to turn with the drum and, therefore, disrupted the relationship between the range of each object and the speed of its image on the retina. This experiment demonstrates that optic-flow cues are crucial to discriminating range. Rotation of the drum does not affect the bees' capacity to perform other kinds of visual tasks, such as color discrimination (Fig. 8b). Therefore, the stimulus created by the rotating drum does not "confuse" the bees in a nonspecific way: it only disrupts cues that rely on image motion. These experiments suggest that bees indeed use cues based on image motion to distinguish among objects at different distances.

Kirchner and Lengler (1994) extended the basic 'meadow' experiment by training bees to distinguish the heights of artificial flowers that carried spiral patterns. Six flowers were presented at the same height, while a seventh was either higher (in one training experiment) or lower (in another experiment). Bees trained in this way were tested with a constellation of three identical spiral-bearing flowers of the same height. One test flower was stationary, one was rotated to stimulate expansion, and the other was rotated to simulate contraction. Bees that had learned to find the higher flower in the training chose the "expanding" flower in the test, whereas bees that had learned to choose the lower flower in the training chose the "contracting" flower. For a bee flying above the flowers and approaching the edge of one of them, the expanding flower produced a higher image motion at its boundary than did the stationary one and was evidently interpreted to be the higher flower. The contracting flower, on the other hand, produced a lower image motion and was therefore taken to be the lower one. This experiment confirms the notion that optic flow is an important cue in establishing the relative distances of objects.

H. Discriminating Objects from Backgrounds

In all the work described earlier, the objects that were being viewed were readily visible to the insects, since they presented a strong contrast—in luminance or color—against a structureless background. What

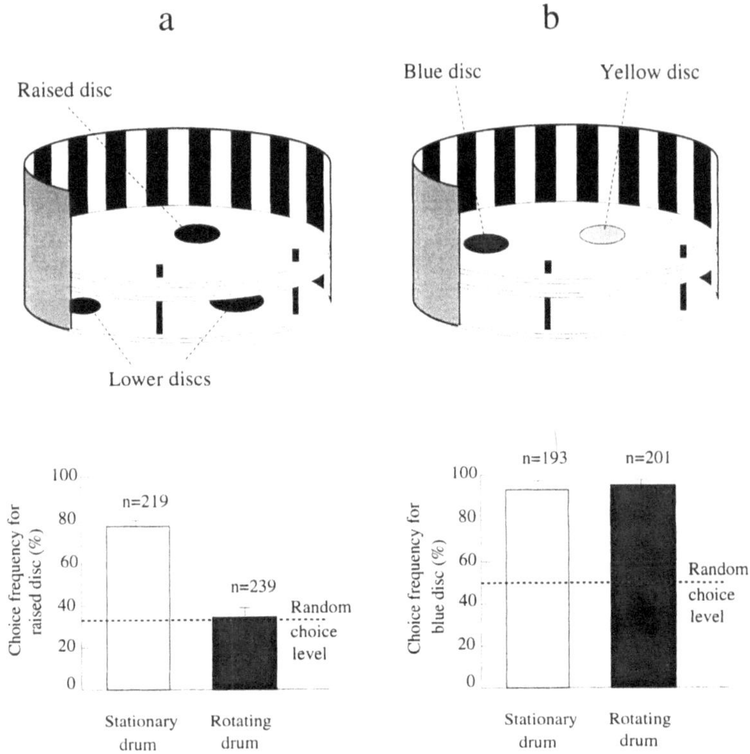

FIG. 8. Experiment demonstrating that bees use image-motion cues to discriminate between objects at different distances. (a) Bees can be trained, by food reward, to discriminate between the raised disc and the two lower ones and always land above the raised disc. Trained bees exhibit a choice frequency of nearly 80% for the raised disc (light bar), which is significantly higher than the expected random-choice level of 33% (dashed line). But when the striped drum is rotated, the resulting optomotor response causes the bees to turn with the drum, thus disrupting the relationship between the depths of the discs and the speeds of their images in the eye. The bees are then unable to discriminate the discs and choose randomly between them (dark bar). (b) Control experiment demonstrating that rotation of the drum selectively disrupts depth discrimination. Bees can be readily trained to distinguish between a blue disc and a yellow disc, both presented at the same height, by associating the blue disc with a reward of sugar water (light bar). This discriminatory capacity is not destroyed when the drum is rotated (dark bar). The choice frequency in each case is significantly greater than the random-choice level of 50% (dashed line). Thus, rotation of the drum selectively disrupts the bees' ability to discriminate depth, indicating that optic-flow information is crucial to this task. *n* denotes the number of choices analyzed. Figure redrawn from *J. Comp. Physiol.*, Freely flying bees discriminate between stationary and moving objects: Performance and possible mechanisms, Lehrer, M., and Srinivasan, M. V., **171,** 457–467, Figs. 8–10, 1992, © Springer-Verlag.

happens if the luminance or color contrast is removed and replaced by 'motion contrast'? To the human eye, a textured figure is invisible when it is presented motionless against a similarly textured background. But the figure "pops out" as soon as it is moved relative to the background. This type of relative motion, termed motion parallax, can be used to distinguish a nearby object from a remote background. Is an insect capable of distinguishing a textured figure from a similarly textured background purely on the basis of motion parallax?

In a series of pioneering experiments, Reichardt and his colleagues in Tübingen showed that a fly is indeed capable of such figure–ground discrimination (Reichardt and Poggio, 1979; Egelhaaf *et al.*, 1988). A tethered, flying fly will show no sign of detecting a textured figure when the figure oscillates in synchrony with a similarly textured background. But it will react to the figure by turning toward it when the figure moves incoherently with respect to the background.

In antipodean Canberra, this question was approached in a different way. Srinivasan *et al.* (1990) examined whether freely flying bees could be trained to find a textured figure when it was presented raised over a background of the same texture. The figure was a disc, bearing a random black-and-white Julesz texture, carried on the underside of a transparent perspex sheet which could be placed at any desired height above the background (Fig. 9a). It was found that bees could indeed be trained to find the figure and land on it, provided the figure was raised at least 1 cm above the background (Fig. 9b). When the figure was placed directly on the background, the bees failed to find it (Srinivasan *et al.*, 1990) demonstrating that the cue used to locate the figure is the relative motion between the images of the figure and the background, caused by the bees' own flight above the setup. Video films of the bees' landings showed that, when the disc was visible to the bees, they did not land at random on it, rather, they landed primarily near the boundary of the disc, facing the visual "cliff" (Fig. 9b). These experiments showed that the boundary has special visual significance, and that bees are capable of detecting it reliably. Recently Kern *et al.* (1997) have shown, through behavioral experiments and modeling, that the detectability of such boundaries can be well accounted for by a neural network which compares optic flow in spatially adjacent receptive fields.

The ability to detect objects through the discontinuities in image motion that occur at the boundaries is likely to be important when an insect attempts to land on a leaf or a shrub. This is a situation where it may be difficult to distinguish individual leaves or establish which leaf is nearest, since cues based on contrast in luminance or color are weak. Visual problems of this nature are not restricted to insects. Over 130

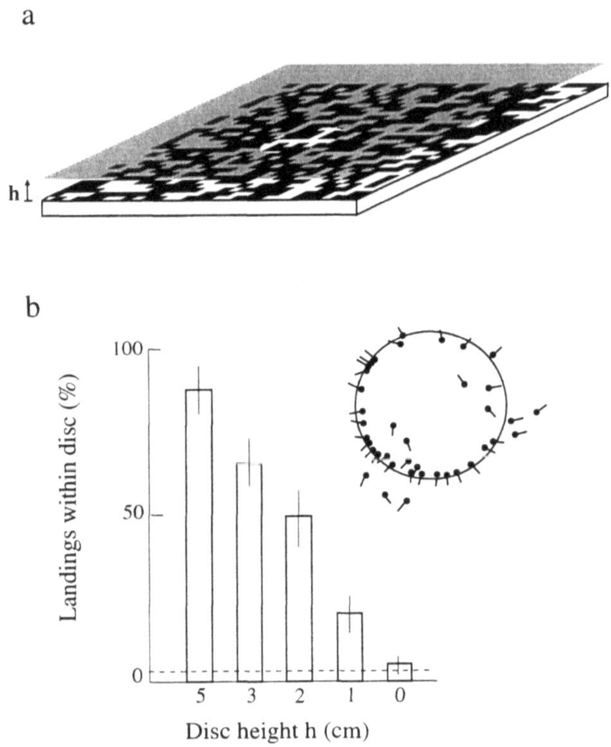

FIG. 9. Experiment investigating the ability of bees to use motion parallax cues to distinguish a figure from a similarly textured background. (a) The apparatus presents a textured disc, 6 cm in diameter, positioned under a sheet of clear perspex at a height h cm above a similarly textured background (42 cm × 30 cm, pixel size 5 mm × 5 mm). The disc is shown as being brighter than the background only for clarity. (b) Bars show percentages of landings occurring within the disc, for various heights (h) of the disc above background. 1121 landings were analyzed. The detectability of the disc reduces as h decreases, reaching the random-hit level (dashed line) when $h = 0$ (i.e., when there is no motion parallax). The inset shows a sample of the distribution of landings of trained bees on the perspex sheet when $h = 5$ cm. Adapted from Srinivasan *et al.* (1990), *Proc. Roy. Soc. Lond. B* **238,** 331–350, The Royal Society.

years ago, Helmholtz (1866) speculated that humans might use optic-flow cues in similar way to distinguish individual trees in a dense forest.

IV. Concluding Remarks

Insects are prime subjects in which to study the ways in which optic flow information can be exploited by visual systems. This is because these creatures, possessing poor or no stereopsis, literally need to move in order to see. We now know that insects use optic flow information in a variety of visually mediated functions and behaviors which extend well beyond the classically studied optomotor response. Flying insects exploit cues derived from image motion to stabilize flight, regulate flight speed, negotiate narrow gaps, infer the ranges to objects, avoid obstacles, orchestrate smooth landings, distinguish objects from backgrounds, and monitor distance traveled. The emerging picture is that there are a number of motion-sensitive pathways in the insect visual system, each with a distinct set of properties and geared to a specific visual function. It would seem unlikely, however, that all these systems (and other, as yet undiscovered ones) are operative all of the time. Obviously, the optomotor system would have to be switched off, or its corrective commands ignored, when the insect makes a "voluntary" turn or chases a target (Srinivasan and Bernard, 1977; Heisenberg and Wolf, 1993; Kirschfeld, 1997). Equally, it would be impossible to make a grazing landing on a surface without first disabling the movement avoidance system! As a third example, there is evidence that the optomotor system is at least partially nonfunctional when insects fly through narrow gaps (Srinivasan *et al.*, 1993). One major challenge for the future, then, is to discover the conditions under which individual systems are called into play or ignored and to understand the ways in which these systems interact to coordinate flight. Another is to uncover the neural mechanisms that underlie these visual capacities.

Acknowledgments

We are grateful to all the friends and colleagues with whom we collaborated at various times to produce the research reviewed here. They are (in alphabetical order) Tom Collett, Adrian Horridge, Wolfgang Kirchner, and Miriam Lehrer. Some of the work described in this review was supported by the International Human Frontier Science Program (Grant RG-84/97).

References

Borst, A., and Bahde, S. (1988). Visual information processing in the fly's landing system. *J. Comp. Physiol. A* **163,** 167–73.

Borst, A., and Egelhaaf, M. (1989). Principles of visual motion detection. *Trends Neurosci.* **12,** 297–306.

Buchner, E. (1984). Behavioral analysis of spatial vision in insects. *In:* "Photoreception and Vision in Invertebrates" (M. A. Ali, Ed.), pp. 561–621. Plenum Press, New York.

Collett, T. S., and Harkness, L. I. K. (1982). Depth vision in animals. *In:* "Analysis of Visual Behavior" (D. J. Ingle, M. A. Goodale, and R. J. W. Mansfield, Eds.), pp. 111–176. MIT Press, Cambridge, MA.

David, C. T. (1982). Compensation for height in the control of groundspeed by *Drosophila* in a new, "Barber's Pole" wind tunnel. *J Comp. Physiol.* **147,** 485–493.

Egelhaaf, M., and Borst, A. (1993). Movement detection in arthropods. *In:* "Visual Motion and Its Role in the Stabilization of Gaze" (F. A. Miles and J. Wallman, Eds.), pp. 203–235. Elsevier, Amsterdam.

Egelhaaf, M., Hausen, K., Reichardt, W., and Wehrhahn, C. (1988). Visual course control in flies relies on neuronal computation of object and background motion. *Trends Neurosci.* **11,** 351–358.

Eriksson, E. S. (1980). Movement parallax and distance perception in the grasshopper (*Phaulacridium vittatum*). *J. Exp. Biol.* **86,** 337–340.

Esch, H., and Burns, J. E. (1995). Honeybees use optic flow to measure the distance of a food source. *Naturwiss.* **82,** 38–40.

Esch, H., and Burns, J. (1996). Distance estimation by foraging honeybees. *J. Exp. Biol.* **199,** 155–162.

Exner, S. (1891). "The Physiology of the Compound Eyes of Insects and Crustaceans." R. C. Hardie, Trans.), pp. 130–131. Springer-Verlag, Berlin, Heidelberg.

Von Frisch, K. (1993). "The Dance Language and Orientation of Bees." Harvard University Press, Cambridge, MA.

Gibson, J. J. (1979). "The Ecological Approach to Visual Perception." Lawrence Erlbaum Associates, Hillsdale, NJ.

Götz, K. G. (1989). Movement discrimination in insects. *In:* "Processing of Optical Data by Organisms and Machines" (W. Reichardt, Ed.), pp. 494–493. Academic Press, New York.

Götz, K. G., and Wandel, U. (1984). Optomotor control of the force of flight in *Drosophila* and *Musca*. II. Covariance of lift and thrust in still air. *Biol. Cybern.* **51,** 135–139.

Götz, K. G., and Wehrhahn, C. (1984). Optomotor control of the force of flight in *Drosophila* and *Musca*. I. Homology of wingbeat-inhibiting movement detectors. *Biol. Cybern.* **51,** 129–134.

Hausen, K. (1993). The decoding of retinal image flow in insects. *In:* "Visual Motion and Its Role in the Stabilization of Gaze" (F. A. Miles and J. Wallman, Eds.), pp. 203–235. Elsevier, Amsterdam.

Hausen, K., and Egelhaaf, M. (1989). Neural mechanisms of visual course control in insects. *In:* "Facets of Vision" (D. G. Stavenga and R. C. Hardie, Eds.), pp. 391–424. Springer-Verlag, Berlin, Heidelberg.

Heisenberg, M., and Wolf, R. (1993). The sensory-motor link in motion-dependent flight control of flies. *In:* "Visual motion and its role in the stabilization of gaze (F. A. Miles and J. Wallman, Eds.), pp. 265–283. Elsevier, Amsterdam.

Helmholtz, H. von (1866). "Handbuch der physiologischen Optik." Voss Verlag, Hamburg (J. P. C. Southall, Trans., 1924; reprinted Dover, New York, 1962).

Horridge, G. A. (1986). A theory of insect vision: Velocity parallax. *Proc. Roy. Soc. Lond. B* **229,** 13–27.

Horridge, G. A. (1987). The evolution of visual processing and the construction of seeing systems. *Proc. Roy. Soc. Lond. B* **230,** 279–292.

Kelber, A., and Zeil, J. (1997). Tetragonsics guard bees interpret expanding and contracting patterns as unintended displacement in space. *J. Comp. Physiol. A* **181,** 257–265.

Kern, R., Egelhaaf, M., and Srinivasan, M. V. (1997). Edge detection by landing honeybees: Behavioural analysis and model simulations of the underlying mechanism. *Vision Res.* **37,** 2103–2117.

Kirchner, W. H., and Lengler, J. (1994). Bees perceive illusionary distance information from rotating spirals. *Naturwiss.* **81,** 42–43.

Kirchner, W. H., and Srinivasan, M. V. (1989). Freely flying honeybees use image motion to estimate object distance. *Naturwiss.* **76,** 281–282.

Kirschfeld, K. (1997). Course control and tracking: Orientation through image stabilization. *In:* "Orientation and Communication in Arthropods" (M. Lehrer, Ed.), pp. 67–93. Birkhäuser Verlag, Basel.

Kral, K. (1998). Side-to-side head movements to obtain motion depth cues: A short review of research on the praying mantis. *Behav. Proc.* **43,**71–77.

Kral, K., and Poteser, M. (1997). Motion parallax as a source of distance information in locusts and mantids. *J. Insect Behav.* **10,** 145–163.

Krapp, H. G., and Hengstenberg, R. (1996). Estimation of self-motion by optic flow processing in single visual interneurons. *Nature (Lond).* **384,** 463–466.

Lehrer, M., and Srinivasan, M. V. (1992). Freely flying bees discriminate between stationary and moving objects; performance and possible mechanisms. *J. Comp. Physiol. A* **171,** 457–467.

Lehrer, M., Srinivasan, M. V., Zhang, S. W., and Horridge, G. A. (1988). Motion cue provide the bee's visual world with a third dimension. *Nature (Lond.)* **332,** 356–357.

Poteser, M., Pabst, M-A., and Kral, K. (1998). Proprioceptive contribution to distance estimation by motion parallax in a praying mantid. *J. Exp. Biol.* **201,** 1483–1491.

Reichardt, W. (1969). Movement perception in insects. *In:* "Processing of Optical Data by Organisms and by Machines" (W. Reichardt, Ed.), pp. 465–493. Academic Press, New York.

Reichardt, W., and Poggio, T. (1979). Figure-ground discrimination by relative movement in the visual system of the fly. Part I: Experimental results. *Biol. Cybern.* **35,** 81–100.

Rossell, S. (1983). Binocular stereopsis in an insect. *Nature (Lond.)* **302,** 821–822.

Sobel, E. C. (1990). The locust's use of motion parallax to measure distance. *J. Comp. Physiol.* **167,** 579–588.

Srinivasan, M. V. (1977). A visually-evoked roll response in the housefly: Open-loop and closed-loop studies. *J. Comp. Physiol.* **119,** 1–14.

Srinivasan, M. V. (1992). How flying bees compute range from optical flow: Behavioral experiments and neural models. *In:* "Nonlinear Vision" (R. B. Pinter, Ed.), pp. 353–375. CRC, Boca Raton.

Srinivasan, M. V. (1993). How insects infer range from visual motion. *In:* "Visual Motion and Its Role in the Stabilization of Gaze" (F. A. Miles and J. Wallman, Eds.), pp. 139–156. Elsevier, Amsterdam.

Srinivasan, M. V., and Bernard, G. D. (1977). The pursuit response of the housefly and its interaction with the optomotor response. *J. Comp. Physiol.* **115,** 101–117.

Srinivasan, M. V., and Lehrer, M. (1984). Temporal acuity of honeybee vision: Behavioural studies using moving stimuli. *J. Comp. Physiol.* **155,** 297–312.

Srinivasan, M. V., and Zhang, S. W. (1997). Visual control of honeybee flight. *In:* "Orientation and Communication in Arthropods" (M. Lehrer, Ed.), pp. 67–93. Birkhäuser Verlag, Basel.

Srinivasan, M. V., Lehrer, M., Zhang, S. W., and Horridge, G. A. (1989). How honeybees measure their distance from objects of unknown size. *J. Comp. Physiol. A* **165,** 605–613.

Srinivasan, M. V., Lehrer, M., and Horridge, G. A. (1990). Visual figure-ground discrimination in the honeybee: The role of motion parallax at boundaries. *Proc. Roy. Soc. Lond. B* **238,** 331–350.

Srinivasan, M. V., Lehrer, M., Kirchner, W., and Zhang, S. W. (1991). Range perception through apparent image speed in freely-flying honeybees. *Visual Neurosci.* **6,** 519–535.

Srinivasan, M. V., Zhang, S. W., and Chandrashekara, K. (1993). Evidence for two distinct movement-detecting mechanisms in insect vision. *Naturwiss.* **80,** 38–41.

Srinivasan, M. V., Zhang, S. W., Lehrer, M., and Collett, T. S. (1996). Honeybee navigation en route to the goal: Visual flight control and odometry. *J. Exp. Biol.* **199,** 237–244.

Srinivasan, M. V., Zhang, S. W., and Bidwell, N. (1997). Visually mediated odometry in honeybees. *J. Exp. Biol.* **200,** 2513–2522.

Srinivasan, M. V., Poteser, M., and Kral, K. (1999). Motion detection in insect orientation and navigation. *Vision Res.* **39,** 2749–2766.

Wagner, H. (1982). Flow-field variables trigger landing in files. *Nature (Lond.)* **297,** 147–148.

Wallace, G. K. (1959). Visual scanning in the desert locust *Schistocerca gregaria,* Forskal. *J. Exp. Biol.* **36,** 512–525.

Wehner, R. (1981). Spatial vision in insects. *In:* "Handbook of Sensory Physiology" (H. Autrum, Ed.) Vol. 7/6C, pp. 287–616. Springer-Verlag, Berlin, Heidelberg, New York.

NEURONAL MATCHED FILTERS FOR OPTIC FLOW PROCESSING IN FLYING INSECTS

Holger G. Krapp

Lehrstuhl für Neurobiologie, Universität Bielefeld, Bielefeld, Germany

I. Introduction

A. Relative Motion and Optic Flow

Relative motions between the eyes and visual structured surroundings always result in retinal image shifts. Moving on a crowded place, for instance, will lead to a complex pattern of retinal image shifts which is induced by a "blend" of two different kinds of relative motions. Self-motion of an observer causes coherent wide field motion patterns covering the entire visual field of the observer. In contrast, external object

0074-7742/00 $30.00

motions within the visual field of an otherwise motionless observer result in image shifts which are usually locally confined. Both kinds of image shifts can be described in terms of velocity vector fields, called 'optic flow fields', where the local vectors indicate the direction and magnitude of the respective relative motion (Gibson, 1950; Nakayama and Loomis, 1974; Koenderink and van Doorn, 1987). In general, the resulting optic flow contains information which may be used to control visually guided behavior. First, the global structure of the optic flow depends on the observer's self-motion—the overall appearance of a flow field induced by a translation defers from a flow field generated during rotation (cf. Figs. 1B and 1C). And second, the magnitude of the translatory optic flow depends not just upon the respective translation speed but also on the distance between the observer and the visual structures of the surroundings. Objects close by result in higher image velocities than more distant ones. Thus, by analyzing the relative velocity differences within translatory optic flow fields, an observer may get information about the distribution of relative distances within the environment presently encountered. Both information about the present self-motion and the 3D layout of the environment is essential for a mobile observer. To control his locomotion adequately (e.g., to stabilize his motion path or to avoid bumping into obstacles), he needs to sense his current self-motion and to estimate the distance to possible obstacles. This holds true for all kinds of observers like humans, animals, and robots—if the latter are equipped with optical sensors. In this chapter, however, I will concentrate on optic flow processing in insects. The significance of optic flow processing for vertebrates, including humans, will be outlined in other chapters in this volume.

II. Visually Guided Behavior and Optic Flow Processing in Flying Insects

A. Trying to Explain Visually Guided Behavior on the Basis of Its Underlying Neuronal Mechanisms

Several species of flying insects show interesting visually guided behavior and have been investigated to find the neuronal basis for processing different aspects of optic flow (general overview: Wehner, 1981). In some cases, the behavioral and the neuronal level were brought together in a rather promising way. Object fixation and figure–ground discrimination (Reichardt and Poggio, 1976; Heisenberg and Wolf, 1984; Egelhaaf, 1985a, b, c; Warzecha et al., 1993; Kimmerle *et al.,* 1996), land-

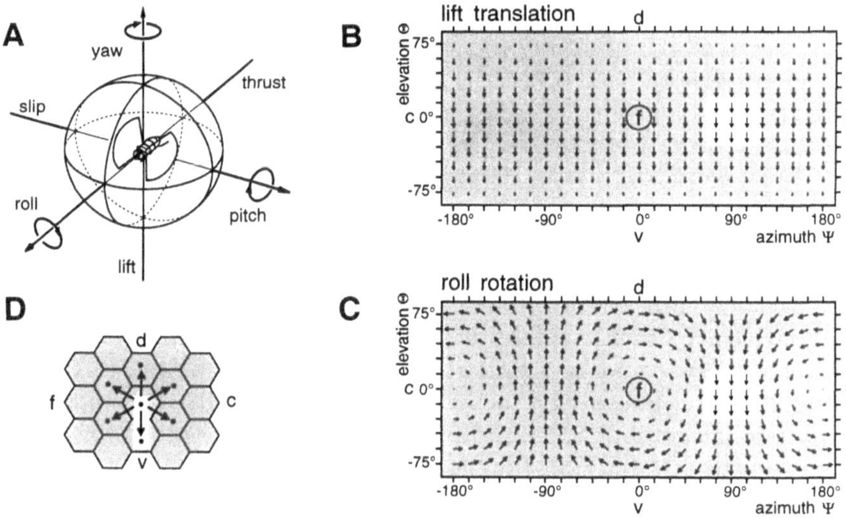

FIG. 1. Self-motion and self-motion-induced optic flow. (A) Self-motion in 3D space can be described in terms of its translation (thrust, slip, lift) and rotation (roll, pitch, yaw) component along and around the animal's three major body axes (body axis, transverse body axis, and vertical body axis). (B) Optic flow field induced by a pure lift translation. It is plotted in a Mercator map of the whole visual field where each location is specified by two angles (i.e., the horizontal azimuth ψ and the vertical elevation θ). The encircled f in the center denotes the direction along the positive body axis of the animal; the left and right halves show the left and right visual hemisphere, respectively (d = dorsal, v = ventral, c = caudal). Due to the Mercator projection, the dorsal and ventral parts of the spherical visual field are highly overemphasized in the map. Each single arrow indicates the direction and velocity of the respective local image shift. In translatory optic flow fields, all velocity vectors are aligned along great circles connecting the focus of expansion (at d) with the focus of contraction (at v). For further explanations, see text. (C) Optic-flow field induced by a pure roll rotation, plotted in the same way as the lift-flow field. In a rotatory flow field, local velocity vectors are aligned along parallel circles centered around the axis of rotation (here: corresponding exactly with the body axis f). Globally, we can easily distinguish the structure of translatory and rotatory flow fields. (D) A set of six local motion detectors (EMDs) analyzing retinal image shifts resulting from relative motion at the same location in the visual field. Arrows connecting the black dots indicate the different preferred directions of the EMDs. Note that the EMD analyzing vertical image shifts (white frame) at ψ = 90° and θ= 0° may be strongly excited in either case—if the animal is performing a lift or roll self-motion (compare the white areas within the respective flow fields). Modified from Krapp *et al.* (1998).

ing (Wagner, 1982; Borst and Bahde, 1988; Borst, 1991) and chasing behavior (Wagner, 1986; Böddeker *et al.*, 1998; Wachenfeld, 1994) in flies, for instance, are controlled by visual cues which rely on particular aspects of optic flow. Dragonflies have developed visual interneurons which are thought to extract particular features from the optic flow to control preying (Mayer, 1957; Olberg, 1981; O'Carroll, 1993; Frye and Olberg, 1995). In the hawk moth, visual interneurons were found to respond to particular self-motion-induced optic flow components which may be involved in their visually controlled feeding behavior (i.e., hovering in front of a flower and sucking nectar like hummingbirds) (Herrera, 1992; Farina *et al.*, 1994; Kern and Varjú, 1998; Kern, 1998). Although it is well known that hymenoptera, first of all honeybees, are capable of solving a great variety of visually controlled tasks based on optic flow processing (Lehrer, 1994, 1997; Srinivansan *et al.*, 1996; Srinivasan and Zhang, this volume), comparatively little is known about the underlying neuronal basis. This rather arbitrary list of examples for different visually guided behaviors in different insects could be continued at length. Nevertheless, in the following I will confine myself to the neuronal basis of the optomotor behavior and gaze stabilization in the fly. In general, the optomotor behavior in flying insects describes the capability of compensating for involuntary course deviations which may be due to gusts of wind or flying through turbulent air, respectively. This capability has been investigated in several insect species including, for instance, flies, locusts and bees, at the behavioral (Wehner, 1981; Lehrer, 1994; Collett *et al.*, 1993; Buchner, 1984; Götz, 1983a,b; Heisenberg and Wolf, 1993; Rowell, 1988; Robert, 1988; Gewecke, 1983; Preiss, 1991) and as well as the neuronal level (Hausen and Egelhaaf, 1989; Gewecke and Hou, 1993; Goodman *et al.*, 1990; Reichert and Rowell, 1986; Rind, 1990; Milde, 1993). In this chapter, gaze stabilization refers to the tendency of insects to keep their eyes aligned with the external horizon by means of rotatory head movements (Hengstenberg, 1991)—a behavior thoroughly investigated in other visually oriented animals and humans (Carpenter, 1988; Dieringer, 1986).

B. The Fly as an Experimental Model System

Among insects and with respect to visual information processing, the fly turned out to be a rewarding experimental animal to address questions of general interest. The fly visual system has been investigated extensively by means of behavioral experiments on the one hand, and by applying neuroanatomical as well as electrophysiological techniques on the other. Often, both the behavior and its underlying neuronal basis can be studied quantitatively in the very same system under similar or even the same stimulus conditions. The combination of the neuronal and the behavioral description level allows us to estimate the system's adaptation regarding the performance in particular behavioral tasks. It is evident, of course, that this kind of neuroethological approach can be ideally pursued in systems where distinct behaviors can be correlated with the activity of identified neuronal circuits or even single nerve cells.

III. How to Gain Self-Motion Information from Optic Flow

A successful performance of both optomotor response and gaze stabilization behavior relies on information about the instantaneous self-motion. In addition to using the mechanosensory signals of the halter system, which measures the velocity of self-rotations (Hengstenberg, 1993), the fly is thought to gain information about the self-motion by exploiting the instantaneous optic flow (Hausen and Egelhaaf, 1989).

A. Features of Rotatory and Translatory Optic Flow

As already mentioned in the introduction, self-motion can be described in terms of translation and rotation. The resulting optic flow field is a linear combination of the translatory and rotatory component induced by the respective motion along and around the three main body axes (Fig. 1A). Translations (thrust, slip, lift) and rotations (roll, pitch, yaw) generate different optic flow fields over the insect's eyes. The local flow vectors in translatory optic flow fields are oriented along meridians connecting the focus of expansion (i.e., the direction point d in the translation is pointing at, Fig. 1B) with the focus of contraction which is the opposite pole of the flow field. Figure 1B shows the optic flow induced by a lift translation plotted along azimuth ψ and elevation θ in a Mercator map of the visual field. The flow field generated during a rotation around the body axis (roll) is shown in Fig. 1C. A general feature of the rotatory flow structure is that all local vectors are aligned along parallel circles centered around the axis of rotation (in this case the axis coincides with point f in Fig. 1C). Within both flow fields (Figs. 1B and 1C), no image shift occurs at the poles. With increasing distance from the poles, the magnitude of the flow vectors increases as well and gets maximum at the equator of translation or rotation (i.e., exactly between the two poles). The local translation vectors additionally depend on the distance between the eyes and the objects in the environment; objects close by generate bigger flow vectors. In the calculated optic flow field shown in Fig. 1B, the same distance was assumed at all location. Rotatory optic flow is distance invariant. In "real" flight situations, the rotatory and translatory components are linearly superimposed and may result in rather complex optic flow fields.

B. How to Get the Global Difference by Local Measurements: The Idea of Neuronal Matched Filters

Globally, we can easily distinguish the flow fields induced by the different self-motions shown here. Visual motion, however, is sensed locally by elementary movement detectors (EMDs; Hassenstein and Reichardt, 1956; Reichardt, 1987; Borst and Egelhaaf, 1989). Each location in the visual field is analyzed by a set of EMDs along at least six different preferred directions. These preferred directions reflect the arrangement of the optical axes of neighboring ommatidia within the fly's compound eye (Fig. 1D; Buchner, 1976; Götz *et al.*, 1979). At the level of a single motion detector, however, it is ambiguous whether the image shift exciting the EMD is due to a translation or a rotation. Such an area where the local optic flow is quite similar for different self-motions is marked in the optic flow fields shown in Figs. 1B and 1C. One way to make the problem less ambiguous is to integrate selectively the outputs of EMDs whose preferred directions correspond to the directions of the local flow vectors at each point in visual space. Figure 2 demonstrates the scheme of

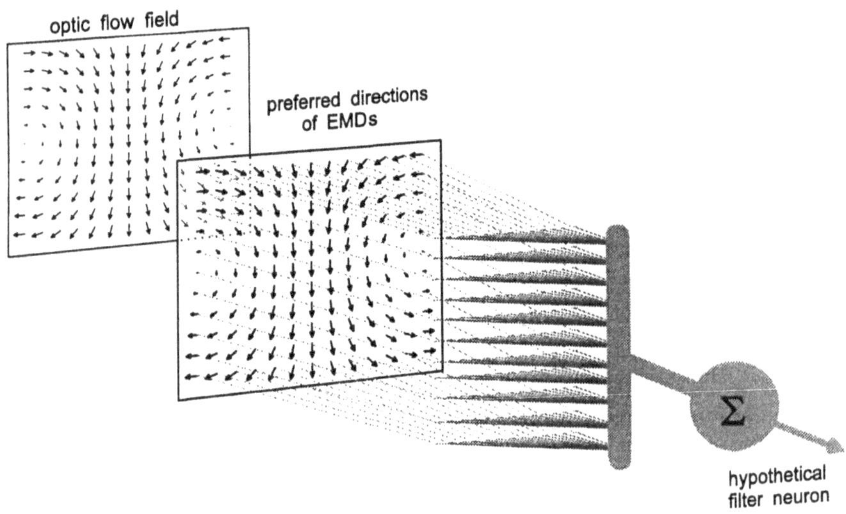

FIG. 2. A hypothetical filter neuron integrates selectively signals of EMDs whose preferred directions correspond to the direction of local flow vectors from a roll flow field over the right eye. This filter neuron with a receptive field that makes up the right visual hemisphere would be strongly excited by a roll rotation around the body axis to the left. Modified from Krapp *et al.* (1998).

a hypothetical filter neuron designed to sense the optic flow over the right eye induced by a rotation to the left around the body axis (roll). Based on this model, some qualitative features of such filter neurons can be expected: these kinds of neurons need to be motion sensitive and directionally selective. They should have extended receptive fields—the larger the receptive field, the better it can be expected to distinguish between different optic flow fields and thus different self-motions. And finally, the distribution of local preferred directions should match the direction distribution of local optic flow vectors.

We have known for a long time that individually identifiable interneurons which respond to wide field motion live within the fly visual system (Bishop and Keehn, 1967; Bishop *et al.*, 1968; MacCann and Dill, 1969; Dvorak *et al.*, 1975; Hausen, 1976). Their big receptive fields and their directionally selective motion response made these so-called tangential neurons good candidates for being involved in the control of optomotor responses and gaze stabilization. After a short introduction into the visual system of the fly, I will try to further explain why some of these neurons are most likely concerned with optic flow processing.

IV. The Fly Visual System

A. Organization of the Visual Neuropils

The fly visual system is organized in retinotopically arranged columns (Strausfeld, 1976, 1989; Bausenwein and Fischbach, 1992) and consists of the retina plus three successive neuropils, called lamina, medulla, and lobula complex. In diptera, the lobula complex is subdivided into the anterior lobula and the posterior lobula plate. With respect to the visual analysis of self-motion, the lobula plate was identified to be the highest processing stage. In the lobula plate, about 60 tangential interneurons have been found so far (Hausen and Egelhaaf, 1989; Hausen, 1984). They are stacked along the anterior-posterior extent of the neuropil within four directional input layers representing horizontal front-to-back, horizontal back-to-front, vertical upward, and vertical downward motion (Buchner and Buchner, 1984). They are thought to integrate the outputs of many movement-detecting small field elements on their dendritic arborizations (Borst and Egelhaaf, 1992). Some tangential neurons are heterolateral spiking elements (Hausen, 1976, 1984). They pick up local visual information within one lobula plate and convey it to the contralateral part of the bilateral symmetric visual system. Other tangential neurons are thought to be pure output elements. In *Calliphora vicina* (previously *Calliphora erythrocephala* Meig.), 13 of these output neurons can be subdivided into two different groups called HS neurons (HS = horizontal system) and VS neurons (VS = vertical system; Hausen, 1976, 1982a; Hengstenberg, 1977; Soohoo and Bishop, 1980; Pierantoni, 1976; Eckert and Bishop, 1978; Hengstenberg *et al.*, 1982).

B. Wide Field Motion-Sensitive Neurons in the Third Visual Neuropil: The Tangential Neurons

Like most of the other tangential neurons, the HS and VS neurons can be individually identified by single-cell-staining methods (Strausfeld *et al.*, 1983; Hengstenberg *et al.*, 1983). Together, the dendritic arborizations of both neuronal subgroups cover the whole lobula plate. The HS divides the visual field into three slightly overlapping areas. The dorsal neuron HSN (N = north) has its receptive field in the upper part of the ipsilateral visual hemisphere, the middle one HSE (E = equatorial) analyzes motion in the equatorial visual field, and the ventral neuron HSS (S = south) monitors the lower part of the visual hemisphere. The dendrites of the HS neurons arborize in the anterior input layers of the lobula plate. Accordingly, on average, these neurons are excited by ipsilateral front-to-back motion. Ipsilateral back-to-front motion inhibits the HS neurons. The HSN and HSE are additionally sensitive to back-to-front motion within the contralateral hemisphere (Hausen, 1982b).

The vertically oriented dendrites of the 10 VS neurons overlap slightly more than those of the HS neurons. Dendritic fields of the VS neurons cover the lobula plate from the distal (VS1) to the proximal margin (VS10) of the neuropil (Hengstenberg *et al.*, 1982). Most of the dendritic arborizations of the VS neurons ramify within the most posterior input layer of the lobula plate, which corresponds with a predominant sensitivity to vertical downward motion (Hengstenberg, 1982). However, some dorsal dendritic branches of VS1 and VS7–VS10 invest the more anterior layers mediating horizontal direction selective inputs (Hengstenberg *et al.*, 1982).

Both, the HS neurons and the VS neurons respond predominantly to visual stimulation with graded membrane potential changes which may be accompanied by irregular superimposed spikes (Hengstenberg, 1977). Motion along their respective preferred direction results in a depolarization of the membrane, whereas motion along the antipreferred, or null direction, causes a hyperpolarization.

C. Behavioral Deficiencies in Flies with Ablated or Degenerated Tangential Neurons

Various kinds of evidence suggest that HS and VS neurons are involved in the control of the optomotor response and gaze stabilization. (i) Detailed neuroanatomical descriptions of the fly brain (Strausfeld, 1976, 1989;) provide insight into the neuronal wiring of the motion pathway. The output regions of the HS and VS neurons are connected via descending neurons to the flight motor centers in the thoracic compound ganglion and, partly, directly to motoneurons of the neck motor system (Gronenberg *et al.*, 1995; Gronenberg and Strausfeld, 1990; Milde *et al.*, 1987; Strausfeld and Gronenberg, 1990; Strausfeld *et al.*, 1987). (ii) Electrophysiological investigations in *Calliphora* showed that the response of these neurons increases with pattern size (Haag *et al.*, 1992; Hausen, 1982b; Hengstenberg, 1982). (iii) Microsurgical lesion experiments in adults or laser ablation of the HS precursor cells resulted in predictable failure of the animal's optomotor response (Geiger and Nässel, 1981; Hausen and Wehrhahn, 1983). (iv) The HS and VS neurons are not developed in the neurological *Drosophila* mutant omb^{H31} (Heisenberg *et al.*, 1978; Pflugfelder and Heisenberg, 1995). This defect selectively affects optomotor responses (Heisenberg *et al.*, 1978; Götz, 1983a,b) and gaze stabilization (Hengstenberg, 1995), whereas the response to small objects is still normal (Bausenwein *et al.*, 1986).

In summary, these findings suggest that the VS and HS neurons in the fly visual system are key elements to process self-motion-induced optic flow and to generate signals for optomotor control and gaze stabilization. However, until recently, it was not known if the receptive field organization of these neurons showed any specialization with respect to this task. In this context, results from early investigations on the VS neurons were quite interesting. In the dorsolateral receptive field, some of these neurons did show responses to horizontal pattern movements (Hengstenberg, 1981). That was a first hint that at least VS neurons may process optic flow in a more specific way rather than extracting only vertical downward motion. To find out if the idea of a neuronal matched filter, proposed in the introduction, may be realized in the fly visual system the local response properties of several tangential neurons were investigated in detail recently.

V. Mapping the Local Response Properties of Tangential Neurons

By applying a fast visual stimulation procedure (Fig. 3), the local directional tuning curves of individually identified tangential neurons

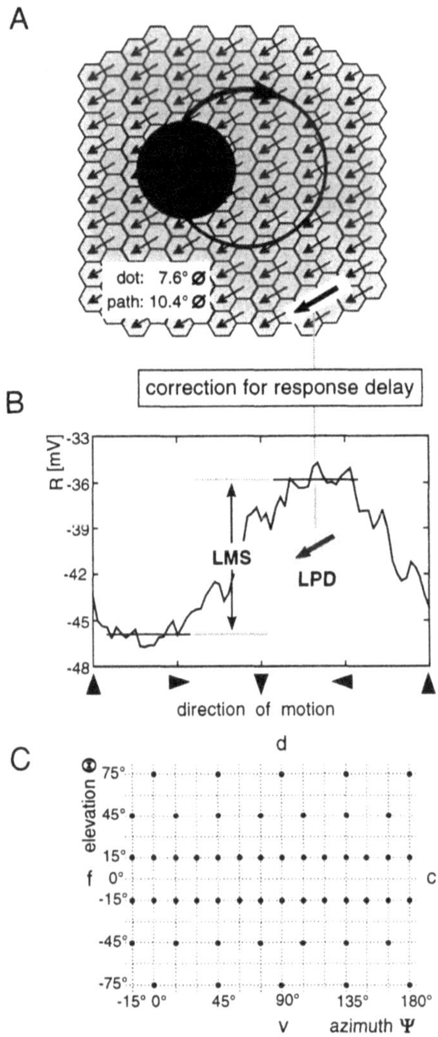

FIG. 3. Stimulation procedure to investigate the receptive field organization of direction-selective wide field neurons. (A) A black dot is moved along a circular path at constant velocity (2 cycles per second). Small arrows within the hexagonal pattern, which schematizes the ommatidium lattice, indicate the preferred directions of EMDs converging on an intracellularly recorded tangential neuron. (B) When the direction of dot motion coincides with the local preferred direction of the EMDs, the response of the recorded neuron becomes maximal; motion in the opposite direction results in a hyperpolarization of the membrane potential. The local preferred direction is determined by comparing the responses to dot motion in clockwise and counterclockwise direction and estimating the response delay. From the corrected tuning curve, the local preferred direction can be obtained by applying circular statistics, or determining the phase shift of the first harmonic of the Fourier transform. The local motion sensitivity is defined by the difference between the LPD quadrant and the opposite quadrant of the tuning curve (thick horizontal lines indicate the quadrants). (C) Measuring positions. The LPDs and LMSs are determined at the positions labeled by small black dots. Data at intermediate positions were obtained by interpolations. Modified from Krapp and Hengstenberg (1997) and Krapp *et al.* (1998).

were measured at about 50 different positions within the visual field (Krapp and Hengstenberg, 1997). From the tuning curves, the local preferred direction (LPD) and the local motion sensitivity (LMS) could be obtained (Fig. 3B). These response parameters were mapped as arrows at the respective measuring positions into a Mercator projection of a little more than one hemisphere of the fly's visual field (Fig. 3C). All positions are determined by their azimuth ψ and elevation θ. The orientation of each arrow gives the local preferred direction, and the length denotes the relative motion sensitivity.

Figure 4 shows the morphology and the response fields of HSN and VS10. The dendritic arborizations of the HSN cover the dorsal part of the neuropil (Fig. 4A). As already shown for the HS neurons by Hausen (1982b), both position and extent of the dendritic branching pattern of the VS neurons are very similar in different individuals. The response fields show the distribution of LPDs and LMSs as determined for the right visual hemisphere and one vertical stripe of the frontal left visual hemisphere (Figs. 4B and 4D; ($\psi = -15°$). Two global features of the response fields are striking. First of all, they extend over wide parts of the ipsilateral hemisphere including the frontal region of binocular overlap at −15° azimuth. Second, the LPD distributions are by no means homogeneous within the response fields. Instead, the LPDs clearly depend on the respective measuring position.

Although, on average, the LPDs within the HSN response field are aligned horizontally, there are considerable deviations from this orientation especially in the frontodorsal and caudal part of the visual field. Purely horizontal LPDs are confined to the lateral and equatorial region of the response field (cf. Fig. 4B, ψ about 90°). In different studies, it was found that in the frontodorsal eye region the LPDs of the HSN and HSE are tilted upward, whereas the LPDs of the HSS and HSE in the frontoventral eye region are tilted downward, relative to the horizontal (cf. Fig: 4B; Hausen, 1982b; Hengstenberg *et al.*, 1997).

The interpretation of the HS response fields is somewhat difficult. On the one hand, as Hausen pointed out (1993), the distribution of LPDs is reminiscent of the dorsal half of a translatory optic flow field. The focus of expansion of such a field would lie at about the frontolateral equator within the contralateral hemisphere. On the other hand, this neuron is thought to receive a rotation-specific input from spiking heterolateral elements (Hausen, 1982a). Thus, it is excited by optic flow generated during rotations around the yaw axis as well. From his investigations and the way tangential neurons in general were thought to integrate ipsilateral and heterolateral wide field motion, Hausen (1993) inferred that these neurons do not specifically discriminate between the

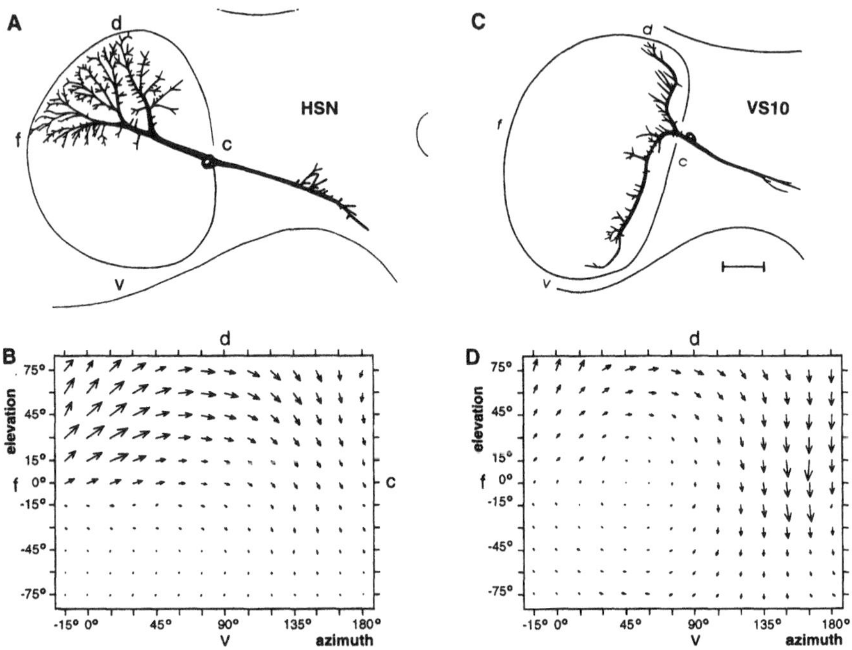

FIG. 4. Morphology and response fields of the HSN neuron and the VS10 neuron. (A) Morphology of the HSN neuron as reconstructed from frontal serial cross sections shown in the contour of the third visual neuropil (lobual plate; the neurons were stained with the intracellular fluorescent dye Lucifer Yellow). The main dendrites of HSN are aligned horizontally and cover the medial to superior part of the lobula plate. The letters f, c, d, and v refer to the retinotopical organization of the neuropil. (C) Morphology of the VS10 neuron, presented in the same way as HSN. Note that the main dendrites of the VS neuron are oriented vertically and are confined to the proximal region of the neuropil. (B) The HSN response field is composed of the whole dorsal hemisphere. In the ventral hemisphere, this neuron does not respond to motion at all. The orientation of each single arrow indicates the local preferred direction and its length gives the local motion sensitivity normalized to the maximum response. The HSN is most sensitive to motion in the frontodorsal visual field. From the global LPD distribution, it is hard to infer which particular self-motion would be sensed most effectively by this neuron. This neuron may contribute to the analysis of different self-motions as proposed by Hausen (1981) (see text). (D) The VS10 response field covers almost the whole visual hemisphere. Even in the frontoventral visual field where the sensitivity is strongly reduced, LPDs can be measured which are consistent with the overall structure of the response field. The neuron is highly sensitivity to vertical downward motion in the caudolateral visual field. However, note that all possible LPDs are present. Within this response field, the LPD distribution shows a strong similarity with an optic flow field induced by a rotation around a horizontally aligned body axis which lies between the pitch and the roll axis (θ about 0° and ψ about 45–60°). As all other VS neurons, VS10 responds more strongly to motion in the dorsal than in the ventral part of the visual field. (C) and (D) are modified from Krapp *et al.* (1998).

rotatory and translatory optic flow components. The new, more detailed data of the local response properties (Fig. 4B; Hengstenberg *et al.*, 1997, Krapp, unpublished results) do not contradict this statement regarding the HS neurons. However, it needs to be reformulated with respect to some other lobula plate neurons (see conclusions).

The dendritic field of the neuron VS10 covers the proximal parts of the neuropil. Its vertical main dendrite arborizes within the most posterior input layer. Some of the dorsal dendrites, however, invade the more anterior input layers which convey signals encoding horizontal motion. A comparison of Figs. 4A and 4C demonstrates the horizontal orientation of the main dendritic branches in HS neurons versus the vertical orientation of the main dendrites of the VS neurons. Another general difference concerns the strong sensitivity to more or less horizontal motion (HS) in contrast to maximum motion responses to vertical downward motion of the VS neurons (cf. Figs. 4B and 4D). Moreover, the VS10 response field shows a marked similarity to a rotatory optic flow field with a presumed axis of rotation at an azimuth of about 45–60° and an elevation of about 0° (Fig. 4D). Even though the HSN response field may not be specialized for sensing a particular optic flow field, the VS10 response field apparently is. Even in the ventral parts of the receptive field where the motion sensitivity is comparatively low, the local preferred directions fit almost perfectly the direction distribution of a global rotatory flow structure. Thus the response field suggests that VS10 is adapted to analyze the momentary optic flow for components induced by a particular self-rotation. Its "preferred axis of rotation" appears to lie between the roll and the pitch axis. The response fields of the other nine VS neurons were also found to mimic rotatory structures induced by self-rotations around horizontally aligned body axes. The different preferred axes of rotation for the ten VS neurons could be estimated from the response fields by applying a least-square algorithm developed by Koenderink and van Doorn (1987). All VS axes are aligned horizontally and scattered along the azimuth (Fig. 5). A slight clustering can be seen for VS8–VS10. The VS4–VS7 axes are more or less aligned with the animal's roll axis, whereas the VS1 and VS2 axes are close to the pitch axis. The preferred rotation axis of VS3 lies between the roll and the pitch axis. The response fields of all VS neurons are highly reliable at the interindividual level (Krapp *et al.*, 1998).

These findings suggest that the VS neurons are adapted to sense self-rotations. However, there is a particular difference between the sensitivity distribution within the response fields and the velocity distribution within rotatory optic flow fields. The velocity distribution within optic flow fields is symmetrical with respect to the equatorial plane. If the roll

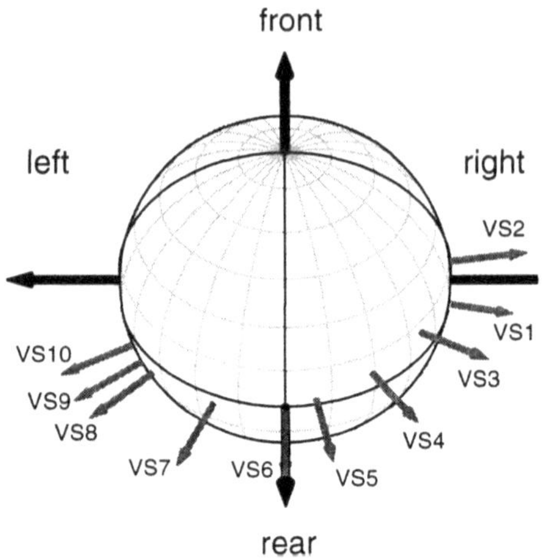

FIG. 5. Preferred axes of rotation of the VS neurons. To calculate the respective axes of rotation from all measured response fields, an iterative least-square algorithm developed by Koenderink and van Doorn (1987) was applied to determine the motion parameters from a "noisy" optic flow field. The plotted arrows represent the mean axes obtained from at least three (up to 12) response fields per neuron type. These preferred axes of rotation are plotted in the visual unit sphere as seen from the rear and above. All rotation axes (gray arrows) are aligned horizontally. The preferred axis of VS6 coincides with the body axis (roll) and the preferred axes of VS1 and VS2 close below and above the transverse body axis (pitch).

flow field shown in Fig. 1C is compared with the VS response fields shown in Figs. 4D and 6A, it appears that the sensitivity within the response fields is asymmetrically distributed. In the ventral part, the sensitivity is smaller than in the dorsal. This observation, which holds true for all VS neurons, was the starting point for a more quantitative approach to understanding the functional significance of the response fields of the VS neurons (see next section).

A different type of response field was measured in another spiking tangential neuron, the so-called neuron Hx (Fig. 6C; Krapp and Hengstenberg, 1996). In contrast to the global rotatory structure within the VS response fields, the Hx response field shows the global structure of a translatory optic flow field. A focus of expansion can be seen at an azimuth of about 135° within the equatorial plane. The results obtained from the Hx show that, within the lobula plate, there are also tangential neurons whose response field is similar to a global translatory structure.

It should be noted that the sensitivity distributions within the translatory Hx response field is inverted with respect to the rotatory response

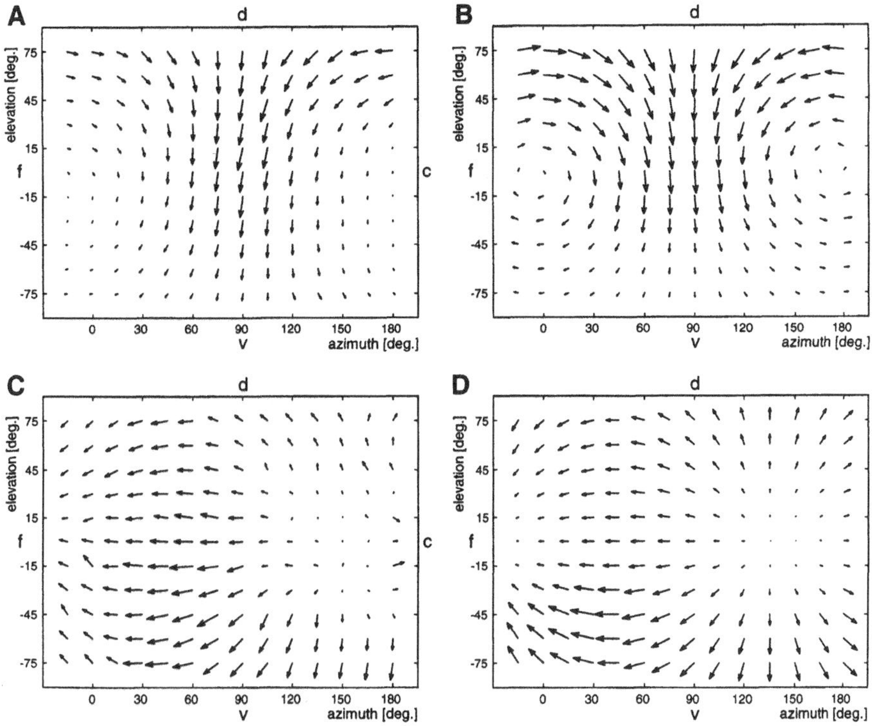

FIG. 6. Response fields and matched filters. (A) Averaged VS6 response field obtained from experiments in five different animals. The preferred axis of the VS6 roughly corresponds to the body axis (roll axis). (B) Weighted direction template corresponding to a particular stage of optic flow processing as derived for an optimal matched filter approach. (C) Response field of the Hx neuron. Note the completely different global structure compared to the VS response field. This response field is highly reminiscent of a translatory optic flow field with a focus of expansion at about $\psi = 135°$ and $\theta = 0°$. Note that Hx is more sensitive to motion in the ventral than in the dorsal visual field. (D) The weighted direction template calculated in the same way as in (B) but for the respective processing stage of a matched filter for translatory self-motion. Note the good correspondence between the measured response field and the weighted direction templates as derived from the theoretical model. See text for further explanations. Modified from Franz and Krapp (submitted).

fields measured in the VS neurons. The Hx responds more strongly to motion stimuli in the ventral than in the dorsal visual field (cf. Fig. 6C).

Until now, not all lobula plate tangential neurons could be investigated with respect to their individual receptive field organization. Only about half of a total of about 60 tangential neurons are characterized so far.

VI. Response Fields and Matched Filters for Optic Flow Processing

A. Matched Filters without Prior Assumptions about Environment and Self-Motion

The matched filter concept originally proposed in the field of image processing (Rosenfeld and Kak, 1982) was later adapted for information processing in biological systems (Wehner, 1987). A "classical" matched filter as applied in image processing is a device whose output is proportional to the cross correlation between the current input and a particularly specified stimulus pattern. Thus a matched filter is not a binary coding device whose output is only different from zero if the input exactly fits the specified stimulus pattern. Instead, input patterns similar but not identical to the specified one will result in a measurable output, too. In this context, the specified stimulus pattern corresponds to a direction template [i.e., a particular distribution of local preferred directions (cf. Fig. 2)], combined with an appropriate set of local weights.

Applying the iterative least-square procedure proposed by Koenderink and van Doorn (1987) Dahmen *et al.* (1997) studied the principal limits of estimating the self-motion parameters from noisy optic flow fields. To extract both the rotation vector and the direction of translation, the authors derived a classical matched filter model. The self-motion estimation is based on a weighted average over the projections of local flow vectors into a direction template. In contrast to the iterative procedure of Koenderink and van Doorn, the approach of Dahmen *et al.* (1997) consists of a "one-shot" mechanism which turned out to be formally equivalent to the first iteration step of the iterative procedure.

The receptive fields of some tangential neurons seems to be organized in a way which is reminiscent of the classical matched filter concept (i.e., to be adapted to sense a specific optic flow field). However, if self-motion is to be estimated, this may be not the optimal strategy. The same combination of rotation and translation in different environments may induce different optic flow fields because the translatory component depends on the distance distribution. Thus many matched filters would be necessary to sense the same self-motion in different 3D layouts.

B. Matched Filters with Prior Assumptions about Environment and Self-Motion

Recently an approach different from the Koenderink and van Doorn procedure and the "one-shot" mechanism was chosen; it was aimed at understanding the particular response field organization of the VS neurons (Franz *et al.*, 1998). In contrast to previous attempts, this approach assumes certain statistics with respect to the distances distribution between the eyes of the fly and the objects in the environment. In addition, assumptions were made about the fly's average flight velocity and the distribution of translation directions occurring during its flight. The resulting type of matched filter was designed to sense particular self-rotations or translations from the momentary optic flow rather than to match a specific optic flow field. In a first processing step, the momentary optic flow is projected into a direction template (e.g., a distribution of EMDs similar to that shown in Fig. 2, but with unit sensitivities). In this case, for instance, the filter would be constructed to sense "roll"-rotations. The local flow projections contain information about the current rotation around the filter axis. However, this information is corrupted by noise and errors of the motion detection process. Furthermore, the local flow projections are contaminated by the current translatory flow, the magnitude of which depends on the respective object distance. Local object distances, however, are not always the same in different environments encountered by the fly but may unpredictably vary around a mean distance. In a second processing stage, therefore, the projections are weighted depending on how much the self-rotation estimation is affected by these factors. The resulting local estimates are subsequently summed in an output stage whose signal then indicates the rotation around the filter axis (Franz and Krapp, submitted).

As derived from the model, local estimations of the rotatory flow component need to be weighted. These local weights were adjusted in such a way as to minimize the variance of the filter output induced by noise and the distance-dependent variability of the translatory flow. The optimal weight distribution was determined under the assumption that during flight the fly's distance toward the ground is closer than toward visual structures in the dorsal visual field. As a consequence, the relative variability of the translatory flow is higher in the ventral than in the dorsal visual field where the translatory flow is reduced because all visual structures are farther away. In addition, it was assumed that directions of voluntary and involuntary translations performed by the fly are broadly distributed with a center of mass that coincides with the fly's body axis. These assumptions were formalized and, including the respective filter axis, used as parameters for the model. From the model, analytic expressions were derived describing matched filter structures which could be compared to the measured response fields. Figures 6B and 6D demonstrate that the global structure of the response fields can be reproduced quite well by the model. Both the local preferred directions and the local motion sensitivities of the VS6 and the Hx neuron are in good agreement with the matched filter models of the respective self-motion sensors. By applying a χ^2 fitting procedure, significant correspondence could be shown between the response fields of VS4–VS6 and the respective theoretical weight distributions (for VS6 see Figs. 7A and 7B).

Although all VS neurons show the dorsoventral anisotropic sensitivity distribution, fitting the matched filter model to the other VS neurons only led to qualitative similarities between model structures and response fields. This is because the weak motion sensitivity in the frontal response field of the neurons VS7–VS10 (for VS10 see Fig. 4D) and in the caudal response field of the neurons VS1–VS3 is not predicted by the model. Nevertheless, neurons of the two groups may complement each other by converging at a later processing stage. Combining the VS1 inverted response field and the VS10 response field, for instance, resulted in a structure perfectly well suited to sense rotations around the transverse body axis ("pitch"-rotation).

Another interesting outcome of the comparison between the response fields and the filter model is that, for fitting the experimental data of the VS neurons, the parameter describing the simplified distance model assumes about the same value. This suggests that all VS neurons make the

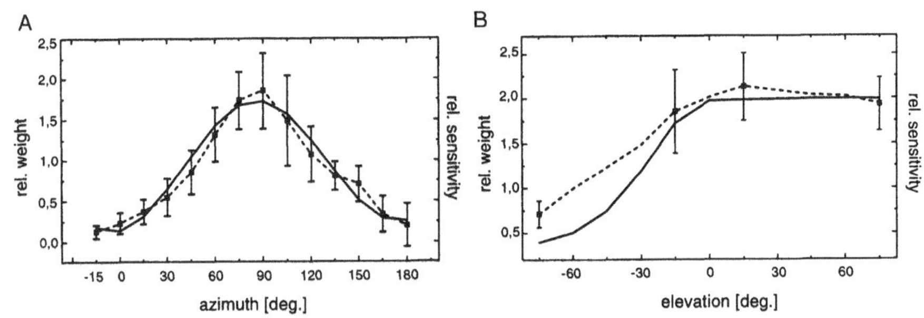

FIG. 7. Cross sections through the optimal weight distribution and the measured sensitivity distribution averaged over five VS6. (A) Section along the azimuth at an elevation of −15°. The measured sensitivities are given by filled squares connected with the dashed line; errorbars denote standard deviations. Theoretical weights are described by the solid line. (B) Cross section along the elevation at an azimuth of 90° through the same weight and sensitivity distributions. The calculated weights are optimized to minimize the filter's output variability given a roll self-rotation in combination with different self-translations. The good accordance of the weights predicted by the model with the experimentally determined sensitivities suggests the dorsoventral anisotropic sensitivity distribution to be a measure of adaptation for processing rotatory self-motions (for further explanations see text). Modified from Franz and Krapp (submitted).

same assumptions about the average distance distributions they are usually encountering.

From these investigations it can be concluded that the tangential neurons apparently extract neither a particular self-motion parameter nor a parameter combination according to the classical matched filter concept. The classical matched filter concept assumes the input organization of the filter to be literally matched to a unique input pattern. The VS neurons, however, can be considered as a generalization of the classical matched filter concept. Although the distribution of the LPDs within their response fields is obviously adapted to a particular rotatory optic flow field, the respective sensitivity distributions are not matched to one unique velocity field. Instead, the sensitivities seems to be adapted to sense an entire class of flow fields, namely those induced by a rotation around a particular axis in combination with a broad distribution of possible translations.

It should be noted that, due to the broad directional tuning of the EMDs, the VS neurons also can be expected to respond, although more weakly, to rotations around axes close to their respective preferred filter axis. In addition, local EMDs do not distinguish between rotatory and translatory optic flow (see Fig. 1). If confronted with a lift translation, for instance, most of the VS neurons may be excited as well. A more specific representation of the momentary self-rotation around a distinct axis could be computed at a later integration stage by subtracting respective correction signals. Such signals may be estimated by other visual wide field neurons and/or by mechanosensory systems like the haltere system, for instance, which senses self-rotations. The processing stage pooling these signals could, in principle, represent the respective self-rotation vector. It may be located at the level of the descending neurons, the interneurons within the motor centers, or the motoneurons.

VII. Conclusion

Rotatory self-motion components are inevitable consequences of locomotion. The resulting optic-flow component, however, does not contain any information about the 3D layout of the environment. This information is only present within translatory optic-flow fields. Thus for all kinds of long-range and short-range distance estimation tasks, a pure translatory optic flow field is desirable (Srinivansan *et al.*, 1996; Land and Collett 1997; Srinivasan, 1993). One possibility to, at least, reduce the rotatory component in the optic flow is to compensate for it by means of stabilizing head movements and steering maneuvers. These measures can be observed in the fly but also in other visually oriented animals, including humans. Of course, to generate the compensatory actions, the respective rotatory self-motion needs to be determined by the sensory systems and transformed into an adequate motor control signal.

The studies reviewed in this chapter suggest that identified interneurons in the fly visual system are adapted to analyze rotatory or translatory optic flow fields (VS neurons and Hx). From their receptive field organization alone, this categorization does not include the HS neurons. It is conceivable that HS neurons are utilized to sense both translation and rotation as has been proposed by Hausen (1981) and will be discussed below.

The VS neurons and the Hx neuron cannot be expected to be insensitive to flow components induced by other than their own respective preferred self-motions. Nevertheless, each neuron is specialized to sense a particular class of optic flow fields. For VS neurons, for instance, each class is defined by its preferred axis of rotation which may be combined with any translations in different environments. The receptive field organization of these neurons shows two adaptations which may have evolved on a phylogenetical time scale: (i) The asymmetric sensitivity distribution within the response fields. This distribution reflects the assumptions about the average distance distributions implemented in the fly visual system (i.e., on average, the distances are closer toward the ground, which makes immediate sense). Furthermore, the local flow estimates are weighted according to their respective reliability, reducing the variability of the neurons response to its respective self-motion axis. (ii) The distribution of the local preferred directions within the response fields reflects the direction distribution of velocity vectors within a class of optic flow fields all induced by a particular self-motion.

The output of a single VS neuron or of combinations of VS neurons needs to be corrected for "apparent rotations." Such apparent rotation may be due to translatory self-motions and to rotations around axes other than the preferred axis of the respective VS neuron. The signals necessary to correct for these erroneous response contributions could be supplied by other wide field neurons. The translation along the body axis and the transverse body axis, for instance, could be estimated by the two Hx neurons living in the left and right lobula plate. Hx has a preferred axis of translation which lies exactly between the body axis and the transverse body axes. Thus, in combination, the two Hx neurons would form a system with two orthogonal measuring axes. The difference of the two Hx outputs could indicate the translation along the transverse body axis, whereas the sum reflects the translation along the body axis. Generally, if the rotatory and translatory correction signals are estimated by other tangential neurons, as proposed for the Hx, reciprocal connections between neurons sensing the respectively different self-motion component need to be assumed. Such connections among the tangential neurons still need to be demonstrated electrophysiologically.

In this context the HS neurons could, in principle, supply correction signals for translations and rotations by appropriately combining their outputs. The HSN in combination with HSS could sense rotations around the transverse body axis (pitch) if the HSS signal is subtracted from the HSN signal (or vice versa). They could also monitor thrust translation if the sum of the outputs is considered. The sum of the HSN and HSS signals, however, may indicate rotations of the animal around the vertical body axis (yaw). The latter possibility is supported by the finding that HSN is also excited by a contralateral spiking wide field element which respond strongly to back-to-front motion. Together, the ipsi- and contralateral inputs to the HSN neuron mediated a high sensitivity to yaw rotation. The HSE neuron may be involved in sensing translatory or rotatory self-motions as well. Like the HSN, it receives contralateral input from spiking elements which are sensitive to back-to-front motion. This input is ineffective during thrust translation because the contralateral element is inhibited in this case, but would facilitate the neurons' response to yaw rotations. Regarding the self-motions presumably sensed by the HS neurons, Hausen (1981) already came to the same conclusions.

Correction signals encoding fast self-rotations may also be supplied by the haltere system (Nalbach, 1994). Because the dynamic range of the haltere system is shifted toward higher angular velocities, it is thought to complement the visual self-motion estimation (Hengstenberg, 1991). By measuring Coriolis forces like a gyroscope, this mechanosensory system is particularly sensitive to fast self-rotations (Nalbach, 1994).

Why are self-rotation and self-translation not represented as a vector at the level of the tangential neurons already? One reason might be that as long as the signals of the VS neurons are kept separately, they could be combined among each other—or with other tangential neurons—to monitor the self-motion along any intermediate motion axis. Such an ensemble coding of self-motion could take place at later processing stages in the nervous system. An advantage of this coding strategy could be to keep the sensory-motor transformation flexible. Information from particular sensory measuring axes could be selected according to the requirements of particular pairs of muscles to be controlled.

Meanwhile the investigations on the visual system of the fly under steady-state conditions did result in a good understanding of some basic aspects regarding optic flow processing. Based on findings with respect to the elementary movement detection and visually controlled stabilization behavior, robots which are capable of autonomously navigating within their respective environments have been designed (Franceschini *et al.*, 1992).

The functional interpretation of the receptive field organization adapting the tangential neurons for optic flow processing is based on the assumption that retinal image shifts are represented in terms of local motion vectors. Part of the theoretical approaches assumed that the output of the local motion analysis will be proportional to the velocity of the respective image shift. In addition, both approaches take for granted that the results of the local motion estimates are summed up in a linear fashion at an integrating processing stage. For insect visual systems, however, it was found that local motion analysis is achieved by elementary motion detectors whose output is not simply proportional to velocity (Egelhaaf and Reichardt, 1987) but also depends on pattern properties like spatial wavelength and contrast (Egelhaaf and Borst, 1993). Hence, it remains unclear how biological sensory systems cope with highly dynamic stimuli as encountered, for instance, by the fly during free flight. It is by no means easy to predict the signals of the tangential neurons under such natural conditions. Moreover, a "gain control" mechanism has been found for the spatial integration properties of the tangential neurons corresponding to the integration stage in the models. The response of the neurons increases with increasing pattern size but saturates at different levels, depending on the respective velocity (Borst *et al.*, 1995; Single *et al.*, 1997). Further experiments using whole field optic flow stimuli will show whether or not the matched filter concept is an appropriate interpretation of the local receptive field organization of the tangential neurons.

Acknowledgments

Many thanks to Bärbel Hengstenberg for doing the histology and reconstructing the tangential neurons. I am also grateful to Karin Bierig for preparing most of the figures. In addition, I thank Roland Hengstenberg, Martin Egelhaaf, and Matthias Franz for critically reading the manuscript and providing helpful discussions. Many thanks to Diane Blaurock for some language corrections. Parts of the work reviewed in this paper were supported by grants from the Max-Planck-Gesellschaft.

References

Bausenwein, B., and Fischbach, K. F. (1992). Activity labeling patterns in the medulla of *Drosophila melanogaster* caused by motion stimuli. *Cell Tissue Res.* **270,** 25–35.

Bausenwein, B., Wolf, R., and Heisenberg, M. (1986). Genetic dissection of optomotor behavior in *Drosophila melanogaster*. Studies on wild-type and the mutant optomotor-blind. *J. Neurogenet.* **3,** 87–109.

Bishop, L. G., and Keehn, D. G. (1967). Neural correlates of optomotor response in the fly. *Kybernetik* **3,** 288–295.

Bishop, L. G., Keehn, D. G., and McCann, G. D. (1968). Studies of motion detection by interneurons of the optic lobes and brain of the flies, *Calliphora phaenicia* and *Musca domestica*. *J. Neurophysiol.* **31,** 509–525.

Böddeker, N., Lutterklas, M., Kern, R., and Egelhaaf, M. (1998). Chasing of free-flying blowflies (*Lucilia* spec.) after a dummy. *In:* "New Neuroethology on the Move" (N. Elsner and R. Wehner, Eds.), Proceeding of the 26th Göttingen Neurobiology Conference 1998, Volume I, p. 138. Thieme, Stuttgart, New York.

Borst, A, and Bahde, S. (1988). Visual information processing in the fly's landing system. *J. Comp. Physiol. A* **163,** 167–173.

Borst, A. (1991). Fly visual interneurons responsive to image expansion. *Zool. Jb. Physiol.* **95,** 305–313.

Borst, A., and Egelhaaf, M. (1989). Principles of visual motion detection. *Trends Neruosci.* **12,** 297–306.

Borst, A., and Egelhaaf, M. (1992). In vivo imaging of calcium accumulation in fly interneurons as elicited by visual motion stimulation. *Proc. Nat. Acad. Sci. USA* **89,** 4139–4143.

Borst, A., Egelhaaf, M., and Haag, J. (1995). Mechanisms of dendritic integration underlying gain control in fly motion-sensitive interneurons. *J. Comput. Neurosci.* **2,** 5–18.

Buchner, E. (1976). Elementary movement detectors in an insect visual system. *Biol. Cybern.* **24,** 85–101.

Buchner, E. (1984). Behavioural analysis of spatial vision in insects. *In:* "Photoreception and Vision in Invertebrates" (Ali, M. A., Ed.), pp. 623–634. Plenum Press, New York, London.

Buchner, E., and Buchner, S. (1984). Neuroanatomical mapping of visually induced nervous activity in insects by ^{3}H-deoxyglucose. *In:* "Photoreception and Vision in Invertebrates" (M. A. Ali, Ed.), pp. 623–634. Plenum Press, New York, London.

Carpenter, R. H. S. (1988). "Movements of the Eye." Pion, London.

Collett, T., Nalbach, H. O., and Wagner, H. (1993). Visual stabilization in arthropods. *In:* "Visual Motion and its Role in the Stabilization of Gaze" (F. A. Miles and J. Wallman, Eds.), pp. 239–263. Elsevier, Amsterdam, London, New York, Tokyo.

Dahmen, H., Wüst, R. W., and Zeil, J. (1997). Extracting egomotion parameters from optic flow: Principal limits for animals and machines. *In:* "From Living Eyes to Seeing Machines" (M. V. Srinivansan and S. Venkatesh, Eds.), pp. 174–198. Oxford University Press, Oxford, New York.

Dieringer, N. (1986). Vergleichende Neurobiologie von blickstabilisierenden Reflexsystemen bei Wirbeltieren. *Naturwiss.* **73,** 299–304.

Dvorak, D. R., Bishop, L. D., and Eckert, H. E. (1975). On the identification of movement detectors in the fly optic lobe. *J. Comp. Physiol.* **100,** 5–23.

Eckert, H., and Bishop, L. G. (1978). Anatomical and physiological properties of the vertical cells in the third optic ganglion of *Phaenicia sericata* (Diptera, Calliphoridae). *J. Comp. Physiol.* **126,** 57–86.

Egelhaaf, M. (1985a). On the neuronal basis of figure-ground discrimination by relative motion in the visual system of the fly. I. Behavioural constraints imposed on the neuronal network and the role of the optomotor system. *Biol. Cybern.* **52,** 123–140.

Egelhaaf, M. (1985b). On the neuronal basis of figure-ground discrimination by relative motion in the visual system of the fly. II. Figure-detection cells, a new class of visual interneurons. *Biol. Cybern.* **52,** 195–209.

Egelhaaf, M. (1985c). On the neuronal basis of figure-ground discrimination by relative motion in the visual system of the fly. III. Possible input circuitries and behavioural significance of the FD-cells. *Biol. Cybern.* **52,** 267–280.

Egelhaaf, M., and Reichardt, W. (1987). Dynamic response properties of movement detectors: Theoretical analysis and electrophysiological investigation in the visual system of the fly. *Biol. Cybern.* **56,** 69–87.

Egelhaaf, M., and Borst, A. (1993). Movement detection in arthropods. *In:* "Visual Motion and its Role in the Stabilization of Gaze" (F. A. Miles and J. Wallman, Eds.), pp. 53–77. Elsevier, Amsterdam, London, New York, Tokyo.

Farina, W. M., Varjú, D., and Zhou, Y. (1994). The regulation of distance to dummy flowers during hovering flight in the hawk moth *Macroglossum stellatarum*. *J. Comp. Physiol. A* **174,** 239–247.

Franceschini, N., Pichon, J. M., Blanes, C., and Brady, J. M. (1992). From insect vision to robot vision. *Phil. Trans. Roy. Soc. Lond. B* **337,** 283–294.

Franz, M. O., and Krapp, H. G. (Submitted) Wide-field, motion-sensitive neurons and optimal matched filters for optic flow.

Franz, M. O., Hengstenberg, R., and Krapp, H. G. (1998). VS-neurons as matched filters for self-motion-induced optic flow fields. *In:* "Göttingen Neurobiology Report 1998" (N. Elsner and R. Wehner, Eds.), Proceeding of the 26th Göttingen Neurobiology Conference 1998, Vol. II, p. 419. Thieme, Stuttgart, New York.

Frye, M. A., and Olberg, R. M. (1995). Visual receptive field properties of feature detecting neurons in the dragonfly. *J. Comp. Physiol. A* **177,** 569–576.

Geiger, G., and Nässel, D. R. (1981). Visual orientation behaviour of flies after selective laser beam ablation of interneurons. *Nature* **293,** 398–399.

Gewecke, M. (1983). Comparative investigations of locust flight in the field and in the laboratory. *In:* "BIONA—Report 2" (W. Nachtigall, Ed.), Akad. Wiss. Mainz, pp. 11–20. G. Fischer, Stuttgart, New York.

Gewecke, M., and Hou, T. (1993). Visual brain neurons in *Locusta migratoria*. *In:* "Sensory Systems of Arthropods" (K. Wiese *et al.*, Eds.), pp. 119–144. Birkhäuser, Basel.

Gibson, J. J. (1950). "The Perception of the Visual World." Houghton Mifflin, Boston.

Goodman, L. J., Ibbotson, M. R., and Pomfrett, C. J. D. (1990). Directional tuning of the motion-sensitive interneurons in the brain of insects. *In:* "Higher Order Sensory Processing" (D. M. Guthrie, Ed.), pp. 27–48. Manchester University Press, Manchester, New York.

Götz, K. G. (1983a). Bewegungssehen und Flugsteuerung bei der Fliege *Drosophila*. *In:* "BIONA—Report 2" (W. Nachtigall, Ed.), Akad. Wiss. Mainz, pp. 21–34. G. Fischer, Stuttgart, New York.

Götz, K. G. (1983b). Genetic defects of visual orientation in *Drosophila*. *Verh. Dtsch. Zool. Ges.* **1983,** 83–99.

Götz, K. G., Hengstenberg, B., and Biesinger, R. (1979). Optomotor control of wing beat and body posture in *Drosophila*. *Biol. Cybern.* **35,** 101–112.

Gronenberg, W., and Strausfeld, N. J. (1990). Descending neurons supplying the neck and flight motor of diptera: Physiological and anatomical characteristics. *J. Comp. Neurol.* **302,** 973–991.

Gronenberg, W., Milde, J. J., and Strausfeld, N. J. (1995). Oculomotor control in calliphorid flies—organization of descending neurons to neck motor-neurons responding to visual-stimuli. *J. Comp. Neurol.* **361,** 267–284.

Haag, J., Egelhaaf, M., and Borst, A. (1992). Dendritic integration of motion information in visual interneurons of the blowfly. *Neurosci. Lett.* **140,** 173–176.

Hassenstein, B., and Reichardt, W. (1956). Systemtheoretische Analyse der Zeit-, Reihenfolgen- und Vorzeichenauswertung bei der Bewegungsperzeption des Rüsselkäfers *Chlorophanus. Z. Naturforsch.* **11,** 513–524.

Hausen, K. (1976). Functional characterization and anatomical identification of motion sensitive neurons in the lobula plate of the blowfly *Calliphora erythrocephala. Z. Naturforsch.* **31**c, 629–633.

Hausen, K. (1981). Monocular and binocular computation of motion in the lobula plate of the fly. *Verh. Dtsch. Zool. Ges.* **1981,** 49–70.

Hausen, K. (1982a). Motion sensitive interneurons in the optomotor system of the fly. I. The horizontal cells: Structure and signals. *Biol. Cybern.* **45,** 143–156.

Hausen, K. (1982b). Motion sensitive interneurons in the optomotor system of the fly. II. The horizontal cells: Receptive field organization and response characteristics. *Biol. Cybern.* **46,** 67–79.

Hausen, K. (1984). The lobula-complex of the fly: Structure, function and significance in visual behaviour. *In:* "Photoreception and Vision in Invertebrates" (M. A. Ali, Ed.), pp. 523–559. Plenum Press, New York, London.

Hausen, K. (1993). Decoding of retinal image flow in insects. *In:* "Visual Motion and Its Role in the Stabilization of Gaze" (F. A. Miles and J. Wallman, Eds.), pp. 203–235. Elsevier, Amsterdam, London, New York, Tokyo.

Hausen, K., and Egelhaaf, M. (1989). Neural mechanisms of visual course control in insects. *In:* "Facets of Vision" (D. G. Stavenga and R. C. Hardie, Eds.), pp. 391–424. Springer, Berlin, Heidelberg.

Hausen, K., and Wehrhahn, C. (1983). Microsurgical lesion of horizontal cells changes optomotor yaw response in the blowfly *Calliphora erythocephala. Proc. Roy. Soc. Lond. B* **219,** 211–216.

Heisenberg, M., and Wolf, R. (1984). "Vision in *Drosophila.*" Springer, Berlin, Heidelberg, New York.

Heisenberg, M., and Wolf, R. (1993). The sensory-motor link in motion-dependent flight control of flies. *In:* "Visual Motion and Its Role in the Stabilization of Gaze" (F. A. Miles and J. Wallman, Eds.), pp. 265–283. Elsevier, Amsterdam, London, New York, Tokyo.

Heisenberg, M., Wonneberger, R., and Wolf, R. (1978). Optomotor-blind^{H31}- a *Drosophila* mutant of the lobula plate giant neurons. *J. Comp. Physiol.* **124,** 287–296.

Hengstenberg, R. (1977). Spike responses in 'non spiking' visual interneurons. *Nature* **270,** 338–340.

Hengstenberg, R. (1981). Rotatory visual responses of vertical cells in the lobula plate of *Calliphora. Verh. Dtsch. Zool. Ges.* **1981,** 180.

Hengstenberg, R. (1982). Common visual response properties of giant vertical cells in the lobula plate of the blowfly *Calliphora. J. Comp. Physiol. A* **149,** 179–193.

Hengstenberg, R. (1991). Gaze control in the blowfly *Calliphora:* A multisensory, two-stage integration process. *Neurosci.* **3,** 19–29.

Hengstenberg, R. (1993). Multisensory control in insect oculomotor systems. *In:* "Visual Motion and its Role in the Stabilization of Gaze" (F. A. Miles and J. Wallman, Eds.), pp. 285–298. Elsevier, Amsterdam, London, New York, Tokyo.

Hengstenberg, R. (1995). Gain differences of gaze-stabilizing head movements, elicited by wide-field pattern motions, demonstrate in wildtype and mutant *Drosophila*, the importance of HS- and VS-neurons in the third visual neuropil for the control of turning behaviour. *In:* "Nervous Systems and Behaviour" (M. Burrows, P. L. Matheson, H. Newland, H. Schuppe, Eds.), Proc. 4th Int. Cong. Neuroethol., p. 264. Thieme, Stuttgart.

Hengstenberg, R., Bülthoff, H., and Hengstenberg, B. (1983). Three-dimensional reconstruction and stereoscopic display of neurons in the fly visual system. *In:* "Functional Neuroanatomy" (N. J. Strausfeld, Ed.), pp. 183–205. Springer, Berlin, Heidelberg, New York Tokyo.

Hengstenberg, R., Hausen, K. and Hengstenberg, B. (1982). The number and structure of giant vertical cells (VS) in the lobula plate of the blowfly *Calliphora erythrocephala*. *J. Comp. Physiol. A* **149,** 163–177.

Hengstenberg, R., Krapp, H. G., and Hengstenberg, B. (1997). Visual sensation of self-motion in the blowfly *Calliphora*. *In:* "Biocybernetics of Vision: Integrative Mechanisms and Cognitive Processes" (C. Taddei-Ferretti., Ed.), World Scientific Publishers, Singapore, London, New York.

Herrera, C. M. (1992). Activity pattern and thermal biology of a day-flying hawkmoth (*Macroglossum stellatarum*) under mediterranean summer conditions. *Ecol. Entomol.* **17,** 52–56.

Kern, R. (1998). Visual position stabilization in the hummingbird hawk moth, *Macroglossum stellatarum* L. II. Electrophysiological analysis of neurons sensitive to wide field image motion. *J. Comp. Physiol. A* **182,** 239–249.

Kern, R., and Varjú, D. (1998). Visual position stabilization in the hummingbird hawk moth, *Macroglossum stellatarum* L. I. Behavioural analysis. *J. Comp. Physiol. A* **182,** 225–237.

Kimmerle, B., Egelhaaf, M., and Srinivansan, M. V. (1996). Object detection by relative motion in freely flying flies. *Naturwiss.* **83,** 380–381.

Koenderink, J. J., and van Doorn, A. J. (1987). Facts on optic flow. *Biol. Cybern.* **56,** 247–254

Krapp, H. G., and Hengstenberg, R. (1996). Estimation of self-motion by optic flow processing in single visual interneurons. *Nature* **384,** 463–466.

Krapp, H. G., and Hengstenberg, R. (1997). A fast stimulus procedure for determining local receptive field properties of motion-sensitive visual interneurons. *Vision Res.* **37,** 225–234.

Krapp, H. G., Hengstenberg, B., and Hengstenberg, R. (1998). Dendritic structure and receptive-field organization of optic flow processing interneurons in the fly. *J. Neurophysiol.* **79,** 1902–1917.

Land, M. F., and Collett, T. S. (1997). A survey of active vision in invertebrates. *In:* From Living Eyes to Seeing Machines" (M. V. Srinivansan and S. Venkatesh, Eds.), pp. 16–36. Oxford University Press, Oxford, New York.

Lehrer, M. (1994). Spatial vision in the honeybee: The use of different cues in different tasks. *Vision Res.* **34,** 2363–2385.

Lehrer, M. (1997). Honeybee's use of spatial parameters for flower discrimination. *Israel J. Plant Sci.* **45,** 157–167.

MacCann, G. D., and Dill, J. C. (1969). Fundamental properties of intensity, form and motion perception in the visual nervous system of *Calliphora phaenicia* and *Musca domestica*. *J. Gen. Pysiol.* **53,** 385–413.

Mayer, G. (1957). Bewegungsweisen der Odonatengattung *Aeschna*. *Östrr Arbeit Jahrb Wildtierforschung* **1957,** 1–4.

Milde, J. J. (1993). Tangential neurons in the moth *Manduca sexta*. Structure and response to optomotor stimuli. *J. Comp. Physiol.* **173,** 783–799.
Milde, J. J., Seyan, H. S., and Strausfeld, N. J. (1987). The neck motor system of the fly *Calliphora erythrocephala*. II. Sensory organization. *J. Comp. Physiol. A* **160,** 225–238.
Nakayama, K. and Loomis, J. M. (1974). Optical velocity patterns, velocity-sensitive neurons, and space perception: a hypothesis. *Perception* **3,** 63–80.
Nalbach, G. (1994). Extremely non-orthogonal axes in a sense organ for rotation: Behavioural analysis of the dipterian haltere system. *Neurosci.* **61,** 149–163.
O'Carroll, D. (1993). Feature-detecting neurons in dragonflies. *Nature* **362,** 541–543.
Olberg, R. M. (1981). Object- and self-movement detectors in the ventral nerve cord of the dragonfly. *J. Comp. Physiol.* **141,** 327–334.
Pflugfelder, G. O., and Heisenberg, M. (1995) Optomotor-blind of *Drosophila-melanogaster*—A neurogenetic approach to optic lobe development and optomotor behavior. *Comp. Biochem. Physiol.* **A110,** 185–202.
Pierantoni, R. (1976). A look into the cockpit of the fly. The architecture of the lobula plate. *Cell Tissue Res.* **171,** 101–122.
Preiss, R. (1991). Separation of translation and rotation by means of eye-region specialization in flying gypsy moths (Lepidoptera: Lymantriidae). *J. Insect Behavior* **4,** 209–219.
Reichardt, W. (1987). Evaluation of optical motion information by movement detectors. *J. Comp. Physiol. A* **161,** 533–547.
Reichardt, W., and Poggio, T. (1976). Visual control of orientation behaviour in the fly. Part I. A quantitative analysis. *Q. Rev. Biophys.* **9,** 311–375.
Reichert, H., and Rowell, C. H. F. (1986). Neuronal circuits controlling flight in locust: How sensory information is processed for motor control. *Trends Neurosci.* **9,** 281–283.
Rind, F. C. (1990). A directionally selective motion-detecting neurone in the brain of the locust: Physiological and morphological characterization. *J. Exp. Biol.* **149,** 1–19.
Robert, D. (1988). Visual steering under closed-loop conditions by flying locusts: Flexibility of optomotor response and mechanisms of correctional steering. *J. Comp. Physiol. A* **164,** 15–24.
Rosenfeld, A., and Kak, A. C. (1982). "Digital Picture Processing." Academic Press, London.
Rowell, C. H. F. (1988). Mechanisms of flight steering in locusts. *Experimentia* **44,** 389–395.
Single, S., Haag, J., and Borst, A. (1997). Dendritic computation of direction selectivity and gain control in visual interneurons. *J. Neurosci.* **17**(16), 6023–6030.
Soohoo, S. L., and Bishop, L. G. (1980). Intensity and motion responses of giant vertical neurons of the fly eye. *J. Neurobiol.* **11,** 159–177.
Srinivansan, M. V., Zhang, S. W., Lehrer, M. and Collett, T. S. (1996). Honeybee navigation en route to the goal: Visual flight control and odometry. *J. Exp. Biol.* **199,** 237–244.
Srinivasan, M. V. (1993). How insects infer range from motion. *In:* "Visual Motion and its Role in the Stabilization of Gaze" (F. A. Miles and J. Wallman, Eds.), pp. 239–263. Elsevier, Amsterdam, London, New York, Tokyo.
Strausfeld, N. J. (1976). "Atlas of an Insect Brain." Springer, Berlin, Heidelberg, New York.
Strausfeld, N. J. (1989). Beneath the compound eye: Neuroanatomical analysis and physiological correlates in the study of insect vision. *In:* "Facets of Vision" (D. G. Stavenga and R. C. Hardie, Eds.), pp. 317–359. Springer, Berlin, Heidelberg.
Strausfeld, N. J., and Gronenberg, W. (1990). Descending neurons supplying the neck and flight motor of diptera: Organization and neuroanatomical relationships with visual pathways. *J. Comp. Neurol.* **302,** 954–972.
Strausfeld, N. J., Seyan, H. S., and Milde, J. J. (1987). The neck motor system of the fly *Calliphora erythrocephala*. I. Muscles and motor neurons. *J. Comp. Physiol. A* **160,** 205–224.

Strausfeld, N. J., Seyan, H. S., Wohlers, D., and Bacon, J. P.(1983). Lucifer yellow histology. *In:* "Functional Neuroanatomy" (N. J. Strausfeld, Ed.), pp. 132–155. Springer, Berlin, Heidelberg, New York Tokyo.

Wachenfeld, A. (1994). Elektrophysiologische Untersuchungen und funktionelle Charakterisierung männchenspezifischer visueller Interneurone der Schmeissfliege *Calliphora erythrocephala* (Meig.). Doctoral thesis, University of Köln.

Wagner, H. (1982). Flow-field variables trigger landing in flies. *Nature* **297,** 147–148.

Wagner, H. (1986). Flight performance and visual control of flight of free-flying housefly (*Musca domestica* L.). II. Pursuit of targets. *Phil. Trans. Roy. Soc. Lond. B* **312,** 553–579.

Warzecha, A. K., Egelhaaf, M., and Borst, A. (1993). Neural circuit tuning fly visual interneurons to motion of small objects. I. Dissection of the circuit by pharmacological and photoinactivation techniques. *J. Neurophysiol.* **69,** 329–339.

Wehner, R. (1981). Spatial vision in insects. *In:* "Handbook of Sensory Physiology," vol. VIIC, (H. Autrum, Ed.), pp. 287–616. Springer, Berlin.

Wehner, R. (1987). Matched filter—Neuronal models of the external world. *J. Comp. Physiol. A* **161,** 511–531.

A COMMON FRAME OF REFERENCE FOR THE ANALYSIS OF OPTIC FLOW AND VESTIBULAR INFORMATION

B.J.Frost* and D.R.W. Wylie[†]

*Department of Psychology, Queen's University, Kingston, Ontario, Canada and
[†]Department of Psychology, University of Alberta, Edmonton, Alberta, Canada

I. Object Motion versus Self-Motion

The movement of animals through the world has exerted two fundamental evolutionary pressures on visual motion detecting systems: first, to evolve mechanisms to detect the presence of other animals from their image motion relative to the images of stationary features in the world and, second, to evolve other separate mechanisms to extract information about an animal's own self-motion in space relative to its stationary world (Gibson, 1979; Frost *et al.*, 1994; Sun and Frost, 1997). In the first case, the detection of animate or object motion in the visual array can provide vital information for recognizing the presence and location of conspecifics—predators and prey—that each require different behavioral sequences for survival. Likewise, optic-flow patterns can potentially inform animals about their current position and trajectory through their environment (e.g., Lee and Aronson, 1974; Owen, 1990; Koenderink, 1986; Wylie *et al.*, 1998a).

Animate or object motion analysis is accomplished through mechanisms specialized for responding to local motion, but not global motion (Frost, 1978; Frost *et al.*, 1981; Allman *et al.*, 1985; Frost *et al.*, 1988). This is accomplished by a double opponent process, directionally specific, center-surround neural mechanism that responds to motion in the preferred direction in the center of the receptive field but is inhibited when the same direction of motion occurs in the very large surround-

0074-7742/00 $30.00

ing inhibitory receptive field. Thus large patterns of optic flow, such as those generated by an animal's own motion will NOT excite these neurons. In contrast, the motion of a small object moving in the preferred direction, relative to backgrounds moving in other directions or at other velocities, will excite these neurons (Frost and Nakayama, 1983).

The visual self-motion system, which is the focus of this volume, extracts information required to stabilize the eyes, head, and gaze in space; information required also for the maintenance of posture; and information about the animal's motion through space. In fact, it is probably even involved in navigation, distance estimation, and "path integration" (Srinivasan *et al.*, 1989; Esch and Burns, 1996; Whishaw and Maaswinkel, 1998; Wylie *et al.*, 1998a). Much of this optic-flow information, that is specifically generated by the animal's own self-motion, is processed in a specialized visual pathway, the accessory optic system (AOS; for reviews see Simpson, 1984; Simpson *et al.*, 1988a; Grasse and Cynader, 1990), which, we will show later, integrates motion vectors from optic flow across very large areas of visual space.

II. The Accessory Optic System

Figure 1 illustrates the AOS pathway in relation to other visual pathways. Figure 1A shows the three major anatomical pathways simplified to illustrate the major components and the equivalent structures in birds (bold letters, above) and mammals (italics, below). This fundamental structural segregation for avian species is shown in more detail in Fig. 1B, where the same color code identifies the same three pathways, with the tectofugal (colliculopulvinar) pathway shown in gray, the thalamofugal (geniculocortical) pathway shown in white, and the AOS in black.

The AOS appears to be highly conserved in evolution with homologous structures (unfortunately with different names) occurring in all vertebrates (Fite, 1985; McKenna and Wallman, 1985; Weber, 1985). Interestingly, invertebrates also appear to segregate motion analysis into object motion and self-motion systems and may even use similar coordinate systems (Egelhaaf and Borst, 1993; Krapp and Hengstenberg, 1996). In birds the AOS consists of two retinal recipient nuclei—the nucleus of the basal optic root (nBOR), whose input arises from the displaced ganglion cells (Karten *et al.*, 1977; Reiner *et al.*, 1979; Fite *et al.*, 1981), and the pretectal nucleus lentiformis mesencephali (Gamlin and Cohen, 1988a). As shown in Fig. 1B, LM and nBOR have extensive connections with many other visual areas (Clarke, 1977; Wylie *et al.*, 1997,

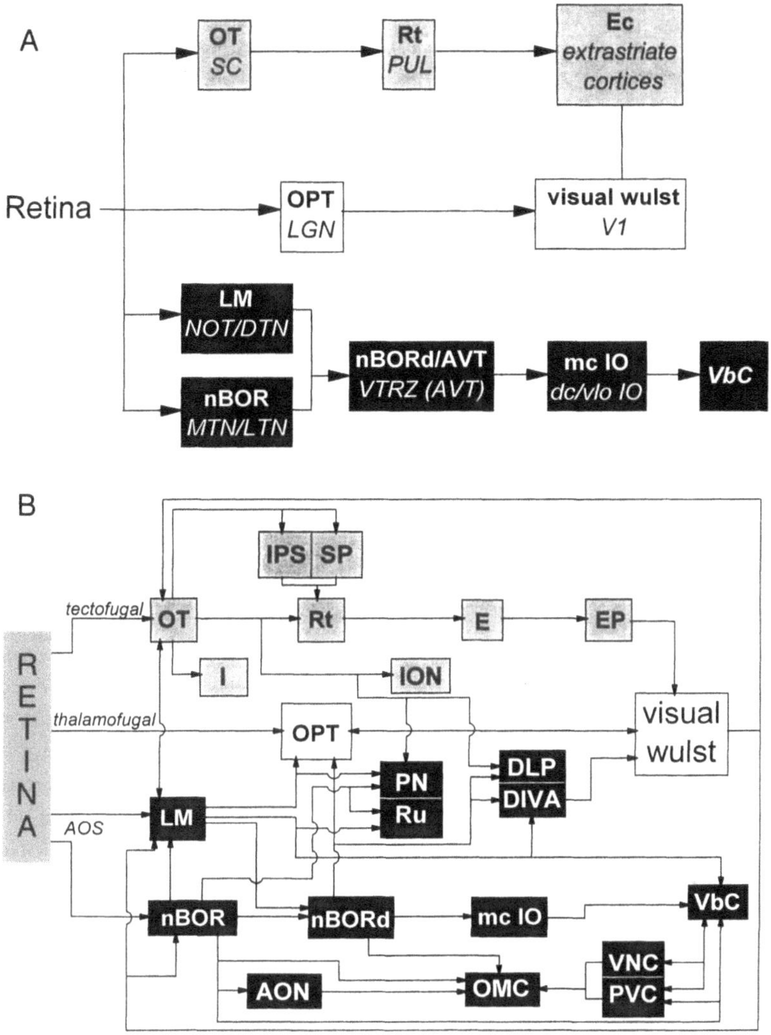

FIG. 1. The three major visual pathways. (A) A very simplified schematic of the visual pathways in birds (bold, above) and the equivalent mammalian structures (italics, below). (B) A more complete illustration of the connectivity within the avian system. Note the abundant interconnections between the three pathways. See text for more details. AVT, area ventralis of Tsai; dc, dorsal cap; DIVA, nucleus dorsalis intermedius ventralis anterior; DLP, nucleus dorsolateralis posterior thalami; DTN, dorsal terminal nucleus; E, Ec, ectostriatum; EP, peri-ectostriatal belt; I, nucleus isthmi; IO, inferior olive; ION, isthmo-optic nucleus; IPS, nucleus interstitio-pretecto-subpretectalis; LGN, lateral geniculate nucelus; LM, nucleus lentiformis mesencephali; LTN, lateral terminal nucleus; mc, medial column; MTN, medial terminal nucleus; nBOR, nucleus of the basal optic root; nBORd, nBOR pars dorsalis; NOT, nucleus of the optic tract; OMC, oculomotor complex; OPT, nucleus opticus principalis thalami; OT, optic tectum; PN, pontine nuclei; PUL, pulvinar; PVC, processus cerebello-vestibularis; Rt, nucleus rotundas; Ru, nucleus ruber; SC, superior colliculus; SP, nucleus subpretectalis; V1, primary visual cortex; VbC, vestibulocerebellum; vlo, ventrolateral outgrowth; VNC, vestibular nuclei complex; VTRZ, visual tegmental relay zone.

1998a; Gamlin and Cohen, 1988b; Arends and Voogd, 1989; Wylie and Linkenhoker, 1996; Lau *et al.,* 1998). Later we will emphasize the pathway from the LM and nBOR to the medial column of the IO (mc IO) which in turn projects as climbing fibers to Purkinje cells in the vestibulocerebellum (VbC). The VbC is a site of visual-vestibular integration: the complex spike (CS) activity of Purkinje cells (which reflects the climbing fiber input) responds to optic-flow stimuli (Graf *et al.,* 1988; Wylie and Frost, 1993, 1999a; Wylie *et al.*, 1993, 1998b) and the simple spike activity of Purkinje cells responds to both visual and vestibular stimulation (e.g., Schwarz and Schwarz, 1983; Ito, 1984; Waespe and Henn, 1987; De Zeeuw *et al.,* 1995).

A. Decomposition of Optic Flow Fields

The AOS in birds and other vertebrates, like other visual pathways, first decomposes visual information from the visual array into elements that are subsequently required for later synthesis and recognition of optic-flow patterns. In the nBOR and LM, therefore, we find neurons are sensitive to large moving patterns, and the preferred directions of movement of these patterns cluster around four cardinal directions. Most birds hold their heads in fairly stereotyped positions during standing, walking, and flying (Erichsen *et al.,* 1989). When this posture is taken into account, nBOR cells have preferred directions that cluster around upward, downward, and backward (i.e., nasal to temporal) directions in their contralateral visual field (Burns and Wallman, 1981; Morgan and Frost, 1981; Gioanni *et al.,* 1984; Wylie and Frost, 1990a), whereas most LM neurons prefer forward motion in the contralateral visual field (Winterson and Brauth, 1985; Wylie and Frost, 1996). Figure 2 illustrates the preferred directions to which nBOR and LM cells respond optimally. It should also be noted that these different preferred directions are represented differentially in different topological zones of nBOR (Burns and Wallman, 1981; Wylie and Frost, 1990a).

These nBOR cells also have several other features that clearly differentiate them from directional cells found in the tectofugal and thalamofugal pathways. These are contrasted in Table I and show that receptive field size, preferred velocity, receptive field structure, and response habituation are quite different from tectal object motion-sensitive neurons (Morgan and Frost, 1981; Frost, 1982, 1985; Frost *et al.,* 1994; Sun and Frost, 1997). However, the functionally important differences are that these AOS directionally specific cells integrate motion over a very large area of visual field, they do not adapt or habitu-

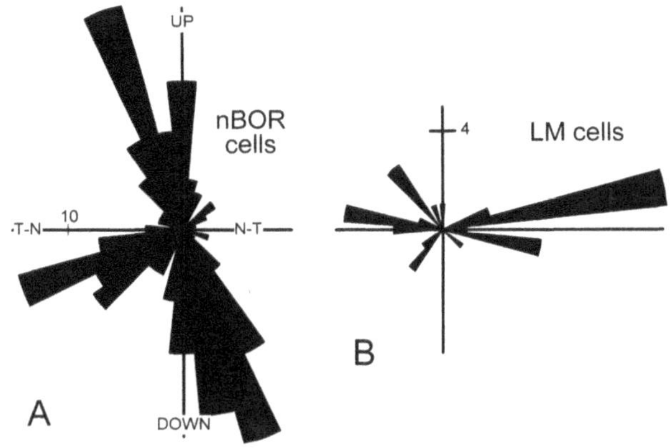

FIG. 2. Preferred directions of neurons in the AOS of pigeons. Polar histograms show the preferred directions of neurons in nBOR (A; adapted from Wylie and Frost, 1990a) and LM (B; adapted from Wylie and Frost, 1996). The preferred direction for each neuron was calculated by a vector analysis of the directional tuning curve in response to large-field stimulus (about 100 × 100°) moving in the contralateral visual field. Note that the majority of neurons prefer one of the cardinal directions: nBOR neurons preferred stimuli moving either upward, downward, or backward (i.e., nasal to temporal; N-T) in the contralateral visual field, whereas most LM neurons preferred forward (T-N) motion.

ate to repeated stimulation, and they are maximally excited by optic flow fields that simulate those produced by an animal's self-motion. In contrast, these very same types of stimuli typically inhibit local motion detectors that are putatively processing animate/object motion like those found in the tectofugal pathway.

B. Binocular Integration

The large majority of cells in the rabbit terminal nuclei and pigeon nBOR have very large monocular receptive fields (Simpson and Alley, 1974; Collewijn, 1975; Burns and Wallman, 1981; Hoffman and Schoppmann, 1981; Morgan and Frost, 1981; Grasse and Cynader, 1982, 1984; Soodak and Simpson, 1988; Rosenberg and Ariel, 1990; Wylie and Frost, 1990a, 1996). However, a small population of binocularly sensitive neurons have been found in the visual tegmental relay zone of rabbits (Simpson *et al.*, 1988b), nBOR of pigeons (Wylie and Frost, 1990b, 1999b), and LM of newts (Manteuffel, 1987). Simpson and his colleagues (Graf *et al.*, 1988; Leonard *et al.*, 1988; Simpson *et al.*, 1988b) showed first in rabbits that some MTN neurons had two spatially separated receptive field zones (bipartite receptive fields) in their monoc-

TABLE I
A COMPARISON OF DIFFERENT FEATURES OF MOTION-SPECIFIC NEURONS FOUND IN THE OPTIC TECTUM AND THE ACCESSORY OPTIC SYSTEM.

	Tectum	AOS
(i) RF structure	Small ERF surrounded by a large IRF with double-opponent process structure	Large ERF, with only single opponent process structure; 2nd-order neurons have bipartite RFs
(ii) Optimal stimulus	Responds best to small moving stimuli	Responds best to large textured patterns; 2nd-order neurons prefer translational or rotational optic flow
(iii) Velocity tuning	Prefers moderate to fast Velocities, 20–90°/s	Prefers slow velocities 1–10°/s
(iv) Spontaneous firing	Normally exhibits no spontaneous firing	Relatively high spontaneous firing
(v) Adaptation	Normally reduced responses to repeated stimulation	Maintains response to repeated stimulation
(vi) Direction preference	Fairly broad tuning curves with well-defined null directions	Sinusoidal tuning curves; 1st-order neurons prefer cardinal directions; 2nd-order neurons aligned with semicircular canals
(vii) Surround modulation	Responses dramatically altered by surround	No surround modulation
(viii) Proposed function	Object/animate motion detection, figure/ground segregation, etc.	Detection of self-motion of head and body through space, etc.

ular visual field, where directional preferences for flow fields were opposite. They also showed that cells in VTRZ, IO, and VbC integrate visual flow information from both eyes and in such a way that rotational flow fields, like those generated with a planetarium projector, optimally stimulate these binocular neurons. It should be stressed that this form of binocular integration is quite distinctively different from that which occurs in cortical and wulst cells for retinal disparity detection for stereopsis (Barlow *et al.,* 1967; Pettigrew *et al.,* 1968; Poggio and Fischer, 1977; Freeman and Ohzawa, 1988; Wagner and Frost, 1994), and the binocular integration that occurs in the AOS of frontal-eyed animals (Hoffman and Schoppman, 1975, 1981; Grasse and Cynader, 1982, 1984, 1990). Here in the rabbit AOS, different directions of motion from different directions of binocular panoramic space are being integrated to represent the binocular flow field. In pigeons, very similar effects have been found (Wylie and Frost, 1990b, 1993, 1999a,b; Wylie *et al.,* 1993, 1998b). At first, using large tangent screens to stimulate each eye separately, it was found that some dorsal nBOR neurons and VbC Purkinje cells were integrating either the same cardinal directions of flow from each eye or opposite directions of flow (Wylie and Frost, 1990b; Wylie *et al.,* 1993). The fact that integration was confined to either one of these two alternatives, and not other possible combinations of the cardinal directions, suggested that binocular integration of the different directions might be associated with rotational flow analysis, whereas integration of the same directions might be associated with translational flow field analysis (Wylie and Frost, 1990b; Wylie *et al.,* 1993).

C. Simulating Rotational and Translation Optic Flow

The possibility therefore exists that flow field vectors from the entire binocular optic flow field generated by an animal's rotation or translation in space might be integrated by these binocularly sensitive AOS neurons. Consequently, visual stimulators that simulate these distinctive patterns of rotation and translation need to be used rather than the spatially restricted patterns presented by conventional tangent screen displays. The planetarium projector (see Fig. 3A) first used by Simpson's group (Simpson *et al.,* 1981) produces a panoramic rotation pattern that is a close approximation of flow fields resulting from an eye or head rotation by the animal. When the planetarium projector, producing rotational flow fields like the one illustrated in Fig. 3B, is placed in gimbals so that the axis of rotation can be positioned in any orientation in 3-D space, then the optimal axis of rotation can be determined for a binocularly sensitive AOS neuron tuned to opposite directions of motion in different parts of its binocular receptive field (RF). However, this device cannot produce panoramic translatory patterns, and so we developed a translating projector, illustrated in Fig. 3C. This device consisted of a small hollow sphere with holes drilled at random positions across its surface. A small filament light was placed inside the sphere and could be moved, under computer control, along a section of the sphere's diameter. Spots of light are projected from the filament through the holes in the sphere onto the ceiling, walls, and floor of the room. When such a translator is placed just above an animal's head, it produces a good approximation of a translatory flow field with expansion of dots at one pole, laminar flow at the equator, and a contracting pattern of dots at the other pole (see Fig. 3D). When placed in gimbals that allow the axis of motion of the translator to be positioned in any direction in 3-D space, the optimal axis for translation-sensitive binocular AOS neurons can be determined.

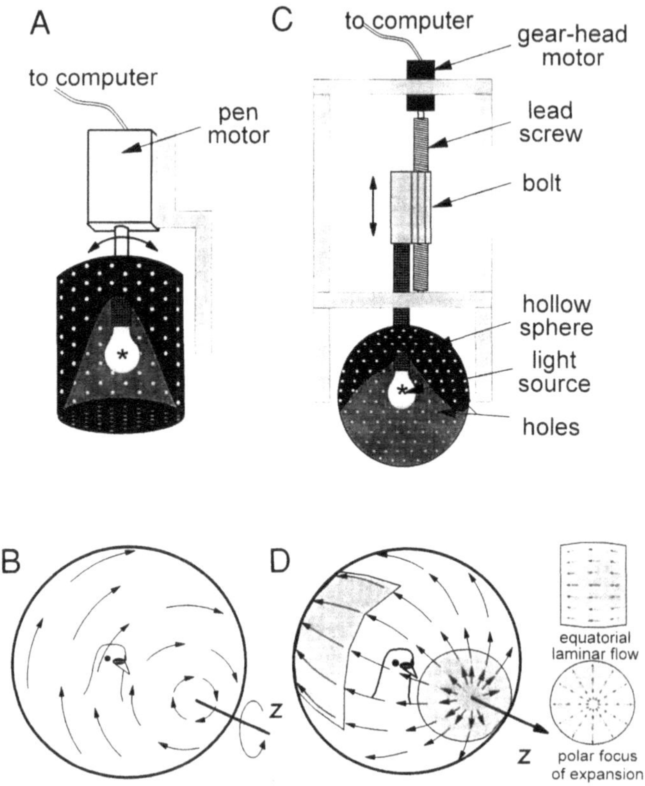

FIG. 3. Simulating rotational and translational optic flow fields. (A) The planetarium we used to stimulate rotation-sensitive neurons. It consisted of a small tin cylinder pierced with numerous small holes. A small filament light source was placed in its center such that a pattern of light dots was projected onto the walls, ceiling and floor of the room. The pen motor oscillated the cylinder about its long axis, effectively producing a rotational flow field. (C) The translator projector used to present translational flow fields. For this device, a light source was moved along a segment of diameter path within a small, hollow, stationary sphere, the surface of which was pierced with holes. This effectively projected a translational flow field onto the floor, walls, and ceiling of the room. (B and D) (with permission, Wylie *et al.*, 1998b) schematics of flow fields resulting from both self-rotation and self-translation. The arrows, as projected onto a sphere, represent local image motion in the flow field. (B) The rotational flow field that would result from a clockwise rotation of the pigeon about the *z*-axis. Note that the flow field is counterclockwise. (D) The translational flow field that would result from a pigeon translating forward (i.e., out of the page) along the *z*-axis. The shaded areas indicate differences in local motion in the translational flow field. At the "pole" in the direction of translation, the arrows radiate outward along "lines of longitude" from "focus of expansion." Likewise, at the pole behind the pigeon's head, the arrows converge to the "focus of contraction." At the "equators" of the sphere, the optic flow is laminar, with all arrows pointing in approximately the same direction.

Using either the planetarium projector or the translation projector, we have been able to show that cells in the dorsal zone of nBOR and the complex spikes of Purkinje cells in the flocculus, nodulus, and uvula of the VbC respond optimally to either rotational or translational flow fields (Wylie and Frost, 1999a,b). Figures 4 and 5 give an example of a rotational and translational cell respectively. In these experiments, we first isolated neurons in the nBOR or VbC of anesthetized pigeons and tested directional sensitivity in the central visual field of each eye with a very large hand-held patterns rich in texture. If the cell preferred the same direction of motion in each eye's visual field, the neuron was most likely a translation cell, and we then proceeded to conduct quantitative observations using the translating projector system. If, on the other hand, the cell preferred the opposite directions of motion of the large hand-held stimulus pattern in each eye, the neuron was most likely a rotation cell, and we then used the planetarium projector to make systematic quantitative observations. Post-stimulus-time-histograms (PSTHs) for different directions of the axes of rotation or translation were collected in order to determine the axis of optimal stimulation for each neuron. Figures 4 and 5 show results typical for rotation-specific neurons and translation-specific neurons, respectively. Note that, typical of all AOS cells, the firing rate of these neurons are modulated by different directions of motion. In other words, AOS cells tend to have a moderate spontaneous firing rate that is either increased by the preferred directions of motion or decreased by the nonpreferred directions of motion of flow fields. Thus these somewhat broad tuning curves are best fitted with a cosine function as shown on the right in Figs. 4 and 5, and the optimal axis for either rotation or translation is determined from the peak of these functions. When tested with rotation or translation axes that are located in an orthogonal plane to the one containing their maximal and minimal firing rates, these cells show little modulation. Thus it seems clear that cells in the AOS pathway ultimately integrate different local vectors of motion so that they specifically respond to either translational or rotational optic flow over their entire binocular visual fields.

D. Coordinate Frame of Reference

When Simpson and his colleagues determined the best axes for rotational flow field stimulation in VTRZ, IO, and VbC neurons in the rab-

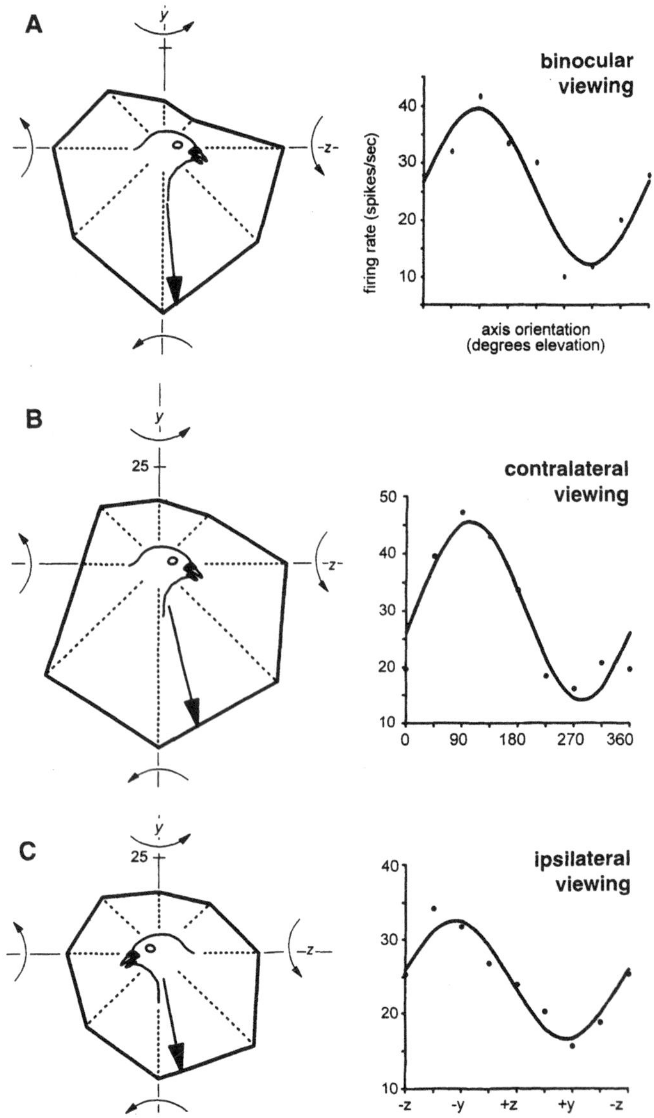

FIG. 4. A rotation-sensitive binocular cell in the nucleus of the basal optic root. The recording was from nBOR on the left side of the head. When tested with a large hand-held stimulus, this cell preferred forward (i.e., temporal to nasal) and backward motion in the contralateral and ipsilateral eyes, respectively. The responses of this cell to rotational flow fields produced by the planetarium projector are shown. Elevation tuning is shown in the sagittal plane under binocular (A), contralateral (B), and ipsilateral (C) viewing conditions. Firing rate is plotted as a function of the axis of rotation in polar coordinates (polar plots). The solid arrows indicate the preferred axis determined from the best cosine fits, which are shown on the right. This cell responded best to the flow field resulting from a leftward rotation about the vertical axis ($-y$ rotation).

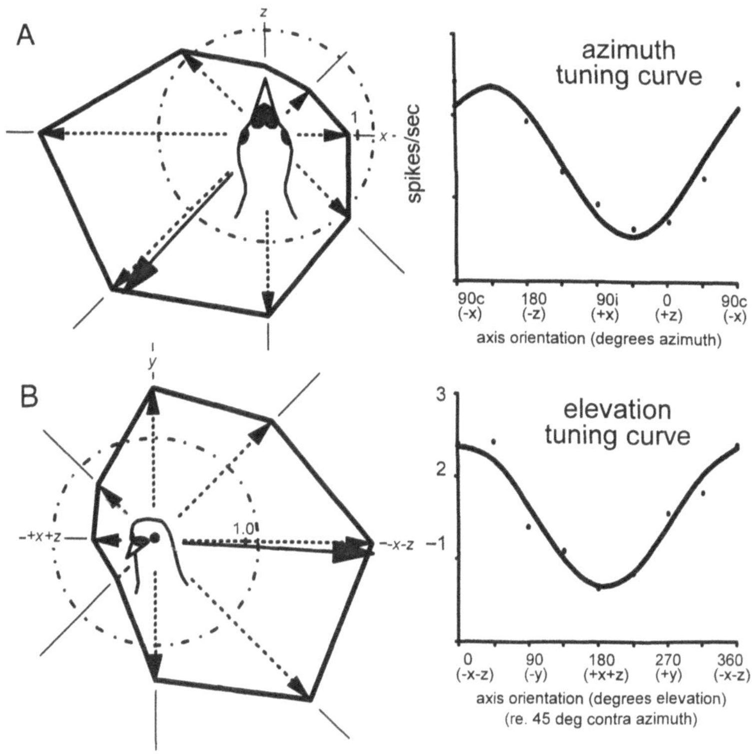

FIG. 5. A translation-sensitive Purkinje cell in the vestibulocerebellum. The recording was from the VbC on the left side of the head. When tested with a large hand-held stimulus, this complex spike activity of this cell responded best to forward motion in both hemifields. The responses of this cell to translational flow fields produced by the translator projector are shown. (A) A polar plot of an azimuthal tuning curve in the horizontal plane. (B) An elevation tuning curve in the vertical plane that intersects the horizontal plane at 45° contralateral azimuth (i.e., elevation tuning is shown in the plane normal to the vector $x-z$). In these curves, the arrowheads point in the direction the animal would move to cause such a flow field. That is, the arrowheads point to the focus of expansion in the translational flow field. The broken circles represent this cell's spontaneous firing rate. Best cosine fits to these tuning curves are on the right. The tuning curves show that this cell responded best to a flowfield with the focus of expansion in the horizontal plane at 135° ipsilateral azimuth. The best vector, indicated by the solid arrows, was approximately $-x-z$.

bit, they found that these axes clustered around the same axes of rotation that optimally stimulate the vestibular semicircular canals (Graf *et al.*, 1988; Leonard *et al.*, 1988; Simpson *et al.*, 1981, 1988b). Likewise, in similar experiments using a planetarium projector to stimulate rotation-sensitive cells in the lateral VbC of pigeons, we also found that the axes of rotation that produced the maximal and minimal firing rates were clustered around the reference axes of the semicircular canals as illustrated in Figs. 6C and 6D (Wylie and Frost, 1993). Moreover, when translation-specific neurons in the medial VbC were similarly tested in a quantitative fashion with the translator projection system and their axes of maximal and minimal firing determined, we also found these to be clustered around the same three orthogonal axes as the semicircular canals and the rotational optic flow system (Wylie *et al.*, 1998b; Wylie and Frost, 1999b). This is shown in Figs. 6A and 6B.

These axes are vertical and 45° to either side of the midline in the azimuthal plane. Thus it appears that the vestibular system and the self-motion optic flow analyzing system share a common frame of reference. Since there is abundant evidence that the visual and vestibular system work synergistically (Waespe and Henn, 1981; Wong and Frost, 1981; Ito, 1984), it is not surprising that visual and vestibular information are processed in a common coordinate system, since this would make integration of information between these two modalities very simple. What is not so obvious is why the frame of reference should be organized with one axis vertical and the other two 45° from the midline.

There are usually many separate forces that exert evolutionary pressure on neural systems to evolve toward optimal solutions of information processing. In the case of detecting information about an animal's position and path through space, some of these forces most likely are (a) neural economy, or the requirement to accomplish the task with as few specialized pattern-specific neurons as possible (Simpson and Graf, 1985); (b) a pressure to develop maximal sensitivity where it is most needed; and (c) to preserve bilateral symmetry in systems as far as possible (Simpson and Graf, 1985).

Since the motion of *any* object through space, including the self-motion of organisms, can be described with reference to six degrees of freedom, rotation about three orthogonal axes, and translation along those three axes, it is consistent with the principle of neural economy that the vestibular system and the visual self-motion systems should possess three rotation-sensitive subsystems organized in orthogonal planes, and three translational sensitive subsystems organized similarly. It may also be argued that whatever coordinate system is the most appropriate organization for one of these sensory systems, the other will most likely share this same coordinate frame to allow efficient neural integration of the inertial and visual optic flow information.

Figures 7A and 7B show two possible arrangements of semicircular canals for coding information about head rotation. Simpson and Graf (1985) have argued that the arrangement shown in Fig. 7A, consisting

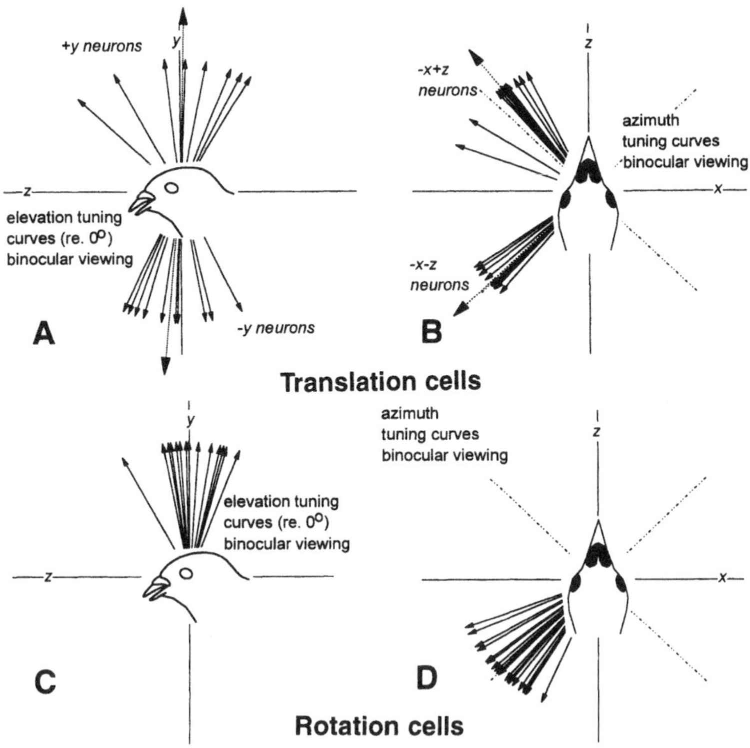

FIG. 6. The best axes of translation and rotation for Purkinje cells in the vestibulocerebellum of pigeons. (A) The distribution of best axes of translation in the sagittal plane for $+y$ and $-y$ translation neurons. (B) The distribution of the best axes in the horizontal plane for $-x+z$ and $-x-z$ translation neurons. (C and D) The best axes of rotation-sensitive cells in the flocculus determined from elevation tuning curves in the sagittal plane and azimuth tuning curves in the horizontal plane, respectively. See text for details. (Adapted from Wylie and Frost, 1993; with permission Wylie *et al.,* 1998b).

of a vertical axis and two horizontal axes oriented 45° to the midline, is more "economical" for the analysis of self-rotation than the arrangement shown in Fig. 7B, which has roll, pitch, and yaw as the principal axes. The arrangement in Fig. 7A satisfies the constraints of bilaterally symmetry and is organized as a "push-pull" or opponent process system. A leftward rotation of the head about the vertical axis excites afferents in the left horizontal canal and inhibits afferents in the right horizontal canal. A head rotation about the horizontal axis 45° to the midline (e.g., the $+x+z$ axis in the direction indicated by the arrows) maximally excites the left anterior canal and inhibits the right posterior canal. With the arrangement in Fig. 7B, consisting of roll, pitch, and yaw axes, al-

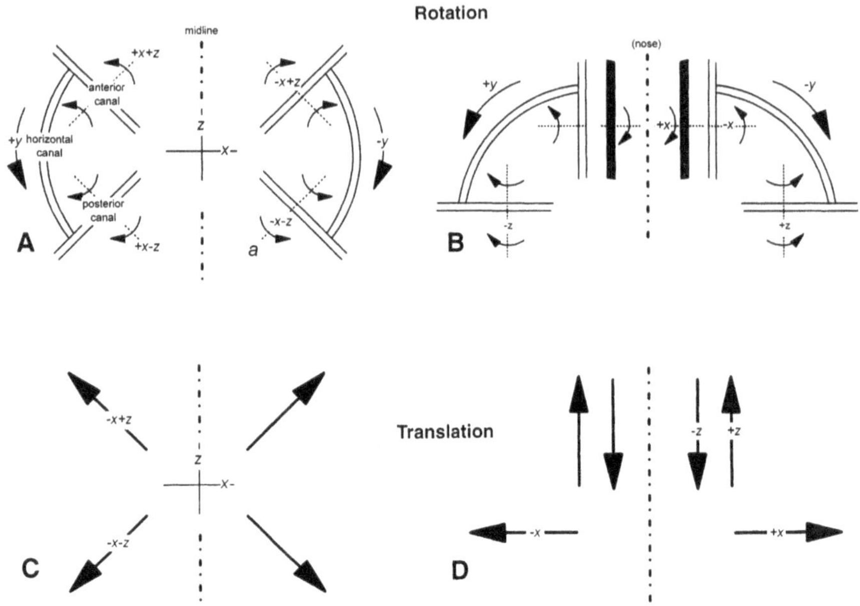

FIG. 7. Efficiency of the common reference frame for self-translation and self-rotation modeled after arguments by Simpson and Graf (1985). (A) The orientation of the vestibular canals in most species: a three axes system that fulfills the requirements of bilateral symmetry and a push-pull organization. (B) A rotation system organized such that the principal axes are the *x*-, *y*- and *z*-axes, where a fourth channel is necessary. (C and D) Diagrams that show that the same argument holds for a self-translation system. See text for a detailed discussion. (From Wylie and Frost, 1999, with permission.)

though bilaterally symmetric, three pairs of canals are insufficient to satisfy the requirement of a push-pull organization. In this arrangement, a rotation about the *y*-axis is not organized in a push-pull fashion. An extra pair of canals (solid black in the figure) would be necessary.

Our findings support a push-pull, bilaterally symmetric reference frame depicted in Fig. 7C for the analysis of self-translation. Like the reference frame for self-rotation shown in Fig. 7A, this system consists of two horizontal axes oriented 45° to the midline as shown in Fig. 7C. This is more economical than the arrangement shown in Fig. 7D, which has the *x*- and *z*-axes as principle axes: as in Fig. 7B, three horizontal axes would be required. Either way, to satisfy the constraints of bilateral symmetry and a push-pull system, both $+y$ and $-y$ *neurons* would be needed on both sides of the brain. Thus, for self-translation, the four-channel (or six-channel, if both sides of brain are considered), three-axes system consisting of $-x+z$, $-x-z$, $+y$, and $-y$ neurons is more economical in satisfying the constraints of bilateral symmetry and a push-pull system than the arrangement shown in Fig. 7D. The conjoint pressures for neural economy and bilateral symmetry would most likely therefore result in a system as depicted in Figs. 7A and 7C rather than Figs. 7B and 7D.

In terms of the optimal axes for translation the point is often made [see, for example, Gibson's (1979) original drawing of flow fields] that the optic flow field for straight-ahead motion consists of an expansion pole directly in front of the animal, a contraction pole behind, and laminar flow vectors at the equator (see Fig. 3D). Since most animals seem

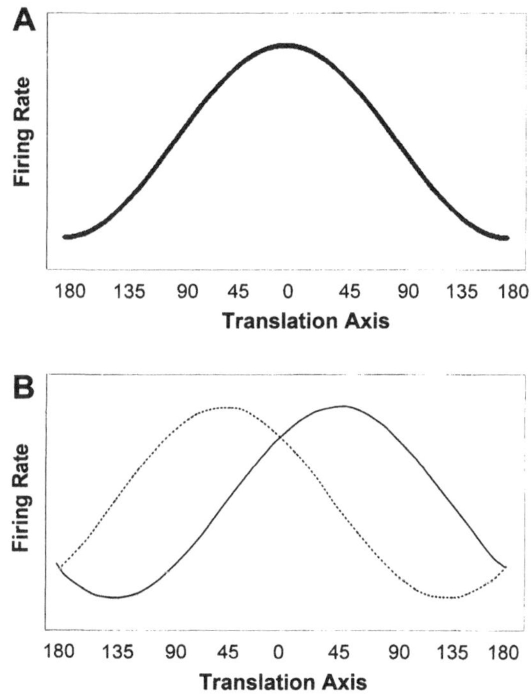

FIG. 8. Two possible arrangements for monitoring forward translation of an animal along a straight ahead path. (A) Translational tuning where the axis of maximal modulation lies on the body midline axis. Note how small departures from the "straight ahead" direction would result in minimal changes in firing rate of neurons tuned in this manner. (B) The arrangement of translation neurons found in pigeons using the translator projector. Note that these are organized in the same frame of reference as the semicircular canals and rotational visual flow field units. With such an arrangement, slight deviations from the straight ahead fall on the more sensitive flanks of the tuning curves, and this scheme allows easy disambiguation of left-right differences. Other axes of maximal sensitivity have been omitted for clarity.

to have adopted a "follow your nose" locomotion strategy where the default direction is "straight ahead," it might be assumed that the most efficient frame of reference for translation might be to have one set of optic flow translation-sensitive neurons tuned to the animals longitudinal body axes, as shown in Fig. 8A, with other orthogonal sets tuned to vertically and horizontally transverse directions. On closer examination, however, this arrangement does not seem optimal. If the directional tuning is approximately sinusoidal, as shown repeatedly in AOS tuning curves, then slight variations from the straight-ahead direction would result in minimal change in firing rate of the forward translation neurons, and left and right variations in direction could be ambiguous, thus compromising steering accuracy required for course maintenance. If however the translation axes were rotated 45° as shown in Fig. 8B, then bilateral symmetry is preserved, and the more sensitive regions of the translational tuning curves fall on the default straight-ahead direction. Moreover, steering would also be facilitated as left/right differences would be disambiguated, and an opponent process system would be readily implemented to accomplish this process. Given all these conjoint constraints, then it seems that the common frame of reference oriented with axes vertical and 45° to either side of the midline is inevitable.

III. Conclusion

Optic flow resulting from self-motion through the world appears to be processed in the accessory optic system where direction-specific neurons have quite different characteristics from directional-specific neurons in the object motion, tectofugal visual pathway. Most importantly, AOS neurons have very large receptive fields, do not habituate to repeated stimulus presentations, prefer large texture patterns of motion, and have preferred directions that cluster around upward, downward, backward, and forward directions of motion. Subsequently, it appears that information from the two eyes is combined in the AOS and VbC, since these higher-order neurons have binocular-receptive fields that prefer either the same or different directions of motion in the central visual field of each eye. When these binocular cells are tested with either a planetarium projector or a translating projector, that produce panoramic rotational or translational flowfields, respectively, six groups of flow-field neurons can be identified. Three of these groups respond maximally to rotation around three separate orthogonal axes, and three groups respond maximally to translation along the same three orthogonal axes. These axes are precisely those used by the semicircular canals, and reasons for this coordinate framework are discussed.

Acknowledgments

We wish to thank Sharon David and Randall Glover for expert technical assistance. The Natural Sciences and Engineering Research Council of Canada (BJF and DRWW), the National Centres of Excellence Program (BJF), the Alberta Heritage Foundation for Medical Research (DRWW), and the Alexander von Humboldt Foundation (BJF) provided generous support for different phases of this work.

References

Allman, J., Miezin, F., and MacGinnes, E. (1985). Stimulus specific responses from beyond the classical receptive field: Neurophysiological mechanisms for local-global comparisons in visual neurons. *Ann. Rev. Neurosci.* **8,** 407–430.

Arends, J. J. A., and Voogd, J. (1989). Topographic aspects of the olivocerebellar system in the pigeon. *Exp. Brain Res. Suppl.* **17,** 52–57.

Barlow, H. B., Blakemore, C., and Pettigrew, J. D. (1967). The neural mechanism of binocular depth discrimination. *J. Physiol.* **193,** 327–342.

Brecha, N., Karten, H. J., and Hunt, S. P. (1980). Projections of the nucleus of basal optic root in the pigeon: An autoradiographic and horseradish peroxidase study. *J. Comp. Neurol.* **189,** 615–670.

Burns, S., and Wallman, J. (1981). Relation of single unit properties to the oculomotor function of the nucleus of the basal optic root (AOS) in chickens. *Exp. Brain Res.* **42,** 171–180.

Clarke, P. G. H. (1977). Some visual and other connections to the cerebellum of the pigeon. *J. Comp. Neurol.* **174,** 535–552.

Collewijn, H. (1975). Direction-selective units in the rabbit's nucleus of the optic tract. *Brain Res.* **100,** 489–508.

De Zeeuw, C. I., Wylie, D. R., Stahl, J. S., and Simpson, J. I. (1995). Phase relations of Purkinje cells in the rabbit flocculus during compensatory eye movements. *J. Neurophysiol.* **74,** 2051–2064.

Egelhaaf, M., and Borst, A. (1993). A look into the cockpit of the fly: Visual orientation, algorithms, and identified neurons. *J. Neurosci.* **13,** 4563–4574.

Erichsen, J. T., Hodos, W., Evinger, C., Bessette, B. B., and Phillips, S. J. (1989). Head orientation in pigeons: Postural, locomotor and visual determinants. *Brain Behav. Evol.* **33,** 268–278.

Esch, H. E., and Burns, J. E. (1996). Distance estimation by foraging honeybees. *J. Exp. Biol.* **199,** 155–162.

Fite, K. V. (1985). Pretectal and accessory-optic visual nuclei of fish, amphibia and reptiles: Theme and variations. *Brain Behav. Evol.* **26,** 192–202.

Fite, K. V., Brecha, N., Karten, H. J., and Hunt, S. P. (1981). Displaced ganglion cells and the accessory optic system of the pigeon. *J. Comp. Neurol.* **195,** 279–288.

Freeman, R. I., and Ohzawa, I. (1988). Monocularly deprived cats: Binocular tests of cortical cells reveal functional connections from the deprived eye. *J. Neurosci.* **8,** 2491–2506.

Frost, B. J. (1978). Moving background patterns alter directionally specific responses of pigeon tectal neurons. *Brain Res.* **151,** 359–365.

Frost, B. J. (1982). Mechanisms for discriminating object motion from self-induced motion in the pigeon. *In:* "Analysis of Visual Behavior" (D. J. Ingle, M. A. Goodale, and J. W. Mansfield, Eds.), pp. 177–196. MIT Press, Cambridge, MA.

Frost, B. J. (1985). Neural mechanisms for detecting object motion and figure-ground boundaries contrasted with self-motion detecting systems. *In:* "Brain Mechanisms of Spatial Vision" (D. Ingle, M. Jeannerod, and D. Lee, Eds.), pp. 415–449. Matrinus Nijhoft, Dordrecht.

Frost, B. J., and Nakayama, K. (1983). Single visual neurons code opposing motion independent of direction. *Science* **220,** 744–745.

Frost, B. J., Scilley, P. L., and Wong, S. C. (1981). Moving background patterns reveal double-opponency of directionally specific pigeon tectal neurons. *Exp. Brain Res.* **43,** 173–185.

Frost, B. J., Cavanaugh, P., and Morgan, B. J. (1988). Deep tectal cells in pigeons respond to kinematograms. *J. Comp. Physiol. A* **162,** 639–647.

Frost, B. J., D. R. Wylie, and Y-C. Wang (1994). The analysis of motion in the visual systems of birds. *In:* "Perception and Motor Control in Birds" (P. Green and M. Davies, Eds.), pp. 249–266. Springer-Verlag, Berlin.

Gamlin, P. D. R ., and Cohen, D. H. (1988a). The retinal projections to the pretectum in the pigeon (*Columba livia*). *J. Comp. Neurol.* **269,** 1–17.

Gamlin, P. D. R., and Cohen, D. H. (1988b). Projections of the retinorecipient pretectal nuclei in the pigeon (*Columba livia*). *J. Comp. Neurol.* **269,** 18–46.

Gibson, J. J. (1979). "The ecological approach to visual perception." Houghton Mifflin, Boston.

Gioanni, H., Rey, J., Villalobos, J., and Dalbera, A. (1984). Single unit activity in the nucleus of the basal optic root (nBOR) during optokinetic, vestibular and visuovestibular stimulations in the alert pigeon (*Columba livia*). *Exp. Brain Res.* **57,** 49–60.

Graf, W., Simpson, J. I., and Leonard, C. S. (1988). Spatial organization of visual messages of the rabbit's cerebellar flocculus. II. Complex and simple spike responses of Purkinje cells. *J. Neurophysiol.* **60,** 2091–2121.

Grasse, K. L., and Cynader, M. S. (1982). Electrophysiology of the medial terminal nucleus of the cat accessory optic system. *J. Neurophysiol.* **48,** 490–504.

Grasse, K. L., and Cynader, M. S. (1984). Electrophysiology of the lateral and dorsal terminal nuclei of the cat accessory optic system. *J. Neurophysiol.* **51,** 276–293.

Grasse, K. L., and Cynader, M. S. (1990). The accessory optic system in frontal-eyed animals. *In:* "Vision and Visual Dysfunction: "The Neuronal Basis of Visual Function," (A. Leventhal, Ed.), vol. IV, pp. 111–139. MacMillan, New York.

Hoffmann, K. P., and Schoppmann, A. (1975). Retinal input to direction selective cells in the nucleus tractus opticus of the cat. *Brain Res.* **99,** 359–366.

Hoffmann, K-P., and Schoppmann, A. (1981). A quantitative analysis of direction-specific response of neurons in the cat's nucleus of the optic tract. *Exp. Brain Res.* **42,** 146–157.

Ito, M. (1984). "The Cerebellum and Motor Control." Raven Press, New York.

Karten, H. J., Fite, K. V., and Brecha, N. (1977). Specific projection of displaced retinal ganglion cells upon the accessory optic system in the pigeon (*Columba livia*). *Proc. Nat. Acad. Sci. USA* **74,** 1752–1756.

Koenderink, J. J. (1986). Optic flow. *Vision Res.* **26,** 161–180.

Krapp, H. G., and Hengstenberg, R. (1996). Estimation of self-motion by optic flow processing in single visual interneurons. *Nature* **384,** 463–466.

Lau, K. L., Glover, R. G., Linkenhoker, B., and Wylie, D. R. W. (1998). Topographical organization of inferior olive cells projecting to translation and rotation zones in the vestibulocerebellum of pigeons. *Neurosci.* **85,** 605–614.

Lee, D. N., and Aronson, E. (1974). Visual proprioceptive control of standing in human infants. *Percept. Psychophys.* **15,** 529–532.

Leonard, C. S., Simpson, J. I., and Graf, W. (1988). Spatial organization of visual messages of the rabbit's cerebellar flocculus. I. Typology of inferior olive neurons of the dorsal cap of Kooy. *J. Neurophysiol.* **60,** 2073–2090.

Manteuffel, G. (1987). Binocular afferents to the salamander pretectum mediate sensitivity of cells selective for visual background motions. *Brain Res.* **422,** 381–383.

McKenna, O., and Wallman, J. (1985). Accessory optic system and pretectum of birds: Comparisons with those of other vertebrates. *Brain Behav. Evol.* **26,** 91–116.

Morgan, B., and Frost, B. J. (1981). Visual response properties of neurons in the nucleus of the basal optic root of pigeons. *Exp. Brain Res.* **42,** 184–188.

Owen, D. H. (1990). Perception & control of changes in self-motion: A functional approach to the study of information and skill. *In:* "Perception & Control of Self-Motion" (R. Warren and A. F. Fuchs, Eds.), pp. 289–322. Lawrence Erlbaum, Hillsdale, NJ.

Pettigrew, J. D., Nikara, T., and Bishop, P. O. (1968). Binocular interaction on single units in cat striate cortex: Simultaneous stimulation by single moving slit with receptive fields in correspondence. *Exp. Brain Res.* **6,** 391–410.

Poggio, G. F., and Fischer, B. (1977). Binocular interactioin and depth sensitivity in striate and prestriate cortex of behaving rhesus monkey. *J. Neurophysiol.* **40,** 1392–1405.

Reiner, A., Brecha, N., and Karten, H. J. (1979). A specific projection of retinal displaced ganglion cells to the nucleus of the basal optic root in the chicken. *Neurosci.* **4,** 1679–1688.

Rosenberg, A. F., and Ariel, M. (1990). Visual-reponse properties of neurons in the turtle basal optic nucleus in vitro. *J. Neurophysiol.* **63,** 1033–1045.

Schwarz, I. E., and Schwarz, D. W. F. (1983). The primary vestibular projection to the cerebellar cortx in the pigeon (*Columba livia*). *J. Comp. Neurol.* **216,** 438–444.

Simpson, J. I. (1984). The accessory optic system. *Ann. Rev. Neurosci.* **7,** 13–41.

Simpson, J. I., and Alley, K. E. (1974). Visual climbing fibre input to rabbit vestibulocerebellum: a source of direction-specific information. *Brain Res.* **82,** 302–308.

Simpson, J. I., and Graf, W. (1985). The selection of reference frames by nature and its investigators. *In:* "Adaptive Mechanisms in Gaze Control: Facts and Theories" (A. Berthoz and G. Melvill-Jones, Eds.), pp. 3–16. Elsevier, Amsterdam.

Simpson, J. I., Graf, W., and Leonard, C. (1981). The coordinate system of visual climbing fibres to the flocculus. *In:* "Progress in Oculomotor Research" (A. F. Fuchs and W. Becker, Eds.), pp.475–484. Elsevier, Amsterdam.

Simpson, J. I., Giolli, R. A., and Blanks, R. H. I. (1988a). The pretectal nuclear complex and the accessory optic system. *In:* "Neuroanatomy of the Oculomotor System" (J. A. Buttner-Ennerver, Ed.), pp. 335–364. Elsevier, Amsterdam.

Simpson, J. I., Leonard, C. S., and Soodak, R. E. (1988b). The accessory optic system of rabbit. II. Spatial organization of direction selectivity. *J. Neurophysiol.* **60,** 2055–2072.

Soodak, R. E., and Simpson, J. I. (1988). The accessory optic system of rabbit. I. Basic visual response properties. *J. Neurophysiol.* **60,** 2037–2054.

Srinivasan, M. V., Lehrer, M., Zhang, S. W., and Horridge, G. A. (1989). How honeybees measure their distance from objects of unknown size. *J. Comp. Physiol. A* **165,** 605–613.

Sun, H-J., and Frost, B. J. (1997). Motion processing in pigeon tectum: equiluminant chromatic mechanisms. *Exp. Brain Res.* **116,** 434–444.

Waespe, W., and Henn, V. (1981). Visual-vestibular interaction in the flocculus of the alert monkey. II. Purkinje cell activity. *Exp. Brain Res.* **43,** 349–360.

Waespe, W., and Henn, V. (1987). Gaze stabilization in the primate: The interaction of the vestibulo-ocular reflex, optokinetic nystagmus, and smooth pursuit. *Rev. Physiol. Biochem. Pharmacol.* **106,** 37–125.

Wagner, H., and Frost, B. (1994). Disparity-sensitive cells in the owl have a characteristic disparity. *Nature* **364,** 796–798.

Weber, J. T. (1985). Pretectal complex and accessory optic system in alert monkeys. *Brain Behav. Evol.* **26,** 117–140.

Whishaw, I. Q., and Maaswinkel, H. (1998). Rats with fimbria-fornix lesions are impaired in path integration: A role for the hippocampus in "sense of direction." *J. Neurosci.* **18,** 3050–3058.

Winterson, B. J., and Brauth, S. E. (1985). Direction-selective single units in the nucleus lentiformis mesencephali of the pigeon (*Columba livia*). *Exp. Brain Res.* **60,** 215–226.

Wong, S. C. P., and Frost, B. J. (1981). The effect of visual-vestibular conflict on the latency of steady-state visually induced subjective rotation. *Percept. Psychophys.* **30,** 228–236.

Wylie, D. R., and Frost, B. J. (1990a). Visual response properties of neurons in the nucleus of the basal optic root of the pigeon: A quantitative analysis. *Exp. Brain Res.* **82,** 327–336.

Wylie, D. R., and Frost, B. J. (1990b). Binocular neurons in the nucleus of the basal optic root (nBOR) of the pigeon are selective for either translational or rotational visual flow. *Vision Neurosci.* **5,** 489–495.

Wylie, D. R., and Frost, B. J. (1993). Responses of pigeon vestibulocerebellar neurons to optokinetic stimulation: II. The 3-dimensional reference frame of rotation neurons in the flocculus. *J. Neurophysiol.* **70,** 2647–2659.

Wylie, D. R. W., and Frost, B. J. (1996). The pigeon optokinetic system: Visual input in extraocular muscle coordinates. *Vision Neurosci.* **13,** 945–953.

Wylie, D. W. and Frost, B. J. (1999a). Complex spike activity of purkinje cells in the ventral uvula and nodulus of pigeons in response to translational optic flowfields. *J. Neurophysiol.* **81,** 256–266.

Wylie, D. W. and Frost, B. J. (1999b). Responses of neurons in the nucleus of the basal optic root to translational and rotational flowfields. *J. Neurophysiol.* **81,** 267–276.

Wylie, D. R. W., and Linkenhoker, B. (1996). Mossy fibres from the nucleus of the basal optic root project to the vestibular and cerebellar nuclei in pigeons. *Neurosci. Lett.* **219,** 83–86.

Wylie, D. R., Kripalani, T-K., and Frost, B. J. (1993). Responses of pigeon vestibulocerebellar neurons to optokinetic stimulation: I. Functional organization of neurons discriminating between translational and rotational visual flow. *J. Neurophysiol.* **70,** 2632–2646.

Wylie, D. R. W., Linkenhoker, B., and Lau, K. L. (1997). Projections of the nucleus of the basal optic root in pigeons (*Columba livia*) revealed with biotinylated dextran amine. *J. Comp. Neurol.* **384,** 517–536.

Wylie, D. R. W., Glover, R. G., and Lau, K. L. (1998a). Projections from the accessory optic system and pretectum to the dorsolateral thalamus in the pigeon (*Columba livia*): A study using both anterograde and retrograde tracers. *J. Comp. Neurol.* **391,** 456–469.

Wylie, D. R. W., Bischof, W. F., and Frost, B. J. (1998b). Common reference frame for coding translational and rotational optic flow. *Nature* **392,** 278–282.

OPTIC FLOW AND THE VISUAL GUIDANCE OF LOCOMOTION IN THE CAT

Helen Sherk and Garth A. Fowler

Department of Biological Structure, University of Washington, Seattle, Washington

I. Introduction

How do humans and animals use optic flow information during locomotion? What neural machinery is involved in this analysis? Gibson's work, starting in 1950, launched a surge of interest in these questions. Recently, investigation of the underlying neural mechanisms has focused on the use of optic flow to determine heading, the discrepancy between the direction of gaze and the direction of locomotion (for a review see Wurtz, 1998). Neural mechanisms responsible for other aspects of visual processing during locomotion have received less attention, although they may be no less essential. In this chapter we will focus on these other aspects of visual processing, and we will also concentrate on species which, like humans, locomote primarily across a two-dimensional substrate. Such observers face a somewhat simpler challenge to the visual system during locomotion than do animals such as birds (Lee, 1994) or arboreal monkeys (Fleagle, 1978) that locomote in a three-dimensional environment.

Section II looks at the specific uses that locomoting observers might make of visual information. Next, Section III considers another issue essential for understanding the neural basis of visual guidance: what optic flow pattern is actually seen by a locomoting observer? As Gibson originally pointed out (1950), the appearance of an optic flow field varies greatly depending on one's direction of gaze, and so we will consider where loco-

0074-7742/00 $30.00

moting observers look when they are relying on visual cues. Section IV, the final section, discusses neural processing as it may relate to analysis of optic flow fields. Although a number of studies have explored the neural basis of heading determination in monkeys, there has been little investigation of additional kinds of visual analysis during locomotion in any species other than the cat, and thus the last section will focus on this animal.

II. Uses of Vision during Locomotion

What might a locomoting observer use visual information for? In addition to the problem of heading, several possibilities come to mind.

- Objects in the environment might serve as visual landmarks when an observer sets or maintains a particular course.
- Visual cues can inform an observer that he is straying from an intended trajectory. This seems most plausible when an observer is following a straight course and deviates unintentionally. Comparing the optic flow fields in Figs. 1A and 1B, the reader will see that even modest turns markedly alter the pattern of flow.
- Visual information can be used to avoid colliding with objects.
- The rate of image expansion in an optic flow field might provide feedback about the observer's own speed.
- Visual information can be used to adjust foot placement in order to avoid stepping on small objects or irregularities in terrain.

What is known about each proposed use of visual information during locomotion? Obviously, vision is useful for orienting oneself toward a goal such as a chair or door that one intends to approach. But is optic flow necessary for accurately reaching that goal? Several studies indicate that it is not. These assessed the accuracy of subjects' locomotion to a goal placed in a large empty room or in an open field; the subjects saw the goal only before beginning to walk, and then approached it with their eyes closed (Thomson, 1983; Elliott, 1986; Steenhuis and Goodale, 1988; Rieser *et al.*, 1990). Over distances of about 3–20 m their performance was surprisingly good, implying that visual information during locomotion is not essential to reach a target that was seen only before the approach. This situation, of course, is unusually simple: the observer moved in a straight line through an uncluttered environment across a smooth substrate. When following a more complex course, it seems probable that observers rely on visual landmarks during locomotion.

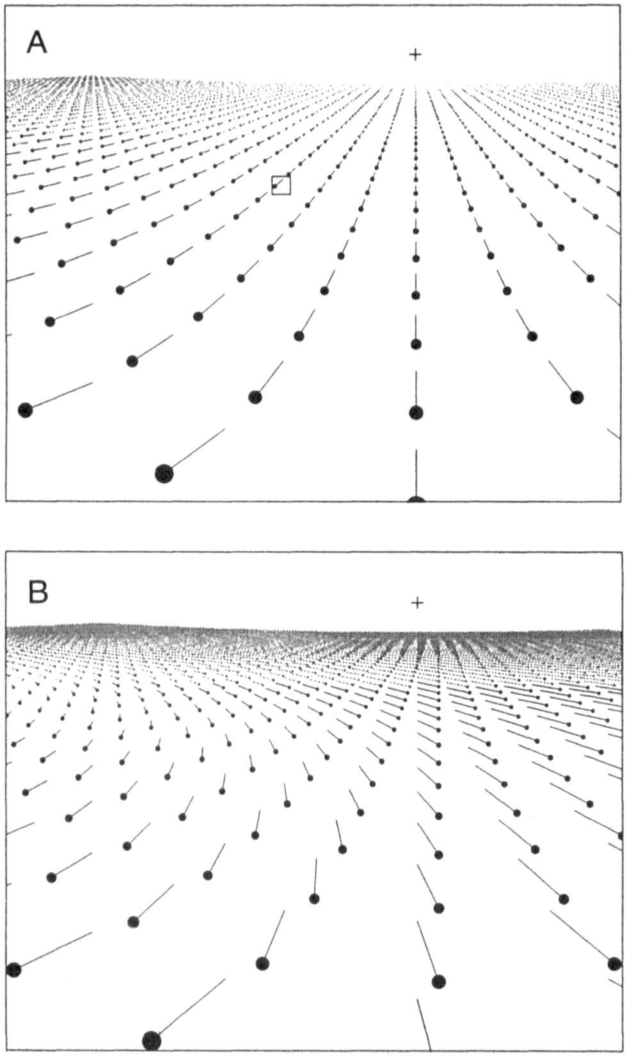

FIG. 1. Vectors of motion seen by an observer locomoting across a substrate covered with small balls. (A) The observer moves in a straight line toward the cross (his heading point). Speed of locomotion is 0.8 m/s, and vector lengths show each ball's motion over a 100 msec interval. Note that if the observer fixates the small square, images move down and to the left across his fovea. (B) The observer moves forward at the same speed, but also turns left at 30°/s.

Whether locomoting observers use visual cues to detect unintended deviations from their course is unknown. That they rely on vision to detect impending collisions, on the other hand, seems intuitively obvious. There has been considerable interest in exactly how visual information is used in this situation. Lee (1976) proposed that moving observers monitor their time to collision by gauging the rate of expansion of an obstacle's image as they approach it. Lee and Reddish (1981) found evidence consistent with this idea in one instance of natural locomotion, gannets' approach to the water surface during diving. In most situations, however, it is difficult to know precisely which cue in an optic flow field an observer is using, or at what time the information is used. Sun *et al.* (1992) solved this problem by constructing a simplified visual environment in which the only salient visual cue was the expansion (or contraction) of a single light disk on a video monitor. They found that when gerbils ran down an alley toward the monitor, their speed varied depending on the disk's behavior. If the disk expanded, the gerbil slowed down. If the disk shrank, the gerbil ran more quickly. Although this experiment could not test whether gerbils compute and use time to collision, it showed that they do use image expansion to gauge their own speed during locomotion. Image expansion might be widely used in this fashion by locomoting observers not only to detect impending collisions, but also in a variety of other situations.

The last item, the use of visual cues to adjust foot placement so as to avoid stepping in holes or on small objects or irregularities, is of considerable interest. Accurate foot placement is crucial in natural environments, as the ground is often uneven and cluttered with small objects such as stones and roots. A walking or running observer must constantly adjust his foot placement, making this a particularly challenging problem for the visual system. One might propose that the observer instead solves this problem using tactile and proprioceptive feedback at foot strike. However, this would appear to be a poor strategy, particularly when running—the information arrives too late to prevent stumbling and sometimes falling.

Little is known about how observers use visual information to solve this problem. Warren *et al.* (1986) tested human observers running on a treadmill in a task in which the runner was required to place his feet on visual targets that were irregularly spaced. They found that runners adjusted step length by modifying the vertical thrust in the stance phase prior to stepping on the target. Walking human subjects in a similar experiment were able to modify their stride with a rather short latency, <400 ms, from the time that a visual cue appeared (Patla *et al.*, 1989). Response latencies when subjects were required to step over a low obstacle were almost as short, <550 ms (Patla *et al.*, 1991).

Patla *et al.* (1996) tested human observers on a similar foot-placement task to explore how much visual sampling is actually required and how often samples need to be taken. Their subjects walked down a path in which they had to place their feet on preexisting footprints, spaced regularly or irregularly on different trials. Subjects wore opaque liquid crystal glasses that could be made transparent by pressing a switch when the subject needed visual information. Subjects generally took a visual sample every step or every other step, with a mean sample duration of about 0.5 s. Overall, they used visual information over only about 10–40% of the time (the longer times occurring with more demanding locomotor situations, as when a hole or barrier was present). Thus even when locomoting observers must place their feet accurately, they do not require continuous visual information.

We have investigated how locomoting cats might use vision to avoid stepping on objects. Cats walked or ran down an alley whose floor was cluttered with small objects. Figure 2 shows top-down views of the alley, which was composed of two legs at right angles to each other. The cat moved from a starting box through a short "tunnel" and into the first leg of the alley, turned right into the second leg, and exited through a door in the far end. The floor of the alley was covered with fine sand graded to a smooth surface, across which were scattered small objects; the sand revealed the cat's footprints.

Cats in this situation avoid stepping on a wide variety of small objects, generally by adjusting stride length. Footprints from two runs by the same cat are shown in Fig. 2. In Run 1, the alley was empty of objects. In Run 2, the alley contained a high density of objects (158 objects/m^2). The cat took the same number of steps in each case and moved at the same speed. The patterns of footprints look similar, but in Run 2 the cat varied its foot placements slightly to avoid stepping on objects Figure 2 illustrates an interesting point about the relative placement of forefeet and hindfeet. In both runs, each hindfoot was placed exactly in the impression left by the forefoot on the same side, so that the pattern of footprints looks bipedal. We think that this strategy lessens the cat's computational problem, since it need only determine where to place its front feet to avoid stepping on objects. Experience with locomotion over difficult terrain probably contributes to this behavior, as not all cats (including the one illustrated in Fig. 2) show this degree of precision when first tested in the alley task.

Do cats use visual cues in this behavior? We suspect that vision makes a major contribution, and possibly is the only cue used most of the time. If the cat instead relied on tactile and proprioceptive feedback, it would need to contact an object with its forefoot at footstrike, retract the foot, and place it either beyond or in front of the object. Unless the cat is mov-

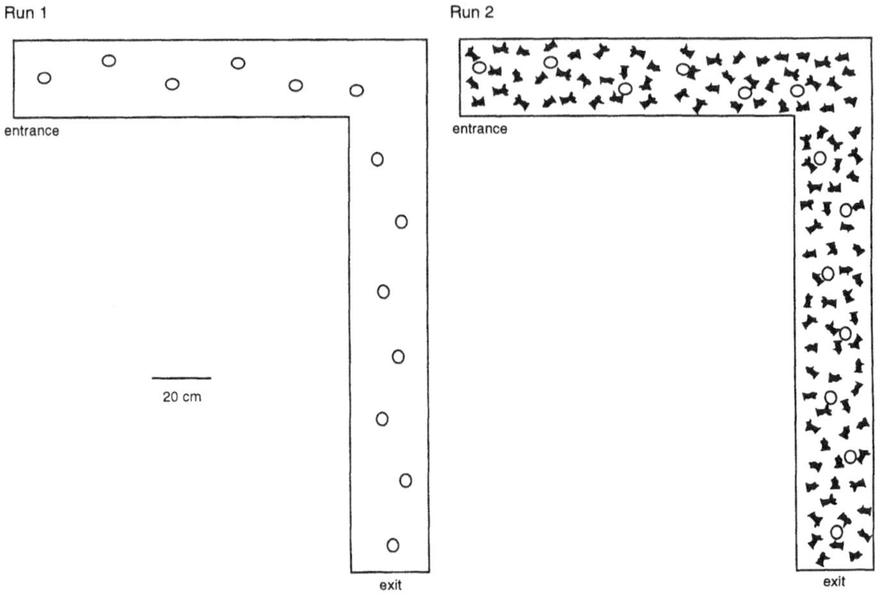

FIG. 2. Top-down view of test alley, together with footprints from two runs by the same cat. In Run 1 the alley was empty. In Run 2 the alley contained a high density of small plastic figures (158 objects/m^2). The cat took the same number of steps in each case, and ran at comparable speeds, 0.6 m/s. However, in Run 2 the cat made slight adjustments of step length so as to avoid stepping on objects.

ing extremely slowly, it seems doubtful that there is time to accomplish this kind of correction. The use of visual cues, on the other hand, is suggested by experiments in which illumination was altered and a higher error rate was found. For example, at scotopic light levels, there was a small but significant increase in error rate. Stroboscopic illumination caused a more severe deterioration in performance.

The cat's accuracy in avoiding small objects during walking is impressive when one considers that this is a predictive behavior. Although the cat's forefoot strikes the ground in front of the body during walking, its neck is extended horizontally so that its head is directly above the location of foot strike (Muybridge, 1957); thus it cannot see its foot approaching the ground without looking directly down. Human observers likewise do not normally see their feet hitting the ground, even when stepping over or onto a low obstacle such as a curb, as readers will realize from their own experience (see also Patla and Vickers, 1997). Thus we too must use cues in optic flow fields to predict where to place our feet, rather than to guide the trajectory of a visible foot.

III. Gaze during Visually Guided Locomotion

A. Gaze in Primates

Observers' directions of gaze during locomotion strongly affect the pattern of optic flow that they see. If an observer fixates his heading point (cross in Fig. 1A), all images move outward from the center of gaze, following approximately radial trajectories, and expanding and accelerating as they move. At the center of the fovea, there is no image motion. If instead he maintains a constant angle of gaze that does not coincide with the heading point (e.g., on the square in Fig. 1A, below and to the left of the heading point) the whole optic flow pattern is shifted on the retina. In this example, images accelerate diagonally downward across the fovea. When the observer turns, a more complex pattern of image motion occurs. In the example in Fig. 1B, an observer is moving forward at 0.8 m/s and turning left at 30°/s. The resulting pattern of image motion is no longer mirror-symmetric about the midline but displays a spiral motion on the left side. Other complex patterns can occur when a locomoting observer tracks a fixed point in the environment (Perrone and Stone, 1994; Royden, 1994; Lappe *et al.*, 1998; Lappe and Rauschecker, 1994).

Little is known about gaze strategy in locomoting observers. The reason is simple: it is technically difficult to measure gaze accurately during locomotion. The use of magnetic fields for inducing current flow in a search-coil attached to the eye (Robinson, 1963), the method of choice in animal experiments, is not feasible for an observer who travels several meters. Solomon and Cohen (1992) circumvented this problem by monitoring gaze in tethered monkeys walking or running in a tight circle. These monkeys often fixated a point in the environment and tracked it with a gain close to unity, and then made a rapid gaze shift to a new fixation point and repeated the process. There was, however, no visual guidance required of the monkeys in this situation, so that it is uncertain how this behavior relates to monkeys' gaze strategy during normal locomotion when using visual information.

Another approach to the problem is to look at the gaze of a stationary observer who views a simulated optic flow field. Lappe *et al.* (1998) found that monkeys in this situation tended to track image motion in the display, but with a rather low gain; thus images continued to move across the fovea at speeds up to about 30°/s during tracking.

Few studies have been performed to assess eye movements or gaze in moving observers who were making use of visual cues. Land and Lee (1994) studied eye movements in subjects driving on winding roads.

Their subjects interspersed saccadic eye movements with smooth movements that tended to track the outer edge of the road when it turned. This result is intriguing but not easy to relate to gaze strategies during natural locomotion (walking or running), which presents different motor challenges and does not usually provide the observer with a consistent visual cue such as a road edge or center line.

Patla and Vickers (1997) have looked at gaze in locomoting human subjects who had to use visual cues to adjust their stride. Subjects walked down a 10-m runway, stepping over a low obstacle whose location on the runway varied from trial to trial. Like Land and Lee's subjects, they shifted their gaze relatively often. Slightly less than 20% of the time was spent fixating the obstacle ahead. Interestingly, they spent a sizable fraction of time (about 33%) with their eyes in a fixed position in the orbits, thus maintaining a constant angle of gaze relative to their heading point (what Patla and Vickers refer to as a travel fixation). The authors point out that this behavior results in an optic flow field similar to that envisioned by Gibson (see Fig. 1A), although since subjects maintained a downward angle of gaze relative to their heading point, images would "flow" downward across the fovea.

When locomoting observers were required to perform a more difficult task, in which they had to shift their foot placements from side to side in a strictly constrained fashion, their gaze strategy was different. Hollands *et al.* (1995) had subjects step on a series of small stepping stones that were offset laterally from each other by irregular amounts. They monitored eye movements (though not gaze), and concluded that subjects fixated a given stone and tracked it until their foot had contacted it or nearly so and then saccaded to the next stone and repeated the process. These authors assumed that their subjects looked directly down, which would be necessary to see the foot as it contacted the substrate (foot contact occurs directly beneath the center of mass in humans during walking). This is not, however, a head position typical of human locomotion, and presumably reflected the very demanding nature of this task.

B. Gaze in Locomoting Cats

One might ask whether cats are likely to show patterns of gaze similar to those of primates during locomotion, given their more limited range of eye movement. Their gaze, however, turns out to be as mobile as that of primates. Collewijn (1977) found that the unrestrained cat made spontaneous saccades at a rate of about 3/s, just as do monkeys and humans. Even when cats are restrained and the head allowed to move only horizontally, they make frequent and rapid gaze shifts up to 70° in size (Guitton *et al.*, 1984, 1990).

We wished to determine what gaze strategy cats follow when walking down our test alley when it is strewn with small objects, a situation in which the animal presumably attends to visual cues on the alley floor. Gaze was estimated from high-resolution videotape images taken at eye level as the cat walked down the first leg of the alley directly toward the camera. A software model was constructed to measure head and eye position in individual frames, taken at a frame rate of 30 Hz.

From these films, it is clear that the cat has a variable gaze behavior during visually guided locomotion. Sometimes only small shifts in gaze occurred. Figure 3 illustrates gaze azimuth and elevation during one run, in which the cat maintained a downward angle of gaze throughout, starting more than 20° below horizontal, and continuing at least 9° below horizontal. The cat's gaze varied from side to side, but at most by 10°, and on average much less. Perhaps the most notable features of this record are the segments during which the cat maintained an approximately constant angle of gaze. The first three frames illustrated in the lower part of Fig. 3 are from such a segment (arrows 1–3). The fourth frame is from a short segment in which it maintained a more eccentric rightward gaze. These segments resemble the frequent episodes of constant gaze-angle ("travel fixations") observed by Patla and Vickers in human subjects.

Does the cat follow a consistent gaze strategy when locomoting across a cluttered substrate? In the test alley, the direction of gaze may be more predictable as the cat crosses the threshold than at later points, as the animal might assess the layout of objects in its path when it first encounters them. The direction of gaze in 35 trials at the entrance of the alley is shown in Fig. 4A. With one exception, the cat's gaze was within 10° of the midline and was directed downward. Indeed, the downward angle was rather pronounced, averaging 17° below horizontal, which corresponds to fixation of a point on the floor 59 cm ahead. About halfway down the alley, the cat's gaze still remained relatively close to the midline (Fig. 4B). However, it was generally not directed so far downward, averaging 9° below horizontal. It was actually looking up in a few trials at this point. Although in the middle of this alley leg its gaze apparently wandered more frequently than at the entrance, its accuracy of foot placement did not deteriorate. Possibly by this point it had acquired the visual information it needed to make the remaining three or four steps before turning the corner.

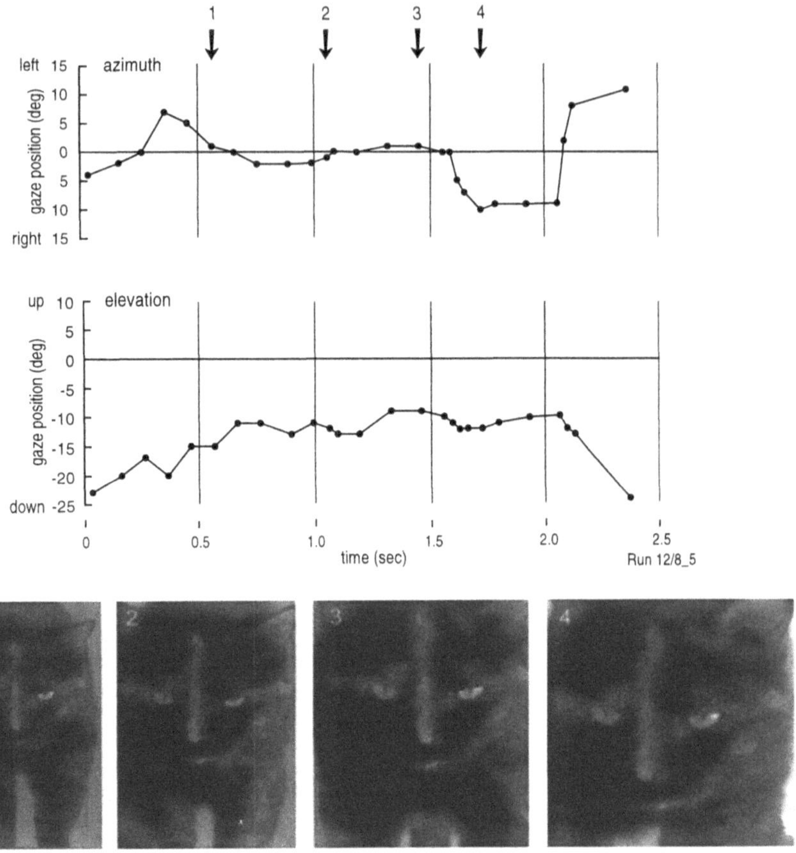

FIG. 3. Cat's direction of gaze as it walked down the first leg of the alley, which contained objects as shown in Fig. 2B. Gaze azimuth is given by the upper plot, and elevation is given by the lower plot. The center of this coordinate system corresponds to fixation of the heading point. The cat consistently looked about 10–24° below this point. Gaze was determined from videotape frames. Sampling rate was 30 Hz, but for clarity, data points for only some frames are illustrated. Four individual frames are shown below, their positions in the whole sequence being given by the correspondingly numbered arrow.

IV. Neural Mechanisms for Analyzing Optic Flow Information

A. Visual Cortex: Area 18

What neural mechanisms does the cat use to analyze optic flow information during locomotion? Given the complexity of optic flow fields, one might suspect that visual cortex must play an essential role. The lay-

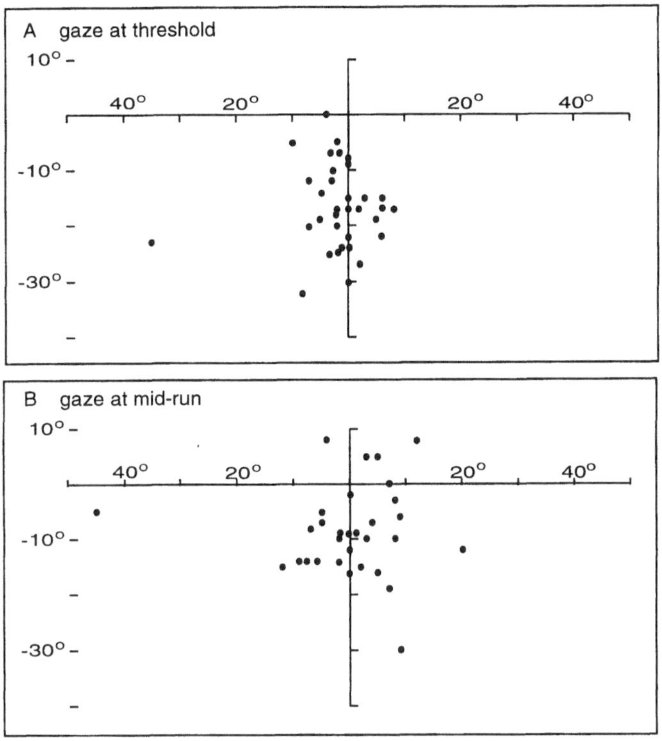

FIG. 4. Cat's direction of gaze in 35 runs down the first leg of the alley, which contained 158 objects/m^2. The coordinate system is centered on the cat's heading point—thus the abscissa represents the horizon, and the ordinate, the cat's midsagittal plane. (A) Gaze was measured in the videotape frame during each run in which the cat's leading forefoot was just crossing the threshold of the alley. (B) Gaze was measured approximately midway down the first leg of the alley.

out of the visual cortex in the cat, insofar as it is known, is relatively simple. Both areas 17 and 18 receive major input from the lateral geniculate nucleus and together function as V1 (Tretter *et al.*, 1975; Ferster, 1981). Beyond them, information processing follows two divergent streams, one via area 19 to area 20, and the other to lateral suprasylvian cortex (see Fig. 5) (Sherk, 1986; Payne, 1993; Lomber *et al.*, 1996). These streams appear similar to those in primates (Ungerleider and Mishkin, 1982). The pathway through area 19 in the cat appears analogous to the primate's ventral stream, as area 20 is concerned with pattern discrimination and object recognition (Lomber *et al.*, 1994, 1996). The lateral suprasylvian pathway resembles the primate's dorsal stream

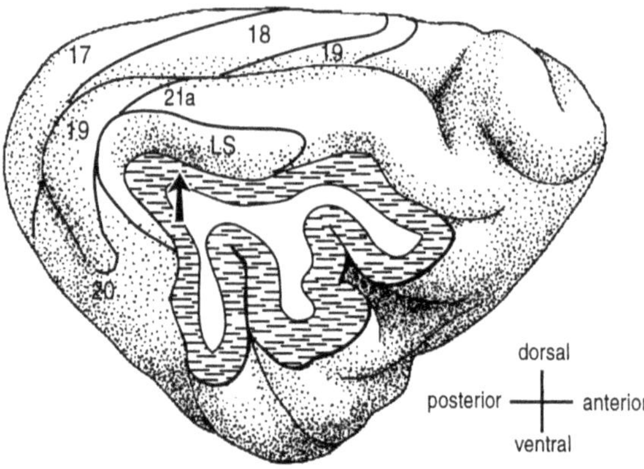

FIG. 5. Right hemisphere of cat brain with the ectosylvian gyri cut away, exposing the medial and posterior walls of the suprasylvian sulcus. Approximate locations of several visual cortical areas are shown. Within area LS, cells anterior to the arrow tend to prefer directions that are radial-outward from the center of gaze through their receptive fields.

in that it is concerned predominately with motion rather than form, texture, or spatial frequency cues, as we will discuss later.

Although it is possible that all areas of visual cortex contribute to visual guidance during locomotion, we think it more likely that cell populations in particular areas are specialized for this function. One cortical area in which cells with potentially appropriate properties have been discovered is area 18. Cynader and Regan (1978) reported that a small number of cells are selective for stimulus motion in depth, principally along an axis through the cat's nose. The authors suggested that such neurons would be useful during locomotion, presumably to signal an impending collision with an object, although it should be noted that the majority of their sample responded preferentially to motion away from the cat rather than toward it.

Area 18 also contains a population of cells that projects to the visual pontine nuclei (Baker *et al.*, 1976). Their response properties are strikingly different from those typical of area 18: they have very large receptive fields, and their preferred stimulus is most commonly a huge field of moving spots (Gibson *et al.*, 1978). Virtually all are direction-selective. As a population, their preferences are strongly biased for downward motion, with some bias also for directions away from the vertical meridian. Since all the receptive fields that Gibson *et al.* studied were centered in the lower hemifield, these fields would see downward motion during locomotion (see Fig. 1). These response properties closely resemble those found by Baker *et al.* (1976) in the pontine visual area, which these authors speculated might have a role in visual guidance of locomotion. This is an attractive idea because the pontine nuclei project to cerebellum, which modulates ongoing motor activity.

B. Visual Cortex: Area LS

1. *Direction Biases in LS*

Beyond areas 17 and 18, can we identify either the area 19 processing stream or the lateral suprasylvian processing stream as a likely site for visual analysis during locomotion? Response properties in area 19 suggest that it is ill-suited for this purpose. Even though images over most of a typical optic-flow field move at moderate to fast speeds, a large majority of cells in area 19 do not respond to stimuli moving faster than 10°/s (Dreher *et al.,* 1980; Duysens *et al.,* 1982; Tanaka *et al.,* 1987). Moreover, direction selectivity is rare in this area (Duysens *et al.,* 1982).

The lateral suprasylvian stream is more promising. Its first stage is a substantial area that we shall refer to as LS (lateral suprasylvian visual area).[1] Following Grant and Shipp (1991), we will define it as the area that receives direct input from V1. Zeki (1974) originally suggested an analogy between LS and the primate "motion" area MT (middle temporal area) based on their similar response properties. Additional similarities in connectivity and retinotopic organization have supported this analogy (Payne, 1993; Sherk and Mulligan, 1993).

The first observation suggesting that LS may be important in optic-flow analysis was made by Baker *et al.* (1976), who noted that, other than area 18, LS is the major source of cortical input to the pontine visual area. Although numerous studies were done subsequently on LS, it was not until 1987 that Rauschecker *et al.* (1987) proposed that this area has a special role in optic flow analysis. They observed that a large fraction of neurons in LS prefer directions of motion that lie along radial lines originating at the center of gaze and passing outward through each neuron's receptive field. If the cat fixates its heading point during locomotion, all images would move from the center of gaze radially outward across the visual field (see Fig. 1A), and thus presumably optimally stimulate this population of neurons.

This was an intriguing proposition, but it ran into several difficulties.

[1]Originally called the Clare-Bishop area by Hubel and Wiesel (1969), LS was subdivided using retinotopic criteria by Palmer *et al.* (1978) but then reconstituted on the basis of cortical connections (Sherk, 1986; Grant and Shipp, 1991). The retinotopic mapping in lateral suprasylvian cortex is complex and variable from one individual to another, so that mapping cannot be used reliably to distinguish one cortical area from another (Sherk and Mulligan, 1993).

First, the evidence from other studies has been mixed. Some investigators have reported a distinct bias for outward and downward directions of stimulus motion, and since in all cases receptive fields were in the lower visual field, these reports would agree with Rauschecker *et al.* (1987) (Toyama *et al.*, 1986a; Hamada, 1987; Weyand and Gafka, 1994). However, Spear and Baumann (1975) found only a weak predominance of preferences for outward over inward directions in LS, and Brenner and Rauschecker (1990) found a radial-outward bias only for peripheral receptive fields (those with eccentricities greater than 20°). One hypothesis that explains these discrepant results is that a radial-outward bias exists only in a part of LS that maps predominately peripheral visual field. This region would lie approximately anterior to the arrow in Fig. 5. It appears to be the region sampled in all studies reporting this bias, and data from our lab also showed such a bias here (Sherk *et al.*, 1995). But we found that neurons located more posteriorly had preferred directions that tended to be orthogonal to radial. Spear and Baumann sampled the full antero-posterior extent of LS and thus would have recorded from both populations. Brenner and Rauschecker probably recorded most of their neurons with central receptive fields in the posterior part of LS since such neurons are scarce more anteriorly.

There is a second difficulty with the hypothesis that this bias in preferred directions indicates a specialization for optic flow analysis. Radial-outward image motion would be seen by each receptive field only if the cat fixated its heading point during locomotion. A given receptive field is likely to see a rather different direction of image motion if the locomoting cat gazes eccentrically or makes a turn or tracks a point in the environment (see Fig. 1). Given the variety of gaze behaviors exhibited by the locomoting cat, about the only general statement we can safely make is that receptive fields in the lower visual field will rarely see upward motion during locomotion. A combination of radial-outward and orthogonal preferences appears optimal for such flow fields (Lappe and Rauschecker, 1995).

A third difficulty concerns the assumption that cells show the same directional preference under all stimulus conditions. This assumption seems so self-evident that it was not explicitly stated in any of the papers cited earlier. If it were true, then a large fraction of cells in LS would be unresponsive to a given optic flow field because image motion through their receptive fields would be in the wrong direction. Surprisingly, it turns out to be untrue for many neurons in LS, as described later. Thus a bias in the preferred directions of cells may be of less consequence for optic flow analysis than one might suppose.

2. *Responses in LS to Solitary Stimuli*

There are, however, additional reasons for thinking that this area may play a role in visual processing during locomotion. First, the response properties of neurons in LS seem appropriate. In any optic flow field, images are in motion throughout the visual field, and cells in LS strongly prefer moving to stationary stimuli, as every investigator has discovered upon recording from this area (see citations that follow). They are generally tolerant of considerable variation in stimulus parameters such as shape, size, and contrast (Hubel and Wiesel, 1969; Wright, 1969; Spear and Baumann, 1975; Camarda and Rizzolatti, 1976; Blakemore and Zumbroich, 1987), but they are selective for motion-related cues. Thus an overwhelming majority of cells are direction-selective (see preceding citations and also Hamada, 1987; Rauschecker *et al.*, 1987; von Grunau *et al.*, 1987; Zumbroich and Blakemore, 1987; Gizzi et al., 1990; Yin *et al.*, 1992). Many show a preference for moderate to fast stimulus speeds (Spear and Baumann, 1975; Rauschecker *et al.*, 1987; Zumbroich and Blakemore, 1987) in a range typical of images that would be seen by a walking or running cat. Rauschecker (1988) observed that cells with more peripheral receptive fields prefer higher speeds than those with central fields and pointed out that this distribution is consistent with detection of looming objects or objects in the path of a locomoting cat.

One might predict that neurons concerned with visual analysis during locomotion would be sensitive to stimulus depth since observers need to know either their distance away from objects in their path or their time to collision with such objects. Binocular vision provides a stationary observer in a static environment with one salient depth cue, positional image disparity, and about a fifth of the cells in LS are reported to be selective for this parameter (Toyama *et al.*, 1986a). If objects move toward a stationary observer, or if the observer instead locomotes, there are two additional binocular disparity cues available—one for direction and the other for speed. Toyama and colleagues have investigated the selectivity of cells in LS for object motion in depth using stimuli that display one or more binocular depth cues; they conclude that motion in depth is important for a large fraction of cells (Toyama and Kozasa, 1982; Toyama *et al.*, 1986a, b; Akase *et al.*, 1998). These results are both intriguing and puzzling. In their most recent study, these authors found that the largest subset of cells (about 40% of the whole sample) was selective for object motion that originated in the cell's receptive field and was directed toward the cat's nose or just above the nose (their "approach" cells). The next largest subset of cells (about 20%) was selective for object motion originating at the nose and directed away toward the receptive field (their "receding" cells). It is possible that approach cells might be activated during locomotion when the cat walks toward an object as tall as itself. For all cells except those with receptive fields in the area centralis, however, optimal image motion would be seen only when the cat maintained an eccentric angle of gaze such that the cell's receptive field was centered on the cat's heading point. Thus to stimulate optimally any receptive field in the lower hemifield (the location sampled by Toyama and colleagues), the cat would need to maintain an upward angle of gaze. Although we do find that cats walking down a cluttered alley glance up at times, we do not find them keeping a constant upward gaze. Even more perplexing is the possible function of receding cells. It is difficult to envision how such cells could see optimal motion during locomotion. It has been suggested that a prey animal might constitute a receding target as seen by a pursuing cat. However, cats hunt largely by stalking and pouncing (Leyhausen, 1979), and if they pursue a running mouse, for example, they move so rapidly that the mouse must constitute an approaching rather than a receding target (unpublished observation). It seems unlikely that 20% of the cells in LS would have evolved to deal with a situation that arises only rarely.

3. *Responses in LS to Whole-Field Stimuli: Surround Effects*

The response properties that we have discussed so far were described using a solitary stimulus moving against a blank background. The situation is quite different during locomotion, when there is motion throughout the visual field. How do cells in LS respond to this sort of stimulus? In 1983 von Grünau and Frost made an intriguing observation using a very large moving stimulus display. This consisted of a huge field of random noise plus a spot stimulus that traveled through the receptive field in the cell's preferred direction. When the noise field moved simultaneously with the spot and in the same direction, cell responses were usually depressed. When the noise field moved in the opposite direction, however, responses were often enhanced. Subsequently, Allman et al. (1985) observed the same behavior among cells in MT in the owl monkey. In both LS and MT, it appears that receptive fields have very large silent surrounds that can modulate responses to stimuli seen by the central receptive field (the region that can be plotted with ordinary moving bars or spots). But the surround's preferred direction is opposite to that of the center.

What is the significance of this behavior for optic flow analysis? Perhaps it helps cells distinguish between two situations in which there is moderate to fast movement throughout the visual field. In the first case, motion is caused by a gaze change and, in the second, by the cat's locomotion. When the cat shifts its gaze, images move at speeds that would drive many cells in LS (Collewijn, 1977; Guitton *et al.*, 1990), but because image motion is in the same direction throughout the visual field, one might expect the silent surrounds described by von Grünau and Frost to suppress responses. An optic flow field, by contrast, includes a broad spectrum of image directions (e.g., Fig. 1); just how it might affect silent surrounds is difficult to predict, but one might at least expect less suppression than during a gaze shift.

We tested this idea using two varieties of large stimulus display (Sherk and Kim, 1997). One simulated an optic flow field seen by a cat walking at 0.8 m/s and maintaining a constant angle of gaze 8° below its heading point. The simulated environment through which the cat moved consisted of small spheres scattered on the ground and hovering in the air (see Fig. 6, top). The second kind of display was made up of the same small gray spheres, but they all moved in a frontal plane (i.e., orthogonal to the cat's line of sight) and had the same speed and direction (Fig. 6, bottom). We would expect this frontoparallel motion display to activate a cell's silent surround when it moved in the cell's preferred direction, and thus suppress responses. If we compared responses to a "whole-field" version of this frontoparallel motion display (about 65 × 65° in size) with responses to this display masked so that it was visible only through a window slightly larger than the excitatory receptive field, we would predict that the full-field, unmasked version would yield a weaker response. For many cells, this was the case. Although suppressive surround effects were less pronounced than von Grünau and Frost had observed, we were able to compare the surround suppression elicited by large frontoparallel motion displays and by large optic flow displays. Few cells showed any significant suppression when tested with large optic flow displays, and more than 20% responded better to the full-field display than to the masked version in which stimulation was confined to the excitatory receptive field center. The upper two poststimulus time histograms (PSTHs) in Fig. 7A show this behavior for one cell. This finding is encouraging if we wish to hypothesize that LS is involved in optic-flow analysis, since obviously this hypothesis requires that cells remain responsive in the presence of optic flow motion throughout the visual field.

4. Responses In LS to Whole-Field Stimuli: Optic Flow versus Frontoparallel Motion

So far we have asked how optic flow stimuli affect receptive field surrounds in LS. What about responses to optic flow movies compared with responses to other kinds of stimuli? We first wanted to compare responses to frontoparallel motion with responses to a simple version of optic flow. These displays incorporated three basic "optic flow" cues—radial-outward motion, expansion of individual image elements, and acceleration of image elements. They simulated a simple situation: the optic flow seen by a cat walking in a straight line across an endless plain

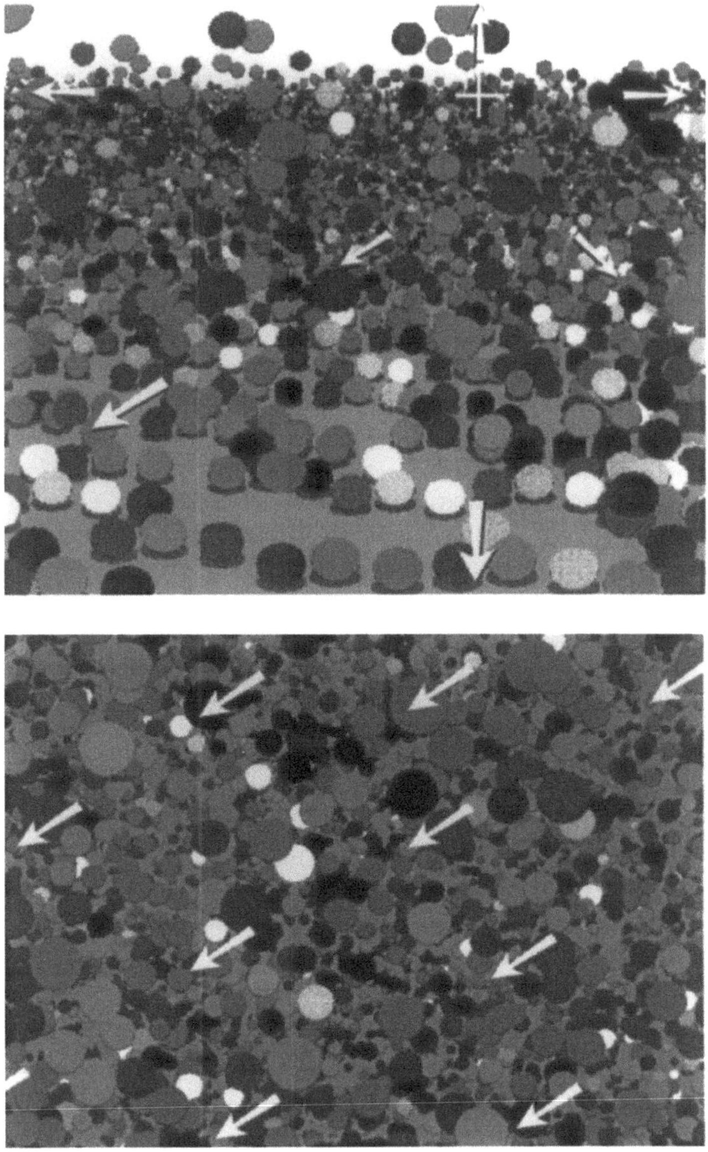

FIG. 6. Frames from two stimulus movies. Upper frame is from an optic flow movie that simulated locomotion through an environment of small balls—these were scattered on the ground and floating in air. The cross shows the heading point, and arrows indicate image motion. Lower frame is from a movie containing strictly frontoparallel motion. The direction and speed of motion were chosen to match the predominate direction and speed of motion seen by a particular receptive field in one optic flow movie. Each frame is about 65 × 65° at the experimental viewing distance.

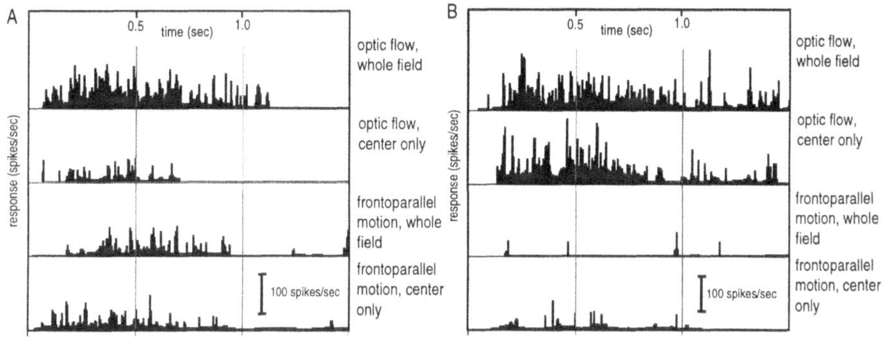

FIG. 7. Responses of two neurons from LS to four different stimuli. (A) This neuron responded strongly to the "full-field" version of an optic flow movie 65 × 65° in size (top PSTH). Its response to a masked version of this movie was much weaker (second PSTH). Responses to a frontoparallel motion movie, whether full-field or masked, were modest (third and fourth PSTHs). (B) This neuron responded about equally strongly to full-field and masked versions of optic flow (first and second PSTHs). Its had little or no response to either version of a frontoparallel motion movie (third and fourth PSTHs).

carpeted with small balls, with the cat keeping a constant angle of gaze 12° below the horizon (Kim *et al.*, 1997). Images moved in a wide range of directions in each optic flow movie. However, because neighboring images had similar directions (e.g., Fig. 1A), a given receptive field center saw predominantly one direction, which depended solely on the receptive field location. Thus for a given cell, it was possible to create a frontoparallel motion display whose direction and speed matched that of the images passing through the cell's receptive field center in the optic flow movie.

A surprising number of cells responded to optic flow movies (about 70% of those tested), and on the whole their responses were better than those to displays of strictly frontoparallel motion. A subset of cells was strongly selective for optic flow, some indeed being unresponsive to any other stimulus tested. Figure 7 illustrates the responses of two cells that were driven well by optic flow movies and more weakly or not at all by frontoparallel motion movies. The cell in Fig. 7A was tested with the displays shown in Fig. 6, whereas the cell in Fig. 7B was tested with movies composed of more complex and naturalistic elements such as leaves and grass (Sherk *et al.*, 1997). We found that the details of the stimulus elements making up the display (shape, texture, and spatial frequency) were generally not critical.

Although such cells suggest that LS is involved in optic flow analysis, it appears that its cell population is quite heterogeneous. Some cells preferred frontoparallel motion to optic flow, some responded more or less equally to both, and a sizable minority responded to neither kind of large display, although they could be driven well by a moving light or dark bar. Thus one might argue that LS is involved in many sorts of motion analysis, and only a handful of its cells are particularly concerned with optic flow. However, an additional finding suggests a stronger link between LS and optic flow analysis: in the sample as a whole, there was a powerful preference for optic flow movies shown in the forward direction compared to the reverse (Kim *et al.,* 1997). These movies simulated the experience of a cat walking forward, whereas movies run in reverse simulated the experience of a cat walking backward. Cats do not normally walk backward, and even if they did so, the visual scene that they would see has no utility for visual guidance, just as the view out the back window of a car is not useful for steering. Thus in a cell population concerned with optic flow analysis, we might expect to see cells responsive to optic flow generated by forward locomotion, but not to optic flow generated by backward locomotion.

Another interesting but quite unexpected finding in these experiments concerned the direction preferences of cells that responded to optic flow movies. As long as an observer travels in a straight line and maintains a fixed direction of gaze during locomotion, the directions of image motion through a receptive field depend only on its location. A receptive field in the lower hemifield, the locus of our sample as well as those in other studies of LS, almost always sees image motion in a downward and/or outward direction during locomotion. We thus anticipated that only cells that preferred downward and/or outward directions would respond well to optic flow movies. Surprisingly, this was not the case (Mulligan *et al.,* 1997). Figure 8A shows data from two cells that preferred directions orthogonal to the direction seen in optic flow movies; nonetheless, they responded vigorously to optic flow. Indeed, most of the cells in this study that responded to optic flow movies had inappropriate preferred directions (compare Figs. 8B and 8C). When we tested optic flow movies containing large salient objects, such as a tall bush in an open field (see Fig. 9), we again found that a number of cells with quite inappropriate preferred directions responded well to such an object (Sherk *et al.,* 1997). These findings suggest that the optic flow stimulus modifies the direction selectivity of many cells. One might speculate that, when the cat is stationary, a cell with a preferred direction inappropriate for optic flow analysis would be useful for signaling the direction of motion of moving objects. When the cat is walking or running, on the other hand, this cell would still be capable of participating in visual analysis because the optic flow field had altered or possibly suppressed its directional preference. The advantage of this behavior is ob-

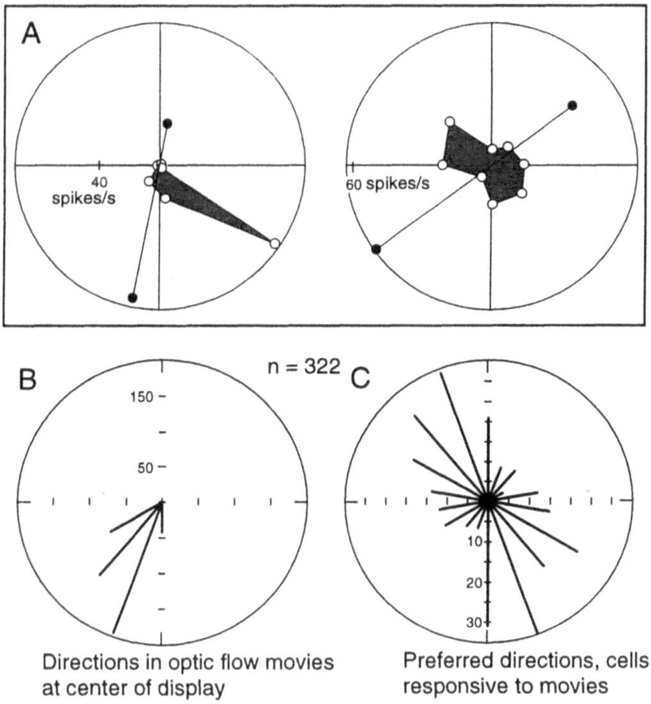

FIG. 8. Relationship between preferred directions of cells in LS and responsiveness to optic flow movies. (A) For two different neurons, polar plots (shaded gray) showing response to a moving bar stimulus as a function of direction. Each cell's responses to optic flow movies are also shown (black dots connected by single line). The angle of this response corresponds to the direction of motion in the optic flow movie at the receptive field center; motion was down and to the left when the movie was run in the forward direction, and up and to the right when the movie was run in reverse Both cells responded strongly to forward-going optic flow movies even though their direction selectivity appeared to be incompatible with such responses. (B) Polar plot of the directions of motion seen by 322 receptive fields in LS in optic flow movies run in the forward direction. (C) Preferred directions of these same neurons when tested with moving bars. All these cells responded to optic flow movies.

vious: a much greater fraction of cells in LS could be used for visual analysis both during locomotion and during stationary viewing than would be possible if cells maintained the same strict directional preference under all conditions.

5. *Responses in LS to Objects Embedded in Optic Flow Displays*

Cells that respond selectively to optic flow fields, like some of those described in the preceding section, might inform the cat that it is locomoting. In humans, optic flow induces a strong sense of self-motion ("ego-motion") even when the observer is in fact at rest (Lishman and

FIG. 9. One frame from an optic flow movie simulating locomotion across an open field carpeted with leaves and containing a single bush. Because of its height, the bush's image moved substantially faster than the images behind it. At a point 8° below the horizon, the bush moved almost three times faster than the adjoining background (relative image speeds are represented by lengths of arrows).

Lee, 1973). Although this is a remarkably robust phenomenon, its function in humans or in cats (supposing that they also experience egomotion) is unknown. Of much more obvious utility would be cells that respond to particular objects in optic flow fields, since these objects might correspond to obstacles to be avoided or landmarks useful for setting a course. Also useful would be cells responsive to small objects on the ground and to bumps and holes in the observer's path; these could cue the observer to adjust his stride and foot placement.

Discrimination of small objects and irregularities from ground during locomotion is a particularly challenging task and might require a substantial amount of neural machinery. With this in mind, we tested the responses of cells in LS to optic flow movies simulating a simple environment: the ground was covered with small balls, and a single black bar lay on top of them, placed so that its image moved through the receptive field of the cell being studied (Sherk *et al.*, 1997). Although we chose cells that responded well to this bar moving against a blank background, when embedded in an optic flow movie the bar failed to drive cells. Most cells, however, responded to a taller object in an optic flow movie such as the bush in Fig. 9; objects that stood at least as high as the cat's simulated eye level were most effective. The images of these objects were larger than that of the black bar, but a more important factor may have been their speed relative to the background. Image speed in an optic flow field depends not only on the observer's speed but also on object distance from the observer. Thus when an observer locomotes across an open space, the image of a tall foreground object will move much faster than the image of the distant substrate that forms its background (compare arrows in Fig. 9). We suspect that it is this difference in speed that makes these objects such effective stimuli.

If LS is crucial for visual guidance of the cat's foot placement when it locomotes across rough terrain, we would expect to find neurons in LS that respond to small, low objects such as stones since behavioral experiments show that cats skillfully avoid stepping on these. Why did we fail to find such cells? One possibility is that such cells are concentrated only in one part of LS, the region mapping the area centralis and perhaps the central few degrees of lower vertical meridian. If cats typically maintain their gaze close to the midline during locomotion, as the videotaped data suggest (e.g., Figs. 3 and 4), then only this part of the visual field will see objects on the ground in the cat's path. This region has an extensive representation in LS, located in the fundus of the suprasylvian sulcus and on its lateral bank at its caudal end (Palmer *et al.,* 1978; Mulligan and Sherk, 1993; Sherk and Mulligan, 1993). However, this part of LS is not commonly explored in recording studies, and little is known of response properties here.

6. *Responses in LS during Turns*

Even though simple detection of optic flow when the cat locomotes in a straight line may not be particularly useful, detection of deviations from this pattern of optic flow could be quite useful for alerting the cat that it had veered away from its intended path. To look for cells sensitive to turns, we tested a set of optic flow movies that again simulated a cat's locomotion through an environment of small balls (Fig. 6A), but now following a path that included a turn either to the left or right (see top of Fig. 10) (Sherk *et al.,* 1996). Different movies simulated turns of different magnitude, ranging from 15 to 60°/s. About a quarter of the cells tested responded well to one or more of these. The cell illustrated in Fig. 10 was selective for left turns of 15°/s. It had no response to a frontoparallel movie with a speed and direction matching that seen during 15°/s left turns (bottom PSTH in Fig. 10), but more commonly cells did respond to this kind of frontoparallel motion.

Interestingly, responses to turns were generally more rapid than to frontoparallel motion. In a sample of about 70 cells, the median latency

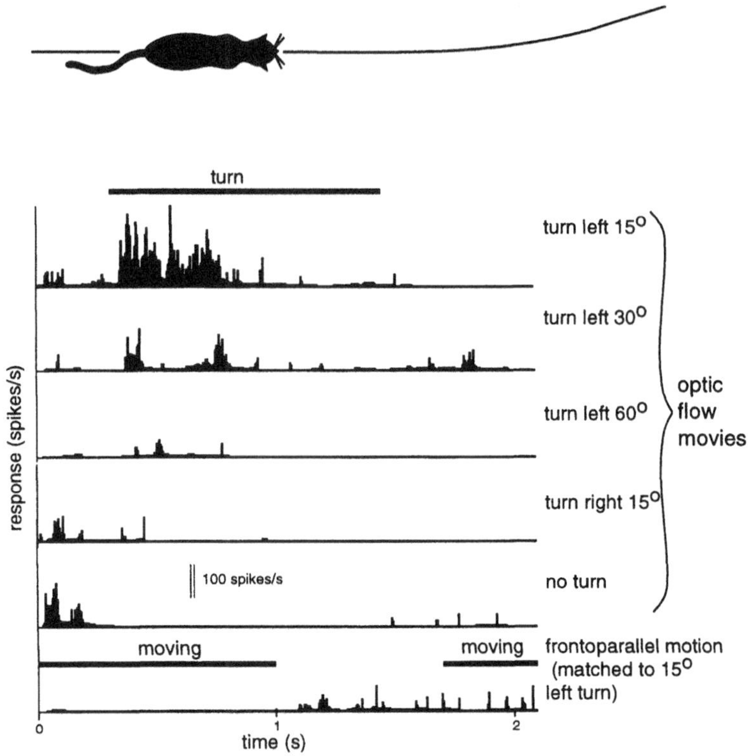

FIG. 10. Responses of one neuron in LS to optic flow movies simulating various courses of locomotion. At top is a view of the path simulated in one such movie, in which the cat starts out moving straight ahead and then makes a leftward turn at 15°/s. The cell had little or no response to the straight-ahead portion of various optic flow movies but responded strongly when a 15° left turn was simulated (top PSTH). Responses to a 30°/s left turn were weak, and there were no responses to any other turn. The cell also failed to respond to frontoparallel motion whose direction and speed matched that seen by the receptive field during the turn portion of the 15° left turn movie (bottom PSTH).

to the turn portion of an optic flow movie was 105 ms, whereas the median latency to frontoparallel motion of comparable speed and direction was 158 ms. Cells that act to detect unintended turns would be useful only if their responses were rapid, which appeared to be the case for most of those in our sample.

7. *Evidence from Behavioral Studies*

It would be of considerable interest to know whether visual guidance of locomotion deteriorates when LS is inactivated, but such experiments have not been done. However, the deficits that have been demonstrated following either damage to LS or inactivation of this region are quite suggestive. First, we might note what sort of visual functions are unaffected: these include pattern discrimination, the capacity to learn reversals of pattern discriminations, the capacity to shift attention to relevant visual cues, and visually triggered orienting (Wood *et al.,* 1974; Spear *et al.,* 1983; Antonini *et al.,* 1985; Hughes and Sprague, 1986; Lomber *et al.,* 1994, 1996). These functions might be more useful to a stationary observer contemplating an environment of objects that are potentially threatening, edible, or otherwise significant than to a moving observer who may be more concerned with the location and size of objects than with their identity.

It is striking that the visual functions that are affected by lesions or inactivation of LS all involve moving visual cues. Simple patterns obscured by fields of moving noise or lines can no longer be discriminated (Kiefer *et al.,* 1989; Lomber *et al.,* 1994, 1996). Although cats can still make gross direction discriminations, for example between left and rightward motion, or between upward and rightward motion (Pasternak *et al.,* 1989; Lomber *et al.,* 1996), fine distinctions between directions of motion are lost (Lomber *et al.,* 1996; Rudolph and Pasternak, 1996). Direction discrimination in lesioned cats is also much more readily disrupted by added noise consisting of randomly moving dots (Rudolph and Pasternak, 1996). Lesions of LS cause deficits in a structure-from-motion task (the ability to see a three-dimensional shape when the motion of dots simulates that of a rotating dot-covered cylinder) (Rudolph and Pasternak, 1996). Finally, cats with such lesions cannot discriminate between the speeds of moving stimuli (Pasternak *et al.,* 1989). These lost functions are ones that might be particularly useful for analyzing optic flow fields. For example, during locomotion it could be helpful to detect a region of similar motion, perhaps analogous to the moving noise fields used by Kiefer *et al.* and Lomber *et al.,* because in an optic flow field such a region corresponds to a contiguous portion of the environment. Speed discrimination might provide information about relative object distance since the images of near objects move faster than the images of more distant objects (e.g., Fig. 9). Finally, images moving in somewhat different directions in an optic flow field might have special significance: when a locomoting observer turns, the images of closer objects move in different directions than do the images of more distant objects at the same visual field location, as shown in Fig. 11.

Although visually guided locomotion has not been tested after lesions of any visual cortical area, one interesting study suggests that motion-sensitive mechanisms are important. Marchand *et al.* (1988) manipulated the temporal pattern of illumination while cats locomoted across horizontally placed ladders with irregularly spaced rung. The use of low-frequency stroboscopic illumination was particularly notable because this eliminated image motion and thus largely abolished optic flow processing

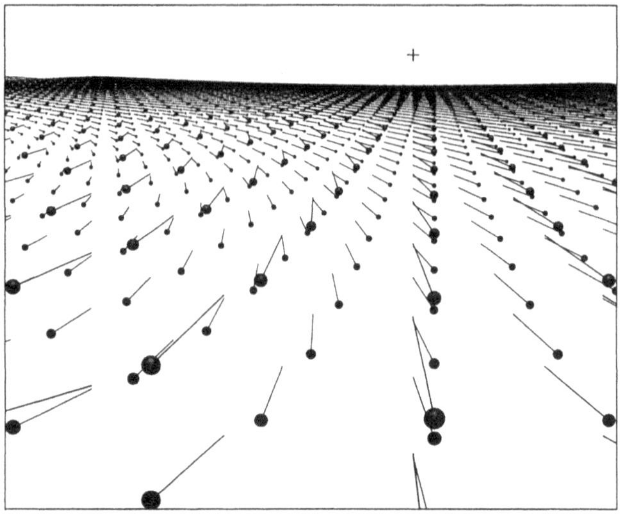

FIG. 11. Vectors of motion seen by an observer locomoting across a substrate dotted with small balls. Additional balls are floating in the air, half way between eye level and the ground. The observer is moving forward at 0.8 m/s and also turning left at 30°/s. Note that the vectors of a ball on the ground and a floating ball with images in close proximity have different lengths, signifying different speeds, because floating balls are closer to the observer's eyes. Interestingly, they frequently also have substantially different directions of motion.

in LS and elsewhere. Cats in this condition moved much more slowly, although they still managed to negotiate the ladders. Similarly, humans have been found to walk more slowly across such a ladder or along a narrow beam under low-frequency stroboscopic illumination than under high-frequency intermittent or continuous illumination (Assaiante *et al.*, 1989).

V. Conclusion

We are still far from understanding how locomoting observers make use of optic flow information. It seems reasonable to think that they use visual cues for several distinct purposes during locomotion—obstacle avoidance, accurate foot placement, and so on. If so, then we might look for neural mechanisms related to these particular tasks. Since the salient characteristic of images in optic flow fields is motion, in the cat the obvious cortical area in which to look is LS, whose cells are almost universally motion-sensitive. Numerous cells there respond selectively to particular optic flow fields or to objects in flow fields. However, its cell population is heterogeneous. We might conclude that this cortical area is important for visual guidance during locomotion but also may play a role in visual analysis performed by the stationary cat.

References

Akase, E., Inokawa, H., and Toyama, K. (1998). Neuronal responsiveness to three-dimensional motion in cat posteromedial lateral suprasylvian cortex. *Exp. Brain Res.* **122,** 214–226.

Allman, J., Miezin, F., and McGuinness, E. (1985). Direction- and velocity-specific responses from beyond the classical receptive field in the middle temporal visual area (MT). *Perception* **14,** 105–126.

Antonini, A., Berlucchi, G., and Sprague, J. M. (1985). Cortical systems for visual pattern discrimination in the cat as analyzed with the lesion method. *In:* "Pattern Recognition Mechanisms" (C. Chagas, R. Gattass, & C. Gross, Eds.), pp. 153–164. Rome, Vatican Press.

Assaiante, C., Marchand, A. R., and Amblard, B. (1989) Discrete visual samples may control locomotor equilibrium and foot positioning in man. *J. Motor Behav.* **21,** 72–91.

Baker, J., Gibson, A., Glickstein, and Stein, J. (1976). Visual cells in the pontine nuclei of the cat. *J. Physiol.* **255,** 415–433.

Blakemore, C., and Zumbroich, T. J. (1987). Stimulus selectivity and functional organization in the lateral suprasylvian visual cortex of the cat. *J. Physiol.* **389,** 569–603.

Brenner, E., and Rauschecker, J. P. (1990). Centrifugal motion bias in the cat's lateral suprasylvian visual cortex is independent of early flow field exposure. *J. Physiol.* **423,** 641–660.

Camarda, R., and Rizzolatti, G. (1976). Visual receptive fields in the lateral suprasylvian area (Clare-Bishop area) of the cat. *Brain Res.* **101,** 427–443.

Collewijn, H. (1977). Gaze in freely moving subjects. *In:* "Control of Gaze by Brain Stem Neurons. Developments in Neuroscience" (R. Baker and A. Berthoz., Eds.) vol. 1, pp. 13–22. Elsevier/North-Holland Biomedical Press, New York.

Cynader, M., and Regan, D. (1978). Neurones in cat parastriate cortex sensitive to the direction of motion in three-dimensional space. *J. Physiol.* **274,** 549–569.

Dreher, B., Leventhal, A. G., and Hale, P. T. (1980). Geniculate input to cat visual cortex: a comparison of area 19 with areas 17 and 18. *J. Neurophysiol.* **44,** 804–826.

Duysens, J., Orban, G. A., Van der Glas, H. W., and de Zegher, F. E. (1982). Functional properties of area 19 as compared to area 17 of the cat. *Brain Res.* **231,** 279–291.

Elliot, D. (1986). Continuous visual information may be important after all: A failure to replicate Thomson (1983). *J. Exp. Psych. Hum. Percept. Perform.* **12,** 388–391.

Ferster, D. (1981). A comparison of binocular depth mechanisms in areas 17 and 18 of the cat visual cortex. *J. Physiol.* **311,** 623–655.

Fleagle, J. G. (1978). Locomotion, posture, and habitat utilization in two sympatric, Malaysian leaf-monkeys (*Presbytis obscura and Presbytis melalophos*). *In:* "The Ecology of Arboreal Folivores" (G.G. Montgomery, Ed.). Smithsonian Inst. Press, Washington, DC.

Gibson, J. J. (1950). "The Perception of the Visual World." Houghton-Mifflin, Boston.

Gibson, A., Baker, J., Mower, G., and Glickstein, M. (1978). Corticopontine cells in area 18 of the cat. *J. Neurophysiol.* **41,** 484–495.
Gizzi, M. S., Katz, E., and Movshon, J. A. (1990). Spatial and temporal analysis by neurons in the representation of the central visual field in the cat's lateral suprasylvian visual cortex. *Visual Neurosci.* **5,** 463–468.
Grant, S, and Shipp, S. (1991). Visuotopic organization of the lateral suprasylvian area and of an adjacent area of the ectosylvian gyrus of the cat cortex: A physiological and connectional study. *Visual Neurosci.* **6,** 315–338.
Guitton, D., Douglas, R. M., and Volle, M. (1984). Eye-head coordination in cats. *J. Neurophysiol.* **52,** 1030–1050.
Guitton, D., Munoz, D. P., and Galana, H. L. (1990). Gaze control in the cat: Studies and modeling of the coupling between orienting eye and head movements in different behavioral tasks. *J. Neurophysiol.* **64,** 509–531.
Hamada, T. (1987). Neural response to the motion of textures in the lateral suprasylvian area of cats. *Behav. Brain Res.* **25,** 175–185.
Hollands, M. A., Marple-Horvat, D. E., Henkes, S., and Rowan, A. K. (1995). Human eye movements during visually guided stepping. *J. Motor Behav.* **27,** 155–163.
Hubel, D. H., and Wiesel, T. N. (1969). Visual area of the lateral suprasylvian gyrus (Clare-Bishop area) of the cat. *J. Physiol.* **202,** 251–260.
Hughes, H. C., and Sprague, J. M. (1986). Cortical mechanisms for local and global analysis of visual space in the cat. *Exp. Brain Res.* **61,** 332–354.
Kiefer, W., Krüger, K., Straub, G., and Berlucchi, G. (1989). Considerable deficits in the detection performance of the cat after lesion of the suprasylvian visual cortex. *Exp. Brain Res.* **75,** 208–212.
Kim, J-N., Mulligan, K., and Sherk, H. (1997). Simulated optic flow and extrastriate cortex. I. Optic flow versus texture. *J. Neurophysiol.* **77,** 534–561.
Lappe, M., and Rauschecker, J. P. (1994). Heading detection from optic flow. *Nature* **369,** 712–713.
Lappe, M., and Rauschecker, J. P. (1995). Motion anisotropics and heading detection. *Biol. Lyb.* **72,** 261–277.
Lappe, M., Pekel, M., and Hoffmann, K-P. (1998). Optokinetic eye movements elicited by radial optic flow in the macaque monkey. *J. Neurophysiol.* **79,** 1461–1480.
Land, M. F., and Lee, D. N. (1994). Where we look when we steer. *Nature* **369,** 742–744.
Lee, D. N. (1976). A theory of visual control of braking based on information about time-to-collision. *Perception* **5,** 437–459.
Lee, D. N. (1994). An eye or ear for flying. *In:* "Perception and Motor Control in Birds" (M.N.O. Davies and P.R. Green, Eds.). Springer-Verlag, Berlin.
Lee, D. N., and Reddish, P. E. (1981). Plummeting gannets: A paradigm of ecological optics. *Nature* **293,** 293–294.
Leyhausen, P. (1979). "Cat Behavior. The Predatory and Social Behavior of Domestic and Wild Cats" (B.A. Tonkin, Trans.). Garland STPM Press, New York.
Lishman, J. R., and Lee, D. N. (1973). The autonomy of visual kinaesthesis. *Perception* **2,** 287–294.
Lomber, S. G., Cornwell, P., Sun J. S., MacNeil, M. A., and Payne, B. R. (1994). Reversible inactivation of visual processing operations in middle suprasylvian cortex of the behaving cat. *Proc. Nat. Acad. Sci., USA* **91,** 2999–3003.
Lomber, S. G., Payne, B. R., Cornwell, P., and Long, K. D. (1996). Perceptual and cognitive visual functions of parietal and temporal cortices in the cat. *Cereb. Cortex* **6,** 673–695.
Marchand, A. R., Amblard, B., and Cremieux, J. (1988). Visual and vestibular control of locomotion in early and late sensory-deprived cats. *Prog. Brain Res.* **76,** 229–238.

Mulligan, K., Kim, J-N., and Sherk, H. (1997). Simulated optic flow and extrastriate cortex. II. Responses to bars versus large-field stimuli. *J. Neurophysiol.* **77,** 562–570.

Mulligan, K., and Sherk, H. (1993). A comparison of magnification functions in area 19 and the lateral suprasylvian visual area in the cat. *Exp. Brain Res.* **97,** 195–208.

Muybridge, E. (1957). "Animals in Motion" (L.S. Brown, Ed.). Dover, New York.

Palmer, L. A., Rosenquist, A. C., and Tusa, R. J. (1978). The retinotopic organization of lateral suprasylvian visual areas in the cat. *J. Comp. Neurol.* **177,** 237–256.

Pasternak, T., Horn, K. M., and Maunsell, J. H. R. (1989). Deficits in speed discrimination following lesions of the lateral suprasylvian cortex in the cat. *Visual Neurosci.* **3,** 365–375.

Patla, A., Adkin, A., Martin, C., Holden, R., and Prentice, S. (1996). Characteristics of voluntary visual sampling of the environment for safe locomotion over different terrains. *Exp. Brain Res.* **112,** 513–522.

Patla, A., Prentice, S. D., Robinson, C., and Neufeld, J. (1991). Visual control of locomotion: strategies for changing direction and for going over obstacles. *J. Exp. Psych. Hum. Percep. Perform.* **17,** 603–634.

Patla, A. E., Robinson, C., Samways, M., and Armstrong, C. J. (1989). Visual control of step length during overground locomotion: task-specific modulation of the locomotor strategy. *J. Exp. Psych. Hum. Percep. Perform.* **15,** 603–617.

Patla, A. E., and Vickers, J. (1997). Where and when do we look as we approach and step over an obstacle in the travel path? *Neuro Report* **8,** 3661–3778.

Payne, B. R. (1993). Evidence for visual cortical area homologues in cat and macaque monkey. *Cereb. Cortex* **3,** 1–25.

Perrone, J. A., and Stone, L. S. (1994). A model of self-motion estimation within primate extrastriate visual cortex. *Vision Res.* **34,** 2917–2938.

Rauschecker, J. P. (1988). Visual function of the cat's LP/LS subsystem in global motion processing. *Prog. Brain Res.* **75,** 95–108.

Rauschecker, J. P., von Grunau, M. W., and Poulin, C. (1987). Centrifugal organization of direction preferences in the cat's lateral suprasylvian visual cortex and its relation to flow field processing. *J. Neurosci.* **7,** 943–958.

Rieser, J. J., Ashmead, D. H., Talor, C. R., and Youngquist, G. A. (1990). Visual perception and the guidance of locomotion without vision to previously seen targets. *Perception* **19,** 675–689.

Robinson, D. A. (1963). A method for measuring eye movements using a scleral search-coil in a magnetic field. *IEEE Trans. Bio-Med. Engng.* **BME-10,** 137–145.

Royden, C. S. (1994). Analysis of misperceived observer motion during simulated eye rotations. *Vision Res.* **23,** 3215–3222.

Rudolph, K. K., and Pasternak, T. (1996). Lesions in cat lateral suprasylvian cortex affect the perception of complex motion. *Cereb. Cortex* **6,** 814–822.

Sherk, H. (1986). Location and connections of visual cortical areas in the cat's suprasylvian sulcus. *J. Comp. Neurol.* **247,** 1–31.

Sherk, H., and Kim, J.-N (1997). Neuronal responses to optic flow in extrastriate cortex: influence of receptive field surrounds. *Soc. Neurosci. Abstr.* **23,** 1126.

Sherk, H., and Mulligan, K. A. (1993). A reassessment of the lower visual field map in striate-recipient lateral suprasylvian cortex. *Visual Neurosci.* **10,** 131–158.

Sherk, H., Mulligan, K., and Kim, J.-N. (1996). Optic flow during simulated turns: Single unit responses in extrastriate cortex. *Soc. Neurosci. Abstr.* **22,** 1617.

Sherk, H., Mulligan, K., and Kim, J.-N. (1997). Neuronal responses in extrastriate cortex to objects in optic flow fields. *Visual Neurosci.* **14,** 879–895.

Sherk, H., Kim, J.-N., and Mulligan, K. (1995). Are the preferred directions of neurons in cat extrastriate cortex related to optic flow? *Visual Neurosci.* **12,** 887–894.

Solomon, D., and Cohen, B. (1992). Stabilization of gaze during circular locomotion in light. I. Compensatory head and eye nystagmus in the running monkey. *J. Neurophysiol.* **67,** 1146–1157.

Spear, P. D., and Baumann, T. P. (1975). Receptive-field characteristics of single neurons in lateral suprasylvian visual area of the cat. *J. Neurophysiol.* **38,** 1403–1420.

Spear, P. D., Miller, S., and Ohman, L. (1983). Effects of lateral suprasylvian visual cortex lesions on visual localization, discrimination, and attention in cats. *Behav. Brain Res.* **10,** 339–359.

Steenhuis, R. E., and Goodale, M. A. (1988). The effects of time and distance on accuracy of target-directed locomotion: Does an accurate short-term memory for spatial locomotion exist? *J. Motor Behav.* **20,** 399–415.

Sun, H.-J., Carey, D. P., and Goodale, M. A. (1992). A mammalian model of optic-flow utilization in the control of locomotion. *Exp. Brain Res.* **91,** 171–175.

Tanaka, K., Ohzawa, I., Ramoa, A. S., and Freeman, R. D. (1987). Receptive field properties of cells in area 19 of the cat. *Exp. Brain Res.* **65,** 549–558.

Thomson, J. A. (1983). Is continuous visual monitoring necessary in visually guided locomotion? *J. Exp. Psych. Hum. Percept. Perform.* **9,** 427–443.

Toyama, K., Komatsu, Y., and Kozasa, T. (1986a). The responsiveness of Clare-Bishop neurons to motion cues for motion stereopsis. *Neurosci. Res.* **4,** 83–109.

Toyama, K., and Kozasa, T. (1982). Responses of Clare-Bishop neurones to three dimensional movement of a light stimulus. *Vision Res.* **22,** 571–574.

Toyama, K., Fujii, K., Kasai, S., and Maeda, K. (1986b). The responsiveness of Clare-Bishop neurons to size cues for motion stereopsis. *Neurosci. Res.* **4,** 110–128.

Tretter, F., Cynader, M., and Singer, W. (1975). Cat parastriate cortex: A primary or secondary visual area? *J. Neurophysiol.* **38,** 1099–1113.

Ungerleider, L. G., and Mishkin, M. (1982). Two cortical visual systems. *In* "Analysis of Visual Behavior" (D.J. Ingle, M.A. Goodale, and R.J.W. Mansfield, Eds.), pp. 549–580. MIT Press, Cambridge, MA.

von Grunau, M. W., and Frost, B. J. (1983). Double-opponent-process mechanism underlying RF-structure of directionally specific cells of cat lateral suprasylvian visual area. *Exp. Brain Res.* **49,** 84–92.

von Grunau, M. W., Zumbroich, T. J., and Poulin, C. (1987). Visual receptive field properties in the posterior suprasylvian cortex of the cat: A comparison between the areas PMLS and PLLS. *Vision Res.* **27,** 343–356.

Warren, Jr., W. H., Young, D. S., and Lee, D. N. (1986). Visual control of step length during running over irregular terrain. *J. Exp. Psych. Hum. Percept. Perform.* **12,** 259–266.

Weyand, T. G., and Gafka, A. C. (1994) Corticotectal cells in areas 17 and PMLS of the cat. *Soc. Neurosci. Abstr.* **20,** 1740.

Wood, C. C., Spear, P. D., and Braun, J. J. (1974). Effects of sequential lesions of suprasylvian gyri and visual cortex on pattern discrimination in the cat. *Brain Res.* **66,** 443–466.

Wright, M. M. (1969). Visual receptive fields of cells in a cortical area remote from the striate cortex in the cat. *Nature* **223,** 973–975.

Wurtz, R. H. (1998). Optic flow: A brain region devoted to optic flow analysis? *Curr. Biol.* **8,** R554–556.

Yin, T. C. T., and Greenwood, M. (1992). Visual response properties of neurons in the middle and lateral suprasylvian cortices of the behaving cat. *Exp. Brain Res.* **88,** 1–14.

Zeki, S. M. (1974). Functional organization of a visual area in the posterior bank of the superior temporal sulcus of the rhesus monkey. *J. Physiol.* **236,** 549–573.

Zumbroich, T. J., and Blakemore, C. (1987). Spatial and temporal selectivity in the suprasylvian visual cortex of the cat. *J. Neurosci.* **7,** 482–500.

PART IV

CORTICAL MECHANISMS

STAGES OF SELF-MOTION PROCESSING IN PRIMATE POSTERIOR PARIETAL CORTEX

F. Bremmer,[*,†] , J.-R. Duhamel,[†,‡] S Ben Hamed,[†,‡] and Werner Graf[†]

[*]Department of Zoology and Neurobiology, Ruhr University Bochum, Bochum, Germany;
[†]Laboratoire de Physiologie de la Perception et de L'Action (LPPA)—Collège de France, Paris, France; and [‡]Institut des Sciences Cognitives (ISC), Centre National de la Recherche Scientifique (CNRS), Bron, France

I. Introduction

Self-motion through the environment requires the processing of a variety of incoming sensory signals, such as visual optical flow on the retina but also sensations from receptors in the skin, induced, e.g., by air flow or brushing leaves or grass. Furthermore, the vestibular organ also signals self-motion: the otolith system in the case of linear movement and the semicircular canals in the case of rotations of the head in space. Visual motion information is processed in parallel by a retino-subcortical and a retino-thalamo-cortical pathway. The latter is thought to be dominant in humans and nonhuman primates. In such cases, visual information arriving from the retina via the dLGN is sent into the dorsal pathway of the visual cortical system that predominantly deals with the processing of spatial information including the encoding of self-motion and object-motion. In this chapter we focus on the cortical processing of self-motion information, and we will give evidence for a parallel processing of motion information in several distinct areas in humans and nonhuman primates.

0074-7742/00 $30.00

II. Motion-Sensitive Areas in the Macaque Visual Cortical System

Self-motion through the environment generates a variety of sensory input signals. In the macaque cortical system, more than half of the neuronal tissue is dedicated to the processing of visual sensory information. This already suggests the importance of the incoming visual information for the processing of self-motion information and indicates its dominance compared to the processing of tactile and vestibular signals. Thus, in the following, we first describe in detail the different stages of cortical visual motion processing. We then will describe briefly cortical areas involved in the processing of self-motion information originating from other sensory modalities (tactile and vestibular).

If self-motion occurs on a straight path, the visual optical flow arriving at the retina is radially symmetric, and the singularity within this flow field indicates the direction of heading (Gibson, 1950). Evidence from single-cell recording experiments suggests that the macaque medial superior temporal area (MST) plays a prominent role in the process of self-motion perception and navigation. Three contributions to this current book describe in great detail this functional role of area MST (see Duffy, this volume; Lappe, this volume; Andersen *et al.*, this volume), which, however, is already a higher-order area in the visual motion-processing pathway.

A. Area MT

Visual information arrives via the thalamus in area V1 of the macaque visual cortex (Fig. 1). Information is sent directly from there or via a further processing stage (area V2) to the middle temporal (MT) area, located in the posterior bank of the superior temporal sulcus (STS) (Ungerleider and Desimone, 1986a, b; Shipp and Zeki, 1989). The vast majority of cells in area MT (or V5) is tuned for the direction and speed of a moving visual stimulus (Albright, 1984; Lagae *et al.*, 1993; Mikami *et al.*, 1986a, b). Cells with similar preferred directions (PDs) are clustered in columns (Albright *et al.*, 1984). The PDs are uniformly distributed, yet centrifugal directions are overrepresented (Albright, 1989). Area MT is topographically organized (i.e., neighboring cells within area MT usually represent neighboring parts within the visual field). Although RFs are on average larger than those in striate cortex and increase in size with increasing eccentricity (Tanaka *et al.*, 1993), they are still small compared to the large field motion across the whole visual field typically

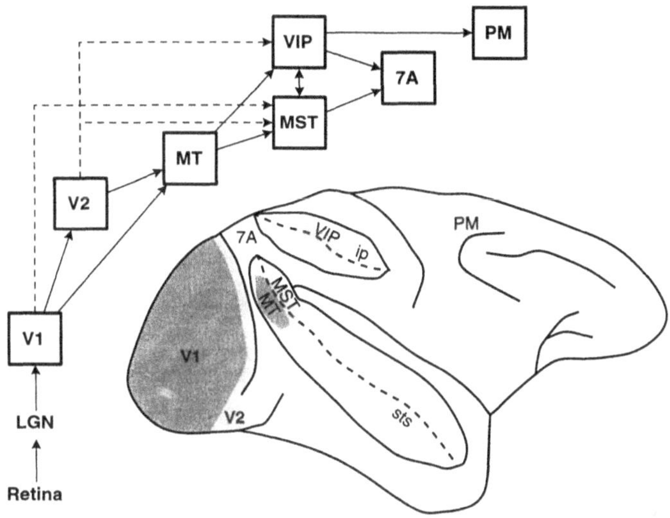

FIG. 1. Motion sensitive areas in the dorsal part of the visual cortical system in monkeys. The panel in the lower right shows a lateral view of a right cerebral hemisphere. The superior temporal sulcus (STS) and the intraparietal sulcus (IP) are unfolded to show the location of areas buried in the depth of the sulcus. The panel in the upper left depicts schematically the information flow from the retina via the lateral geniculate nucleus (LGN) and areas V1 and V2 into the higher-order motion-sensitive areas. Arrows indicate existing connections between areas. Dashed lines indicate weak connections.

occurring during self-motion. Area MT thus can be considered as a relay station for visual motion processing not specifically dedicated to the analysis of visual motion pattern occurring during self-motion.

B. Area MST

1. *Optic-Flow Responses*

Since the functional role of area MST is described in detail in other chapters of this book (see: Lappe, this volume; Andersen *et al.*, this volume; Duffy, this volume), it will be reviewed only briefly here. In contrast to neurons in area MT, those in area MST have large receptive fields (Tanaka *et al.*, 1993), some of them covering the whole visual field. Most neurons, like those in area MT, are tuned for the direction and speed of moving visual stimuli (Saito *et al.*, 1986). In addition, many MST neurons also respond selectively to optic-flow stimuli mimicking either forward (expansion) or backward (contraction) motion as well as rotation or motion resulting from a combination of these canonical movement types (e.g., spiral stimuli) (Saito *et al.*, 1986; Lagae *et al.*, 1994; Tanaka *et al.*, 1986, 1989; Tanaka and Saito, 1989; Graziano *et al.*, 1994; Lappe *et al.*, 1996; Duffy and Wurtz, 1991a, b, 1995; see also Lappe this volume, and Duffy this volume). Area MST shows no clear topography, yet another kind of ordered structure seems to exist: neurons responding to canonical movement types are organized in a columnar fashion (Geesaman *et al.*, 1997; Britten and van Wezel, 1998).

Response strength to optic flow stimuli is often influenced by the location of the singularity of the optic flow (SOF; i.e., the very single point in the image with zero velocity of local motion), corresponding to the apparent point of origin of a given radial optic flow pattern (Lappe *et al.*, 1996; Duffy and Wurtz, 1995; see also Duffy, this volume). Such a shift of the SOF corresponds to a constant angular difference between the direction of heading and the direction of gaze. Tuning to the focus location often is very broad and can be modeled by 2-D linear or sigmoidal functions. Since heading cannot be deduced from the activity profile of a single neuron, it was suggested that the direction of self-motion might be encoded by a population of neurons whose discharge is modulated by the location of the SOF (Lappe *et al.*, 1996).

The retinal flow pattern during actual self-motion becomes even more complex when eye movements are considered. Eye movements are performed naturally (Lappe *et al.*, 1998) in order to track targets or obstacles. In such case heading direction is not indicated by the singularity of the optic flow (Regan and Beverly, 1982; Warren and Hannon, 1990; Lappe and Rauschecker, 1995). Psychophysical experiments showed that human subjects could still detect the direction of heading in the presence of actual or simulated eye movements. In the latter case, however, this was possible only if the optic flow mimicked a movement across a ground plane (see also van den Berg, this volume). Neurophysiological studies have shown that single neurons from area MST can compensate for the disturbance of the retinal flow caused by pursuit eye movements while the animal viewed a scene simulating a self-motion toward a vertical 2-D plane of dots (Bradley *et al.*, 1996). The neurons did not compensate if the retinal flow only simulated eye movements across an optic flow field. However, recent studies from our laboratory (Bremmer *et al.*, 1998b) have shown that MST neurons (and also VIP neurons, see later discussion) can compensate for retinal flow simulating eye movements provided that the optic flow simulates a self-motion across a ground plane.

Taken together, all data observed from purely visual studies therefore have established the view of an involvement of macaque area MST in heading perception.

2. *Responses to Translational Movements*

Further evidence for the involvement of area MST in the process of self-motion perception comes from studies showing responses of single neurons to real physical motion. In these studies, animals were moved either on a parallel swing (Fig. 2) or on a moveable chair. Movement occurred only along the anterior-posterior axis (Bremmer *et al.,* 1999) or in the whole horizontal plane (Duffy, 1998; see also Duffy, this volume). Vestibular responses were observed during movement in light (visual vestibular interaction) as well as in darkness (pure vestibular sensory information). Usually, vestibular responses were smaller than visual responses. Finally, there exists only a weak, if any, correlation between preferred directions for real movement (vestibular stimulation) and simulated movement (optic flow stimuli).

C. Area VIP

Area MST, however, is not the only major output structure of area MT. Based on anatomical data (Ungerleider and Desimone, 1986a; Maunsell and Van Essen, 1983), the ventral intraparietal area (VIP) was originally defined as the MT projection zone in the intraparietal sulcus. These results suggested neurons in area VIP to be responsive for the direction and speed of moving visual stimuli, which could be confirmed in later physiological studies (Colby *et al.,* 1993; Duhamel *et al.,* 1991). This functional behavior makes area VIP very different from the neighboring lateral intraparietal (LIP) and medial intraparietal (MIP) areas. In the following, after having contrasted it with response features of areas LIP and MIP, we will review in detail response characteristics of neurons in area VIP.

Most neurons in area LIP respond to flashed, static stimuli (Colby *et al.,* 1996). These visual responses can be modulated by the attentional or intentional state of the animal (Bracewell *et al.,* 1996; Mazzoni *et al.,* 1996; Gottlieb *et al.,* 1998; Colby *et al.,* 1996). Furthermore, many neurons in area LIP show saccade-related activity (Barash *et al.,* 1991a, b; Colby *et al.,* 1996; Thier and Andersen, 1998; Platt and Glimcher, 1998), some of them also show pursuit related activity (Bremmer *et al.,* 1997b) or anticipate the "visual consequence" of an upcoming saccade (Duhamel *et al.,* 1992; Colby *et al.,* 1995). Spontaneous activity, visual responses as well as oculomotor activity often are influenced by the position of the eyes in the orbit. (Andersen *et al.,* 1990; Bremmer *et al.,* 1997b, 1998a).

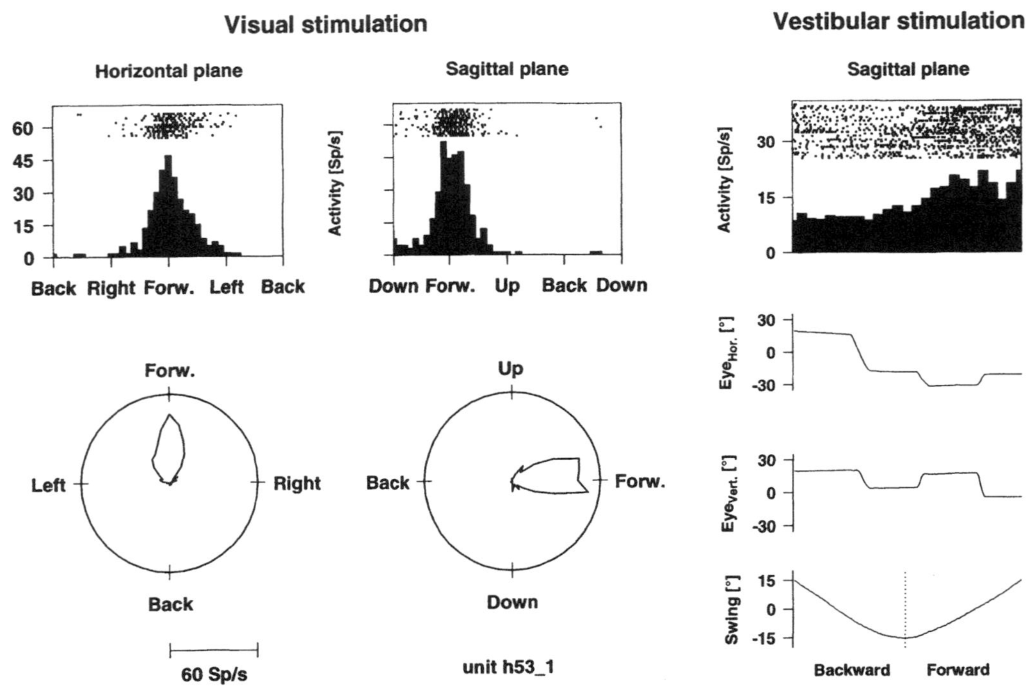

FIG. 2. Response to real and simulated self-motion in area MST. The left and middle columns show the response of a neuron from area MST to optic flow fields simulating self-motion. The left column indicates the responses for simulated movement in the horizontal plane as response histogram (upper row) and as polar plot (bottom row). The middle column shows responses for simulated movement in the sagittal plane. Movement directions changed continuously throughout the trial in both stimulus conditions. Movement directions indicated specifically: F, forward; L, left; B, backward; R, rightward; U, upward; D, downward. The right column shows the responses of this neuron to vestibular stimulation. The top histogram indicates the firing rate of the neuron during sinusoidal backward and forward motion. The lower panels indicate sample traces of the horizontal (second row) and vertical (third row) eye position and the position of the parallel swing (fourth row) during the trial.

Area MIP, on the other hand, can be considered to be part of a parietofrontal cortical network subserving the control of reaching movements (Battaglia *et al.*, 1998; Caminiti *et al.*, 1998; Johnson *et al.*, 1996; Wise *et al.*, 1997). Neurons respond to static visual stimuli dictating a reaching movement. In addition, many neurons show set-related activity during the wait period preceding a movement as well as responses to tactile stimuli. None of the neurons in areas LIP and MIP show directionally selective responses to moving visual stimuli.

1. *Optic-Flow Responses*

Recent studies suggested an involvement of area VIP in the processing of self-motion information (Schaafsma and Duysens, 1996; Schaafsma *et al.*, 1997; Bremmer *et al.*, 1995, 1997c). In the following, we will review response features of neurons in area VIP related to optic-flow stimuli and present, in addition, some new findings. Furthermore, we will unveil possible similarities and/or differences between response characteristics of neurons in areas MST and VIP.

In the experiments reported here, we tested neurons in area VIP for their responsiveness to a basic optic-flow pattern. Optic-flow stimuli simulating forward (expansion) or backward (contraction) motion or a rotation [clockwise (CW) or counterclockwise (CCW)] was presented with the singularity at the screen center. Expansion, contraction, and rotation stimuli were presented interleaved in pseudorandom order.

About two thirds of the neurons in area VIP respond selectively to optic flow stimuli (Fig. 3). Activity often encompasses strong phasic responses to the onset of the simulated movement, which then decreases to a much weaker tonic discharge level. The majority of cells prefers expansion over contraction stimuli (72%). In addition, the average response of the population of neurons for an expansion stimulus is significantly stronger compared to the response for a contraction stimulus (Rank sum test, $p < .01$).

2. *Compatibility of Optic Flow Responses with Sensitivity to Frontoparallel Motion*

We were interested, in particular, whether optic flow responses could be easily explained by a cell's directional selectivity to frontoparallel stimulus motion, or whether optic flow responses were unique and could not be accounted for by other response properties. Thus, we determined for a subset of cells not only the responses to optic flow stimuli but also mapped a given neuron's receptive field with a technique described previously by Duhamel *et al.* (1997a).

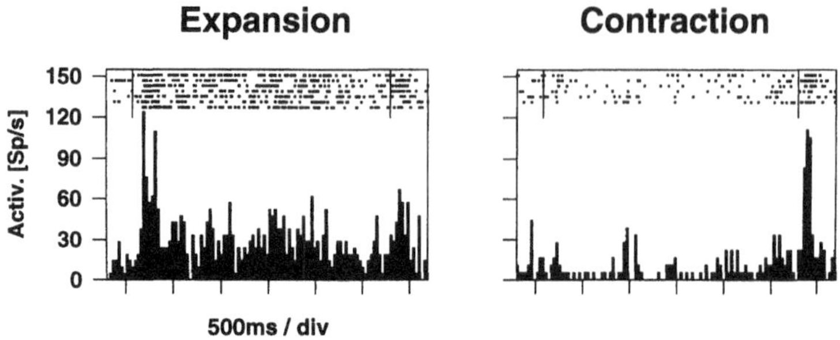

FIG. 3. Optic flow responses in area VIP. During the experiment, the head fixed animal was facing a translucent screen subtending the central 80° by 70° of the visual field. Computer-generated visual stimuli as well as a fixation target were back-projected by a liquid crystal display system. During visual stimulation, the monkey had to keep the eyes for 3500 ms within the tolerance window always at straight-ahead position [$(x, y) = (0°, 0°)$] to receive a liquid reward. Visual stimuli were random dot patterns, consisting of 240 dots, each individual dot 0.5° in size. A single dot leaving the display area at its outer borders was replaced at a randomly chosen new location. This replacement guaranteed a constant density of dots across the entire screen throughout the stimulation period. The two histograms show the responses of a single VIP neuron to an expansion stimulus (left) and a contraction stimulus (right). Raster displays indicate the response on a single-trial basis. Vertical lines in the raster displays indicate the onset (left) and the offset (right) of visual stimulus motion.

A prototypical example is shown in Fig. 4. The polar plot in the upper-left corner depicts the directional selectivity of this cell to frontoparallel motion as determined with the circular pathway paradigm (see legend for details). This cell preferred visual stimulus motion to the right. The upper-right panel indicates the size of the receptive field as plotted with a bar moving rightward (i.e., into the cell's preferred direction). The map represents the central 70° by 70° of the monkey's visual field. Neuronal activity is gray shaded with white corresponding to high and black corresponding to low discharge rates. The white contour line indicates the outer border of the receptive field (RF). We determined this border by connecting all points in the RF map where the neuron discharged 30% of its maximum firing rate. The lower two panels show superpositions of the cells outer RF border onto the optic flow patterns (expansion and contraction). It is obvious that this cell would be optimally stimulated with an expansion stimulus, whereas a contraction stimulus would be in the directions of the cell's nonpreferred direction. This cell thus should prefer an expansion over a contraction stimulus, as was indeed the case (Fig. 5). The two PSTHs show the neuronal response to expansion (left) and contraction (right). The cell responded with a phasic response to the onset of the pattern motion and decreased its discharge

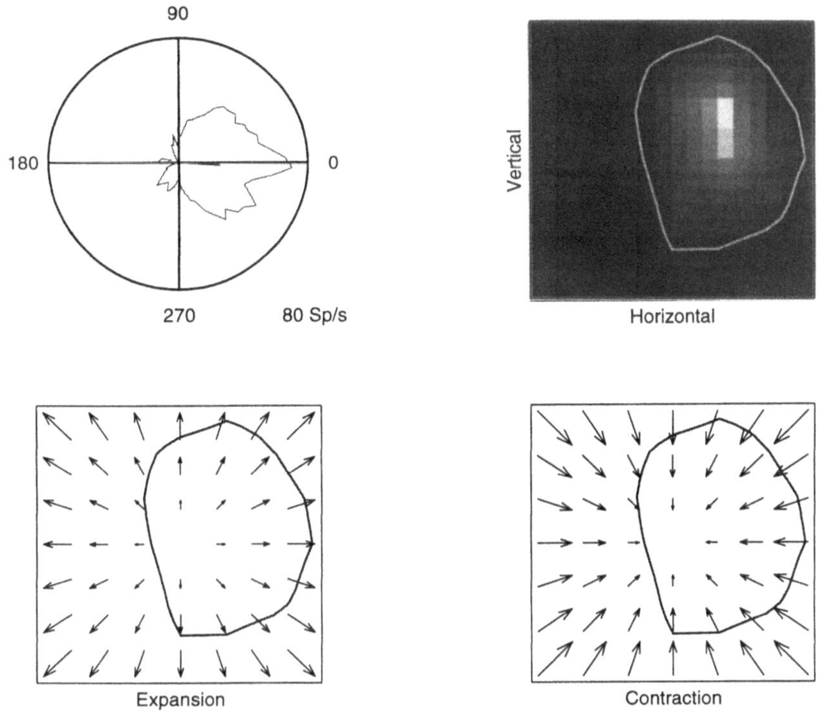

FIG. 4. Directional selectivity and RF location. The "circular pathway" paradigm was used to map directional selectivity. In this paradigm the speed of the stimulus is kept constant throughout a stimulus trial (cycle), but stimulus direction changes continuously (0–360°) within a complete stimulus cycle. Thus, each pattern element moves with the same speed (typically 27 or 40°/s) around its own center of motion (the radius was typically 5 or 10°). This paradigm allows for covering the full 2-D stimulus space during a single trial. The upper left panel indicates the directional selectivity of a cell to such kind of frontoparallel stimulus motion. This cell preferred stimulus motion to the right. The panel in the upper right shows the location and size of this neuron's receptive field. The displayed area represents the central 70° by 70° of the visual field. Neuronal activity is gray shaded with white corresponding to high and black corresponding to low activity. The RF of this cell was mostly located in the right part of the visual field. The white line demarcates a 30% activity line (i.e., it connects all points which fired at 30% of the cell's maximum discharge rate). The lower two panels superimpose the applied optic flow stimuli (expansion: left panel; contraction: right panel) onto the outer border of the cell's RF in order to visualize the distribution of the optic flow fields across the cell's RF.

to a tonic response level. Presentation of a contraction stimulus resulted in a slight inhibition of the cell's activity, as shown in the right PSTH.

Thus, for this cell, the information of its directional selectivity to frontoparallel motion and the location of its RF could easily predict re-

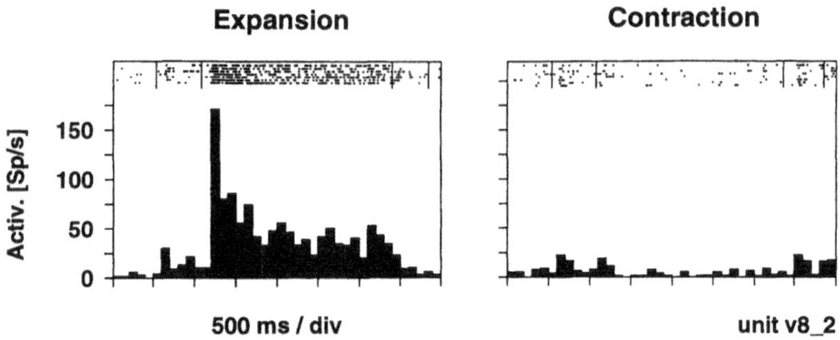

FIG. 5. Responses to expansion and contraction. Responses of the same cell as in Fig. 4 to expansion and contraction. The tick marks in the spike trains indicate stimulus onset (first tickmark), motion onset (second tickmark), motion offset (third tickmark), and stimulus offset (fourth tickmark). The monkey had to fixate throughout the trial. Expansion and contraction stimuli were presented interleaved in randomized order. This cell responded to the expansion stimulus but was inhibited by the contraction stimulus. Optic flow responses could be deduced from the cell's directional preference to frontoparallel stimulus motion and its RF location.

sponses to optic flow stimuli. Such predictive behavior was found in one half of the cells tested. For the other half of the cells, optic flow responses could not be deduced from information about directional preferences and RF location. An example is shown in Figs. 6 and 7.

Figure 6 shows a cell whose directional preference to frontoparallel motion is directed predominantly to the right (upper-left panel). The RF covers large parts of the central part of the visual field and is largely oriented vertically (upper-right panel). Superposition of the RF onto the visual stimulation pattern (in this case clockwise and counterclockwise rotation) clearly indicates that both visual stimuli contained about the same amount of visual motion components moving into the cells PD as well as into the cell's NPD. Thus, no prediction concerning the best response to any of the two stimuli is possible. Yet, as shown in Fig. 7, this cell clearly preferred CCW stimuli over CW stimuli.

Our finding concerning response predictability is similar to data from area MST (Lagae *et al.*, 1994) where also only about 50% of the optic flow responses could be explained by the directional selectivity of the cells to frontoparallel motion.

3. *Shifting the Singularity of the Optic Flow*

Optic flow stimuli with central singularities mimic a particular situation: gaze direction and movement direction (or the direction of the axis of rotation) are codirected. During natural orienting behavior,

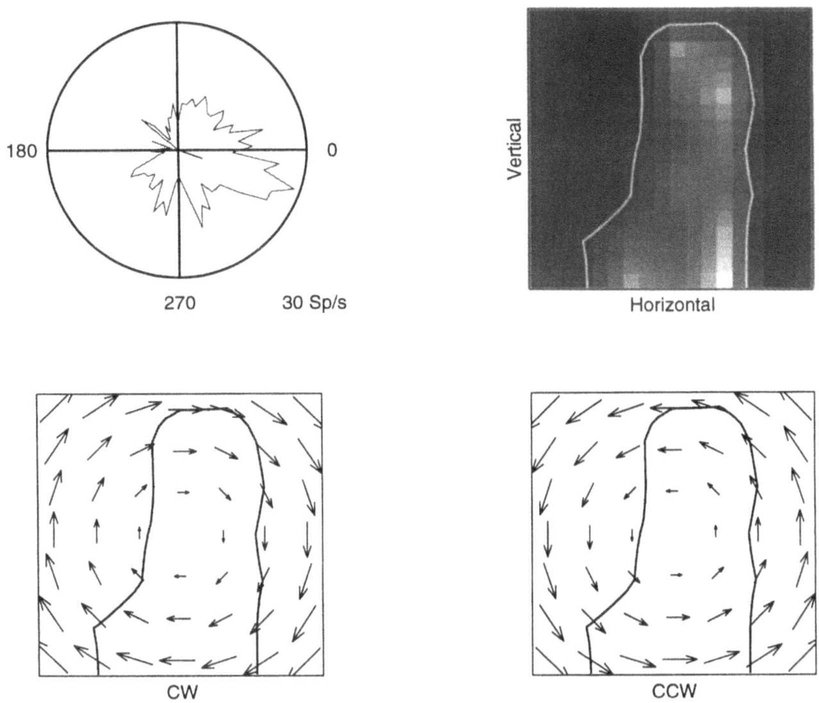

FIG. 6. Directional selectivity and RF location. The upper-left panel indicates (as in Fig. 4) the directional selectivity of the cell to frontoparallel stimulus motion. This cell preferred stimulus motion predominantly to the right. The panel on the right shows the location and size of this neuron's receptive field. The RF of this cell was located in the center of the visual field. The white line demarcates a 30% activity line. The lower two panels superimpose the applied optic flow stimuli (clockwise rotation: left panel; counterclockwise rotation: right panel) onto the outer border of the cell's RF.

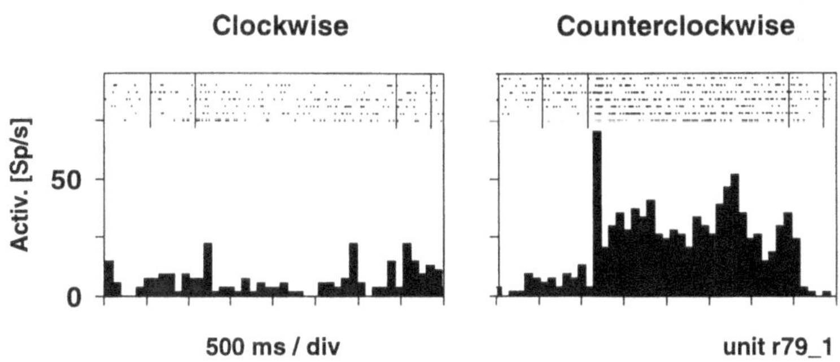

FIG. 7. Responses to rotational optic-flow stimuli. Responses of the same cell as in Fig. 6 to clockwise and counterclockwise rotation. The tick marks in the spike trains indicate again stimulus onset (first tickmark), motion onset (second tickmark), motion offset (third tickmark), and stimulus offset (fourth tickmark). This cell clearly preferred CCW rotation over CW rotation, although this could not be deduced from the directional selectivity and the RF location.

gaze direction and movement direction most often do not coincide, leading to an offset between the retinal projection of both points. Thus, we were also interested in the question, whether response strength, as in area MST, might be influenced by the location of the SOF on the retina. In order to investigate this question, we tested a subset of neurons for their response to nine different focus locations (expansion, contraction, clockwise and counterclockwise rotation), one central focus and eight foci shifted 25° into the periphery. The vast majority of neurons (95%) showed a significant influence of the location of the SOF with regard to their responses. An example is shown in Fig. 8. This neuron was tested with clockwise and counterclockwise rotation stimuli. Mean discharges (plus sd) for the nine different focus locations are shown in the upper two histograms. Variation of the focus location had a significant influence on the neuronal discharges ($p < .0005$ for CW and CCW rotation).

We compared qualitatively different statistical models in order to quantify the modulatory influence of the SOF location on the neuronal responses. Previous studies had suggested to use Gaussian tunings (Perrone and Stone, 1994) or sigmoidal tunings (Lappe *et al.*, 1996) to describe analogue response characteristics. From pure visual inspection, most neurons did not reveal a single-response peak, as would be the case for a Gaussian tuning. Sigmoidal tunings, on the other hand, need to have specified saturation values. These were obviously not available from our data. We thus decided on a different statistical model, namely the two-dimensional linear regression analysis to quantify the modulatory influence. A two-dimensional regression plane could be approximated significantly to the discharges of the neuron to CW ($p < .001$) and CCW rotation ($p < .002$).

At the population level, this modulatory influence was balanced out, as shown in Fig. 9. This histogram shows the average response of all neurons tested in this paradigm ($n = 13$). Statistical tests revealed no significant difference between the responses for the different SOFs. This result is similar to our data on expansion versus contraction responses (Duhamel *et al.*, 1997b) and to data obtained previously in area MST (Lappe *et al.*, 1996).

Equilibrium of the response strength for all focus locations together with a modulatory influence at the single-cell level can be considered as prerequisite for a population code concerning the focus location within the visual field. We could indeed show that a population of neurons is capable of encoding the location of the SOF within the visual field (Duhamel *et al.*, 1997b).

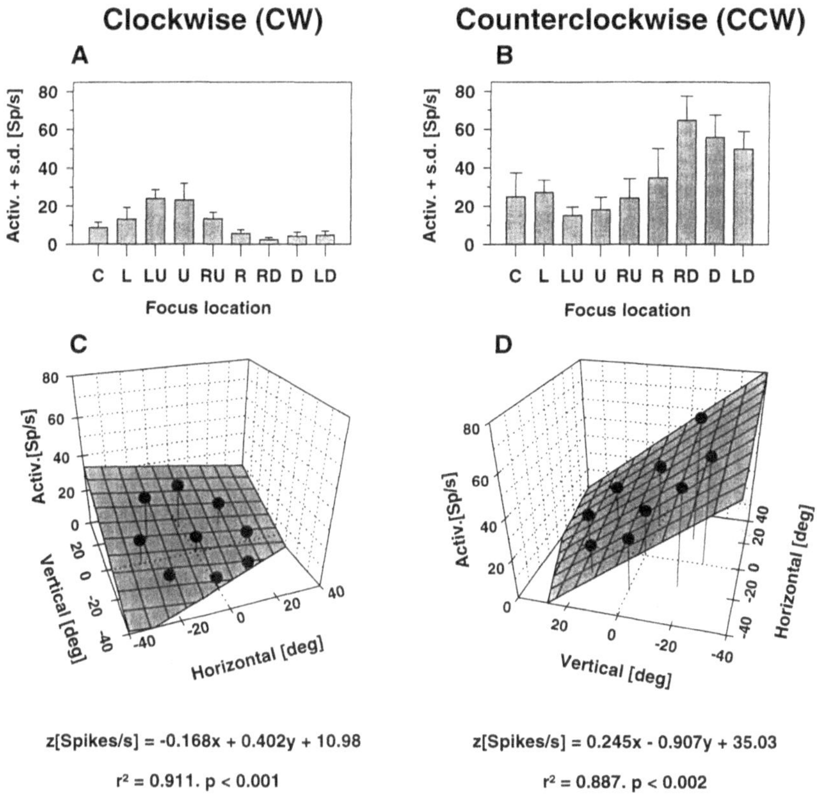

FIG. 8. Modulation of cell activity by shifting the location of the singularities of the optic flow. The figure shows an example of a single cell whose activity was profoundly influenced by the location of the singularity ($p < .0005$) for clockwise (A and C) and counterclockwise (B and D) rotation. (A, B) Each histogram shows the average response (± sd) of the cell for stimuli with the SOF at different locations in the visual field (C, Center; LU, Left Up; U, Up; RU, Right Up; R, Right; RD, Right Down; D, Down; LD, Left Down; L, Left). (C, D) Each shaded plane represents the two-dimensional linear regression to the mean discharge. The x-y plane in these plots represents the central 80° by 80° of the visual field. The base point of each drop line depicts the SOF location on the screen, and the height of each line depicts the mean activity value for stimulation with the SOF at this location. Approximation of a linear regression planes was statistically significant (CW: $r^2 = 0.911$, $p < .001$; CCW: $r^2 = 0.887$, $p < .002$).

4. *Tactile Stimulation*

In addition to visual information, somatosensory signals also can be used as to signal self-motion. Many neurons in area VIP are responsive to tactile stimulation (Fig. 10) (Colby *et al.*, 1993; Duhamel *et al.*, 1991, 1998).

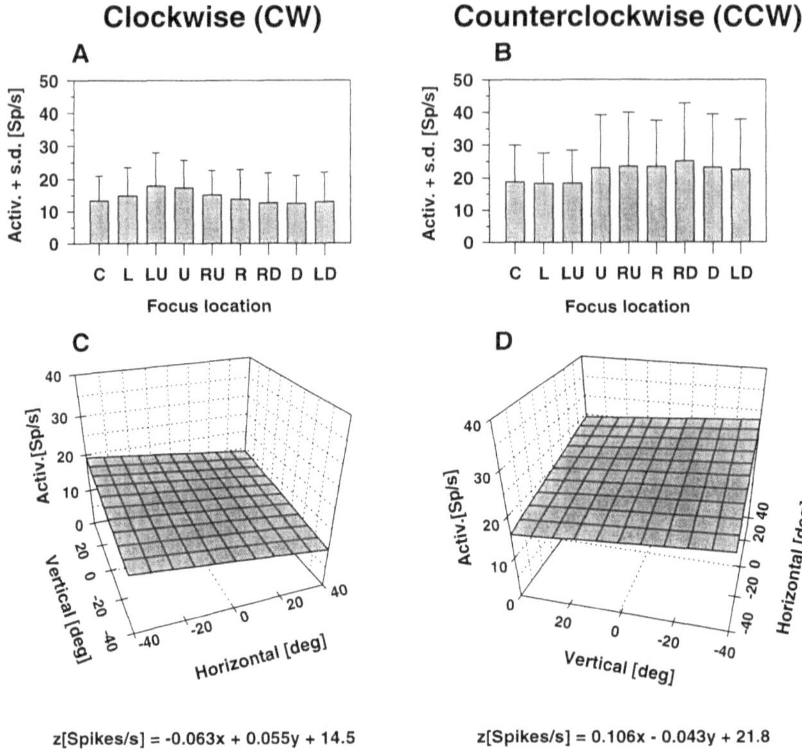

FIG. 9. Population response to optic-flow stimuli with shifted SOFs. The histograms and response planes represent the population response (n = 13) for the CW and CCW stimuli with shifted SOFs: on average there was no significant bias for any focus location in the visual field (distribution-free ANOVA).

These data were established while the monkey was blindfolded, by evoking responses with mechanical stimuli including air-puffs, light pressure applied with the tip of a stationary or moving cotton applicator and individual joint rotations. Most VIP cells that have a somatosensory RF respond well to passive superficial stimulation of restricted portions of the head, with the upper and lower face areas being represented in approximately equivalent manner. Less often, neurons have an RF located on the body or upper limbs and are driven by joint rotation of the arm or the shoulder. The body representation in VIP appears to be organized to include a higher-resolution region around the muzzle, where many small RF fields are found, and a coarser peripheral region extending to the back of the head, the upper body, and the limbs, where RFs are much larger. Somatic and visual RFs are organized in an orderly manner with tactile RF

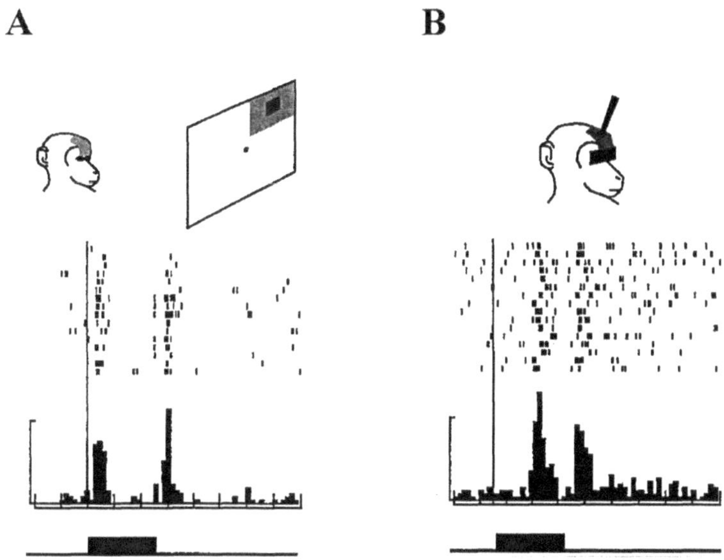

FIG. 10. Tactile responses in area VIP. Bimodal tactile and visual responses in area VIP. The neuron shows the same biphasic (ON/OFF) response pattern to a stimulation of (A) the visual RF on the top left side of a tangent screen and (B) the tactile RF on the top left side of the head (Duhamel *et al.*, 1998).

showing a systematic relation to the main axes of the visual field. Central visual RFs are matched to small tactile RFs around the lips and nose, and large peripheral visual RFs are associated with large tactile RFs on the side of the head or body, suggesting that the center of the face constitutes a somatosensory "fovea" in area VIP. Importantly, the matched tactile and visual RFs often demonstrate coaligned direction selectivity, and in some cases this parallelism was found to extend to orientation selectivity. The function of such multisensory responses is only partially understood. There may be an advantage of representing together sensory patterns that are strongly correlated because they are likely to have a common origin in the external world. In a natural environment, congruent directional stimulation across the visual and tactile modalities occurs in two types of circumstances—during a movement of an object (or the animal's own hand) occurring close enough to the face as to produce a concomitant tactile stimulation and during self-motion when head displacements produce congruent optical and tactile flow, as is the case when the animal navigates in dense vegetation. Under such conditions, multisensory input may allow us to encode the movement pattern of a stimulus moving in one sensory system using the reference frame provided by another sensory system.

5. *Vestibular Stimulation*

Vestibular sensory signals can indicate rotational and translational self-motion. In recent studies, we could show that many neurons in area VIP are not only bimodal but also trimodal [i.e., they respond to visual, tactile, and vestibular (vertical axis) stimulation] (Bremmer *et al.*, 1995; Graf *et al.*, 1996; Bremmer *et al.*, 1997c) (Fig. 11). Neurons usually were tested for vestibular responsiveness in light (first column of Fig. 11) and in total darkness (i.e., with all lights shut off and the animals' eyes covered with light-tight material) (second column of Fig. 11). During these experiments, animals perform reflexive, compensatory eye movements [optokinetic reflex (only in light) and vestibulo ocular reflex]. In order to test for possible influences of these compensatory eye movements, some neurons were tested during VOR suppression. To this end, animals fixated a chair-mounted LED while being rotated in otherwise darkness. All neurons also

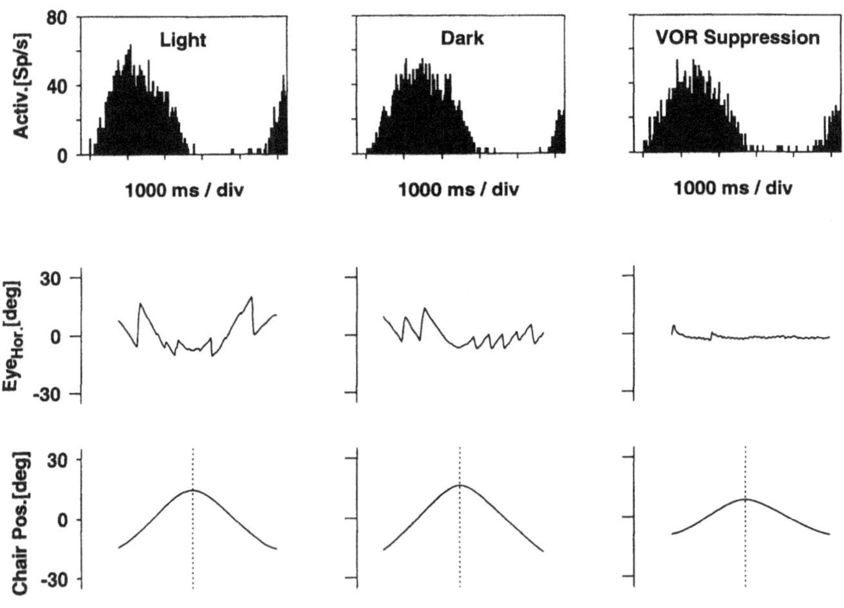

FIG. 11. Vestibular responses in area VIP. The top row shows the response histograms of a single VIP neuron to vertical axis rotation in light (left), in darkness (middle), and during VOR suppression (right). Sample eye position traces are shown in the middle row for the respective conditions. The chair positions during these movements are shown in the bottom row. This cell preferred rightward movement in all three conditions.

tested kept their selectivity for vestibular stimulation during this task (i.e., while no systematic eye movements occurred) (third column of Fig. 11).

All neurons with vestibular responses also showed directionally selective visual responses. Interestingly, preferred directions for visual and for vestibular stimulation were codirectional, i.e., nonsynergistic or noncomplementary. This is opposite to what one would expect and what one finds, e.g. in vestibular brain stem neurons. This is because the net retinal flow during vestibular-driven compensatory eye movements for rotations in light is opposite to the rotation direction. Yet, all VIP neurons without exception showed this nonsynergistic response characteristic. The functional role for this unexpected feature is only partially understood. It might be a hint for the role of area VIP in the coding of motion in near-extra personal space (Bremmer *et al.,* 1997c).

6. *Areas MST and VIP: Similarities and Differences in the Processing of Self-Motion Information*

Neurons in areas MST and VIP seem to have very similar response properties: they have large receptive fields, respond selectively to either expansion or contraction, clockwise or counterclockwise rotation. Shifting of the SOF results in a modulation of discharges for the majority of cells, whereas no bias exists for the population of cells. Cells in both areas also respond to spiral stimuli (Graziano *et al.,* 1994; Schaafsma and Duysens, 1996). Population responses in both areas allow the retrieval of the location of the SOF in the visual field (Lappe *et al.,* 1996; Duhamel *et al.,* 1997b). Neurons in both areas furthermore constantly encode the direction of heading, irrespective of superimposed, simulated eye movements (Bremmer *et al.,* 1998b).

The question thus arises about the particularities of optic flow processing in these two areas, and about what the specific role of one or the other area might be. As mentioned earlier, the multimodality of the neuronal responses makes area VIP different from area MST. The VIP neurons respond to visual, vestibular, and tactile stimuli. In particular, the tactile responses make area VIP different from area MST. The impressive coincidences of locations of visual and tactile RFs strongly support the idea of an important functional role of this response characteristic. We suggest that area VIP might be especially dedicated to the encoding of movement in near-extrapersonal space. Near objects approaching the animal during locomotion are potential candidates for becoming tactile stimuli in the very near future. Thus both visual and tactile responses are well suited to encode self-motion.

Furthermore, contrary to neurons in area MST, many neurons in area VIP encode visual information *explicitly* in a nonretinocentric reference frame (Duhamel *et al.*, 1997a). Thus, sensory information from all three sensory modalities can be encoded in a common frame of reference. This is different from area MST, where only eye-centered RFs are described (Bremmer *et al.*, 1997d). Instead, the occurrence of modulatory influences of eye position on neuronal discharges (Bremmer *et al.*, 1997a, 1998a) was taken as indication for an *implicit* encoding of visual sensory information in a nonretinocentric frame of reference.

Furthermore, the anatomical connections between these parietal areas and their target structures are different. Both areas have interconnections within the posterior parietal cortex. But it is area VIP that also makes direct connections with a specific part of the premotor cortex, a region representing movement of the animal's head and neck (Rizzolatti *et al.*, 1998). Thus, area VIP may be involved in head movement control, either when heading toward a target or when avoiding an object on collision course.

D. Area 7A

Areas MST and VIP are not the final stages in the motion-sensitive part of the dorsal stream of the macaque visual cortical system. As already shown in Fig. 1, both areas send feed-forward connections to area 7A. Two recent studies have shown that neurons in area 7A respond selectively to optic flow stimuli (Read and Siegel, 1997; Siegel and Read, 1997). In this study, monkeys had to detect changes in translational, rotational, or radially structured optic-flow fields. Some neurons showed the classical tuning for the direction of the optic flow (e.g., they preferred expansion over contraction stimuli). However, others were tuned for classes of optic flow [e.g., they preferred radial optic flow (expansion and contraction) over rotations optic flow (clockwise and counterclockwise)]. There was no strong correlation between cells classified as opponent vector cells (e.g., *cells preferring* centrifugal motion for all parts of the visual field) (Motter and Mountcastle, 1981) and those responding to optic flow fields (e.g., expansion cells). Furthermore it was shown, that optic-flow responses were modulated by the retinal position of a stimulus and by the position of the eyes in the orbit, as previously had been shown for area MST (Bremmer *et al.*, 1994).

E. Area STPa

Visual information processing is not strictly kept segregated between the ventral and the dorsal pathway of the monkey visual cortical system. The anterior part of the macaque superior temporal polysensory area

(STPa) is an example of converging signals from both cortical streams. Area STPa receives input from ventral stream area TEO and dorsal stream areas MST and 7A (Distler *et al.*, 1993; Boussaoud*et al.*, 1990).

According to this input, neurons have been described solely as sensitive to form (Perrett *et al.*, 1991) or motion (Oram *et al.*, 1993). A recent study could show specific, form-invariant sensitivity to optic flow fields (Anderson and Siegel, 1999) in more than half of the cells. Interestingly, preferred directions of visual motion were strongly biased toward expansion stimuli and upward and downward motion. It was suggested from this study that area STPa may be specifically involved in the processing of forward motion and/or looming stimuli. The latter hypothesis is corroborated by previous studies on selectivity of neurons in area STPa for object- and self-motion (Hietanen and Perrett, 1996a, b). Using real environmental objects, these authors could show that the majority of neurons responds only to visual motion originating from movements of external objects rather than to retinal stimulus motion caused by a movement of the animal. This response feature makes area STPa functionally very different from areas MST and VIP where neurons respond to simulated (optic flow) and real translational self-motion (Bremmer *et al.*, 1999, and unpublished observations).

Finally, some neurons in area STPa combine both response selectivity for form and motion (Oram and Perrett, 1996). The majority of these neurons responds selectively to only one combination of form, usually a head or a body, and movement direction (viewer-centered). In addition, a small fraction of cells (2%) responds in an object-centered manner in the sense that such cells respond to all directions of motion where the body moves in a direction compatible with the direction it faces.

III. Cortical Vestibular Areas

As mentioned earlier, self-motion induces not only visual sensory signals but also characteristic tactile and vestibular sensations. In addition to the previously described vestibular responses in areas MST and VIP, responses to vestibular stimulation have been reported also in other regions of the macaque parietotemporal cortex (Guldin and Grusser, 1998; Fukushima, 1997). Ödkvist *et al.* (1974) described such responses in the "neck region" of somatosensory cortical area 3a. Two other regions considered "vestibular" are area 2v at the anterior tip of the intraparietal sulcus (Buttner and Buettner, 1978; Graf *et al.*, 1995), and the parietoinsular vestibular cortex (PIVC), located in the parietal operculum of the sulcus lateralis and the retroinsular region (Grusser *et al.*, 1990a, b). Sakata *et al.* (1994) reported vestibular responses in some neurons "localized in the posterolateral part of area PG (area 7a of Vogt), on the anterior bank of the caudal superior temporal sulcus (STS), in the region partly overlapping the medial superior temporal (MST) area" (p.183). These latter neurons shared the nonsynergistic response characteristics of area VIP neurons regarding the preferred directions for visual and vestibular stimulation. The majority of neurons in the other areas respond in a synergistic fashion as it is also typical for vestibular brainstem neurons.

Further evidence for vestibular input to cortical areas of the dorsal stream of the macaque visual system comes from two other studies. Thier and Erickson (1992) showed a direction-selective increase in activity during VOR suppression in neurons of area MST. Brotchie *et al.* (1995) found an influence of head position on the activity of parietal neurons (areas 7A and LIP). In the latter case, an integrated vestibular head velocity signal could be the origin for this head position signal.

Area VIP is reciprocally connected to areas 7A, LIP, and MST. Area 7A in turn is connected to area PIVC which itself is part of a cortical network of vestibular sensitive areas (2v and 3a). Yet, area VIP differs remarkably from the other cortical areas, which are considered vestibular, in the sense that essentially all neurons show the nonsynergistic response characteristic of coaligned preferred directions for visual and vestibular stimulation.

IV. Human Brain Areas Involved in the Processing of Self-Motion Information

New imaging techniques like PET and fMRI allow also for the noninvasive investigation of the human brain for responses to moving visual stimuli. The human visual cortical system also has been shown to contain two functionally specialized parallel pathways each dealing with the encoding of specific stimulus features—PET (Haxby *et al.*, 1991; Zeki *et al.*, 1991) and fMRI (Tootell *et al.*, 1996). As in the monkey visual system, these streams are also separated topographically into a more ventral and a more dorsal part, and the existence of human homologues of area V4 and area MT has been hypothesized. In these studies, area MT proved to be more responsive to moving than to stationary stimuli, indicating that it is well suited for the encoding of motion. In addition, coherent motion could drive area MT (or V5), whereas, incoherent motion could not (Cheng *et al.*, 1995). Interestingly, however, the human homologue of area MT was not selectively active during the presentation of optic flow stimuli (De Jong *et al.*, 1994). Instead, three other brain regions showed selective activity: the right area V3, a region in the right superior parietal lobule, and the bilateral occipitotemporo ventral surface. This was taken as indication of a specific optic flow processing in higher parietal areas in humans.

V. Conclusion

There exist several stages of optic flow processing in the macaque visual cortical system. Area MT serves mostly as input stage. Neurons there respond to local motion vectors but not selectively to optic flow fields. Area 7A might be seen as a generalizing stage already providing abstract parameters as classes of motion (e.g., expansion and contraction). Two areas are localized in between with (at first glance) very similar response properties—areas MST and VIP. A more detailed analysis, however, reveals profound differences between the two, suggesting that area VIP might be especially dedicated to the encoding and control of motion in near extrapersonal space. New imaging techniques revealed a similar pattern of processing of visual information in the human brain: different stimulus features are processed in functionally different streams. The human homologues of areas MT and MST seem to be identified. The existence of the human homologue of area VIP, however, is still not clear. Part of our future work will concentrate on this issue.

Acknowledgments

This work was supported by the HCM program of the European Union (CHRXCT 930267) and the Human Frontier Science Program (RG 71/96B).

References

Albright, T. D. (1984). Direction and orientation selectivity in visual area MT of the macaque. *J. Neurophysiol.* **52,** 1106–1130.

Albright, T. D. (1989). Centrifugal directional bias in the middle temporal visual area (MT) of the macaque. *Vis. Neurosci.* **2,** 177–188.

Albright, T. D., Desimone, R., and Gross, C. G. (1984). Columnar organization of directionally selective cells in visual area MT of the macaque. *J. Neurophysiol.* **51**(1), 16–31.

Anderson, K. C., and Siegel, R. M. (1999). Optic flow sensitivity in the anterior superior temporal polysensory area. STPa, of the behaving monkey. *J. Neurosci.* **19,** 2681—2692.

Andersen, R. A., Bracewell, R. M., Barash, S., Gnadt, J. W., and Fogassi, L. (1990). Eye position effect on visual, memory, and saccade-related activity in areas LIP and 7A of macaque. *J. Neurosci.* **10**(4), 1176–1196.

Barash, S., Bracewell, R. M., Fogassi, L., Gnadt, J. W., and Andersen, R. A. (1991a). Saccade-related activity in the lateral intraparietal area I: Temporal properties; comparison with area 7A. *J. Neurophysiol.* **66**(3), 1095–1108.

Barash, S., Bracewell, R. M., Fogassi, L., Gnadt, J. W., and Andersen, R. A. (1991b). Saccade-related activity in the lateral intraparietal area II: spatial properties. *J. Neurophysiol.* **66**(3), 1109–1124.

Battaglia, M. A., Ferraina, S., Marconi, B., Bullis, J. B., Lacquaniti, F., Burnod, Y., Baraduc, P., and Caminiti, R. (1998). Early motor influences on visuomotor transformations for reaching: A positive image of optic ataxia. *Exp. Brain Res.* **123,** 172–189.

Boussaoud, D., Ungerleider, L. G., and Desimone, R. (1990). Pathways for motion analysis: cortical connections of the medial superior temporal and fundus of the superior temporal visual areas in the macaque. *J. Comp. Neurol.* **296,** 462–495.

Bracewell, R. M., Mazzoni, P., Barash, S., and Andersen, R. A. (1996). Motor intention activity in the macaque's lateral intraparietal area. II. Changes of motor plan. *J. Neurophysiol.* **76,** 1457–1464.

Bradley, D. C., Maxwell, M., Andersen, R. A., Banks, M. S., and Shenoy, K. V. (1996). Mechanisms of heading perception in primate visual cortex. *Science* **273,** 1544–1547.

Bremmer, F., Distler, C., and Hoffmann, K.-P. (1997a). Eye position effects in monkey cortex. II: Pursuit and fixation related activity in posterior parietal areas LIP and 7A. *J. Neurophysiol.* **77**(2), 962–977.

Bremmer, F., Duhamel, J.-R., Ben Hamed, S., and Graf, W. (1995). Supramodal encoding of movement space in the ventral intraparietal area of macaque monkeys. *Soc. Neurosci. Abstr.* **21,** 282.

Bremmer, F., Duhamel, J.-R., Ben Hamed, S., and Graf, W. (1997c). The representation of movement in near extra-personal space in the macaque ventral intraparietal area (VIP). *In:* "Parietal Lobe Contributions to Orientation in 3D Space" (P. Thier and H.-O. Karnath, Eds.), pp. 619–630. Springer Verlag, Heidelberg.

Bremmer, F., Ilg, U. J., Thiele, A., Distler, C., and Hoffmann, K.-P (1997b). Eye position effects in monkey cortex. I: Visual and pursuit related activity in extrastriate areas MT and MST. *J. Neurophysiol.* **77**(2), 944–961.

Bremmer, F., Kubischik, M., Pekel, M., and Lappe, M. (1998b). Selectivity for heading direction during simulated eye movements in macaque extra-striate cortex. *Eur. J. Neurosci. (Suppl.)* **10,** 244.

Bremmer, F., Kubischik, M., Pekel, M., Lappe, M., and Hoffmann, K.-P. (1999). Linear vestibular self-motion signals in monkey medial superior temporal area. *Ann. N.Y. Acad. Sci.* **871,** 272–281.

Bremmer, F., Lappe, M., Pekel, M., and Hoffmann, K.-P. (1994). Representation of gaze direction during egomotion in macaque visual cortical area MSTd. *Eur. J. Neurosci. (Suppl.)* **17,** 150.

Bremmer, F., Pouget, A., and Hoffmann, K.-P. (1998a). Eye position encoding in the macaque posterior parietal cortex. *Eur. J. Neurosci.* **10,** 153–160.

Bremmer, F., Thiele, A., and Hoffmann, K.-P. (1997d). Encoding of spatial information in monkey visual cortical areas MT and MST. *Soc. Neurosci. Abstr.* **23,** 459.

Britten, K. H., and van Wezel, R. J. A. (1998). Electrical microstimulation of cortical area MST biases heading preception in monkeys. *Nat. Neurosci.* **1,** 59–64.

Brotchie, P. R., Andersen, R. A., Snyder, L.H., and Goodman, S.J. (1995). Head position signals used by parietal neurons to encode locations of visual stimuli. *Nature* **375,** 232–235.

Büttner, U. and Buettner, U. W. (1978). Parietal cortex (2v) neuronal activity in the alert monkey during natural vestibular and optokinetic stimulation. *Brain Res.* **153,** 392–397.

Caminiti, R., Ferraina, S., and Mayer, A. B. (1998). Visuomotor transformations: Early cortical mechanisms of reaching. *Curr. Opin. Neurobiol.* **8,** 753–761.

Cheng, K., Fujita, H., Kanno, I., Miura, S., and Tanaka, K. (1995). Human cortical regions activated by wide-field visual motion: An $H_2{}^{15}O$ PET study. *J. Neurophysiol.* **74,** 413–427.

Colby, C. L., Duhamel, J.-R., and Goldberg, M. E. (1993). Ventral intraparietal Area of the macaque: Anatomical location and visual response properties. *J. Neurophysiol.* **69,** 902–914.

Colby, C. L., Duhamel, J. R., and Goldberg, M. E. (1995). Oculocentric spatial representation in parietal cortex. *Cereb. Cortex.* **5,** 470–481.

Colby, C. L., Duhamel, J. R., and Goldberg, M. E. (1996). Visual, presaccadic, and cognitive activation of single neurons in monkey lateral intraparietal area. *J. Neurophysiol.* **76,** 2841–2852.

De Jong, B. M., Shipp, S., Skidmore, B., Frackowiak, R. S. J., and Zeki, S. (1994). The cerebral activity related to the visual perception of forward motion in depth. *Brain* **117,** 1039–1054.

Distler, C., Boussaoud, D., Desimore, R., and Ungerleider, L. G. (1993). Cortical connections of inferior temporal area TEO in macaque monkeys. *J. Comp. Neurol.* **334,** 125–150.

Duffy, C. J. (1998). MST neurons respond to optic flow and translational movement. *J. Neurophysiol.* **80,** 1816–1827.

Duffy, C. J., and Wurtz, R. H. (1991a). Sensitivity of MST neurons to optic flow stimuli. I: A continuum of response selectivity to large-field stimuli. *J. Neurophysiol.* **65**(6), 1329–1345.

Duffy, C. J., and Wurtz, R. H. (1991b). Sensitivity of MST neurons to optic flow stimuli. II: Mechanisms of response selectivity revealed by small-field stimuli. *J. Neurophysiol.* **65**(6), 1346–1359.

Duffy, C. J., and Wurtz, R. H. (1995). Response of monkey MST neurons to optic flow stimuli with shifted centers of motion. *J. Neurosci.* **15,** 5192–5208.

Duhamel, J.-R., Bremmer, F., Ben Hamed, S., and Graf, W. (1997b). Encoding of heading direction during simulated self-motion in the macaque ventral intraparietal area (VIP). *Soc. Neurosci. Abstr.* **23,** 1545.

Duhamel, J.-R., Bremmer, F., Ben Hamed, S., and Graf, W. (1997a). Spatial invariance of visual receptive fields in parietal cortex neurons. *Nature* **389,** 845–848.

Duhamel, J.-R., Colby, C. C., and Goldberg, M. E. (1998). Ventral intraparietal area of the macaque: congruent visual and somatic response properties. *J. Neurophysiol.* **79,** 126–136.

Duhamel, J.-R., Colby, C. L., and Goldberg, M. E. (1991). Congruent representations of visual and somatosensory space in single neurons of monkey ventral intra-parietal cortex (area VIP). *In:* "Brain and Sapce," (J. Paillard, Ed.), pp. 223–236. Oxford University Press, Oxford.

Duhamel, J.-R., Colby, C. L., and Goldberg, M. E. (1992). The updating of the representation of visual space in parietal cortex by intended eye movements. *Science* **255,** 90–92.

Duhamel, J.-R., Colby, C. L., and Goldberg, M. E. (1998). Ventral intraparietal area of the macaque: congruent visual and somatic response properties. *J. Neurophysiol.* **79,** 126–136.

Fukushima, K. (1997). Corticovestibular interactions: Anatomy, electrophysiology, and functional considerations. *Exp. Brain Res.* **117,** 1–16.

Geesaman, B. J., Born, R. T., Andersen, R. A., and Tootell, R. B. H. (1997). Maps of complex motion selectivity in the superior temporal cortex of the alert macaque monkey: A double label 2-deoxyglucose study. *Cereb. Cortex* **7,** 749–757.

Gibson, J. J. (1950). "The Perception of the Visual World." Houghton Mifflin, Boston.

Gottlieb, J. P., Kusunoki, M., and Goldberg, M. E. (1998). The representation of visual salience in monkey parietal cortex. *Nature* **391,** 481–484.

Graf, W., Bremmer, F., Ben Hamed, S., and Duhamel, J.-R. (1996). Visual-vestibular interaction in the ventral intraparietal area of macaque monkeys. *Soc. Neurosci. Abstr.* **22,** 7. [Abstract]

Graf, W., Bremmer, F., Ben Hamed, S., Sammaritano, M., and Duhamel, J.-R. (1995). Oculomotor, vestibular and visual response properties of neurons in the anterior inferior parietal lobule of macaque monkeys. *Soc. Neurosci. Abstr.* **21,** 665.

Graziano, M. S. A., Andersen, R. A., and Snowden, R. (1994). Tuning of MST neurons to spiral motions. *J. Neurosci.* **14**(1), 54–67.

Grüsser, O.-J., Pause, M., and Schreiter, U. (1990b). Localization and responses of neurons in the parieto-insular vestibular cortex of awake monkeys (macaca fascicularis). *J. Physiol.* **430,** 537–557.

Grüsser, O.-J., Pause, M., and Schreiter, U. (1990a). Vestibular neurons in the parieto-insular cortex of monkeys (macaca fascicularis): Visual and neck receptor responses. *J. Physiol.* **430,** 559–583.

Guldin, W. O. and Grüsser, O.-J. (1998). Is there a vestibular cortex? *Trends Neurosci.* **21,** 254–259.

Haxby, J. V., Ungerleider, L. G., and Rapoport, S. I. (1991). Dissociation of object and spatial visual processing pathways in human extrastriate cortex. *Proc. Nat. Acad. Sci. USA* **88,** 1621–1625.

Hietanen, J. K., and Perrett, D. I. (1996a). A comparison of visual responses to object- and ego-motion in the macaque superior temporal polysensory area. *Exp. Brain Res.* **108,** 341–345.

Hietanen, J. K., and Perrett, D. I. (1996b). Motion sensitive cells in the macaque superior temporal polysensory area: Response discrimination between self-generated and externally generated pattern motion. *Behav. Brain Res.* **76,** 155–167.

Johnson, P. B., Ferraina, S., Bianchi, L., and Caminiti, R. (1996). Cortical networks for visual reaching: Physiological and anatomical organization of frontal and parietal lobe arm regions. *Cereb. Cortex* **6,** 102–119.

Lagae, L., Maes, H., Raiguel, S., Xiao, D.-K., and Orban, G. A. (1994). Responses of macaque STS neurons to optic flow stimuli: A comparison of areas MT and MST. *J. Neurophysiol.* **71,** 1597–1626.

Lagae, L., Raiguel, S., and Orban, G. A. (1993). Speed and direction selectivity of macaque middle temporal neurons. *J. Neurophysiol.* **69,** 19–39.

Lappe, M., Bremmer, F., Pekel, M., Thiele, A., and Hoffmann, K.-P. (1996). Optic flow processing in monkey STS: A theoretical and experimental approach. *J. Neurosci.* **16,** 6265–6285.

Lappe, M., Pekel, M., and Hoffmann, K.-P. (1998). Optokinetic eye movements elicited by radial optic flow in the macaque monkey. *J. Neurophysiol.* **79,** 1461–1480.

Lappe, M. and Rauschecker, J. P. (1995). Motion Anisotropies and Heading Detection. *Biol. Cybern.* **72,** 261–277.

Maunsell, J. H. R., and Van Essen, D. C. (1983). The connections of the middle temporal visual area (MT) and their relationship to a cortical hierarchy in the macaque monkey. *J. Neurosci.* **3,** 2563–2580.

Mazzoni, P., Bracewell, R. M., Barash, S., and Andersen, R. A. (1996). Motor intention activity in the macaque's lateral intraparietal area. I. Dissociation of motor plan from sensory memory. *J. Neurophysiol.* **76,** 1439–1456.

Mikami, A., Newsome, W. T., and Wurtz, R. H. (1986a). Motion selectivity in macaque visual cortex. I: Mechanisms of direction and speed selectivity in extrastriate area MT. *J. Neurophysiol.* **55**(6), 1308–1327.

Mikami, A., Newsome, W. T., and Wurtz, R. H. (1986b). Motion selectivity in macaque visual cortex. II: Spatiotemporal range of directional interactions in MT and V1. *J. Neurophysiol.* **55**(6), 1328–1339.

Motter, B. C. and Mountcastle, V. B. (1981). The functional properties of the light-sensitive neurons of the posterior parietal cortex studied in waking monkeys: Foveal sparing and opponent vector organization. *J. Neurosci.* **1**(1), 3–26.

Ödkvist, L. M., Schwarz, D. W. F., Fredrickson, J. M., and Hassler, R. (1974). Projection of the vestibular nerve to the area 3a arm field in the squirrel monkey (*Saimiri sciureus*). *Exp. Brain Res.* **21,** 97–105.

Oram, M. W., and Perrett, D. I. (1996). Integration of form and motion in the anterior superior temporal polysensory area (STPa) of the macaque monkey. *J. Neurophysiol.* **76,** 109–129.

Oram, M. W., Perrett, D. I., and Hietanen, J. K. (1993). Directional tuning of motion-sensitive cells in the anterior superior temporal polysensory area of the macaque. *Exp. Brain Res.* **97,** 274–294.

Perrett, D. I., Oram, M. W., Harries, M. H., Bevan, R., Hietanen, J. K., Benson, P. J., and Thomas, S. (1991). Viewer-centered and object-centered coding of heads in the macque temporal cortex. *Exp. Brain Res.* **86,** 159–173.

Perrone, J. A. and Stone, L. S. (1994). A model of self-motion estimation within primate extrastriate visual cortex. *Vision Res.* **34,** 2917–2938.

Platt, M. L. and Glimcher, P. W. (1998). Response fields of intraparietal neurons quantified with multiple saccadic targets. *Exp. Brain Res.* **121,** 65–75.

Read, H. L., and Siegel, R. M. (1997). Modulation of responses to optic flow in area 7a by retinotopic and oculomotor cues in monkey. *Cereb. Cortex* **7,** 647–661.

Regan, D. and Beverly, K. I. (1982). How do we avoid confounding the direction we are looking and the direction we are moving? *Science* **215,** 194–196.

Rizzolatti, G., Luppino, G., and Matelli, M. (1998). The organization of the cortical motor system: New concepts. *Electroencephalogr. Clin. Neurophysiol.* **106,** 283–296.

Saito, H., Yukie, M., Tanaka, K., Hikosaka, K., Fukada, Y., and Iwai, E. (1986). Integration of direction signals of image motion in the superior temporal sulcus of the macaque monkey. *J. Neurosci.* **6**(1), 145–157.

Sakata, H., Shibutani, H., Ito, Y., Tsurugai, K., Mine, S., and Kusunoki, M. (1994). Functional properties of rotation-sensitive neurons in the posterior parietal association cortex of the monkey. *Exp. Brain Res.* **101,** 183–202.

Schaafsma, S. J. and Duysens, J. (1996). Neurons in the ventral intraparietal area of awake macaque monkey closely resemble neurons in the dorsal part of the medial superior temporal area in their responses to optic flow patterns. *J. Neurophysiol.* **76,** 4056–4068.

Schaafsma, S. J., Duysens, J., and Gielen, C. C. (1997). Responses in ventral intraparietal area of awake macaque monkey to optic flow patterns corresponding to rotation of planes in depth can be explained by translation and expansion effects. *Vis. Neurosci.* **14,** 633–646.

Shipp, S., and Zeki, S. (1989). The organization of connections between areas V5 and V1 in macaque monkey visual cortex. *Eur. J. Neurosci.* **1**(4), 309–332.

Siegel, R. M. and Read, H. L. (1997). Analysis of optic flow in the monkey parietal area 7A. *Cereb. Cortex* **7,** 327–346.

Tanaka, K., Fukada, Y., and Saito, H. (1989). Underlying mechanisms of the response specificity of expansion/contraction and rotation cells in the dorsal part of the medial superior temporal area of the macaque monkey. *J. Neurophysiol.* **62,** 642–656.

Tanaka, K., Hikosaka, K., Saito, H., Yukie, M., Fukada, Y., and Iwai, E. (1986). Analysis of local and wide-field movements in the superior temporal visual areas of the macaque monkey. *J. Neurosci.* **6**(1), 134–144.

Tanaka, K., and Saito, H. (1989). Analysis of motion of the visual field by direction, expansion/contraction, and rotation cells clustered in the dorsal part of the medial superior temporal area of the macaque monkey. *J. Neurophysiol.* **62,** 626–641.

Tanaka, K., Sugita, Y., Moriya, M., and Saito, H.-A. (1993). Analysis of object motion in the ventral part of the medial superior temporal area of the macaque visual cortex. *J. Neurophysiol.* **69,** 128–142.

Thier, P., and Andersen, R. A. (1998). Electrical microstimulation distinguishes distinct saccade-related areas in the posterior parietal cortex. *J. Neurophysiol.* **80,** 1713–1735.

Thier, P., and Erickson, R. G. (1992). Vestibular input to visual-tracking neurons in area MST of awake rhesus monkeys. *Ann. NY Acad. Sci.* **656,** 960–963.

Tootell, R. B. H., Dale, A. M., Sereno, M. I., and Malach, R. (1996). New images from human visual cortex. *Trends Neurosci.* **19,** 481–489.

Ungerleider, L. G. and Desimone, R. (1986a). Cortical connections of visual area MT in the macaque. *J. Comp. Neurol.* **248,** 190–222.

Ungerleider, L. G. and Desimone, R. (1986b). Projections to the superior temporal sulcus from the central and peripheral field representations of V1 and V2. *J. Comp. Neurol.* **248,** 147–163.

Warren, W. H. J., and Hannon, D. J. (1990). Eye movements and optical flow. *J. Opt. Soc. Am. [A]* **7,** 160–169.

Wise, S. P., Boussaoud, D., Johnson, P. B., and Caminiti, R. (1997). Premotor and parietal cortex: corticocortical connectivity and combinatorial computations. *Ann. Rev. Neurosci.* **20,** 25–42.

Zeki, S., Watson, J. D., Lueck, C. J., Friston, K. J., Kennard, C., and Frackowiak, R. S. (1991). A direct demonstration of functional specialization in human visual cortex. *J. Neurosci.* **11,** 641–649.

OPTIC FLOW ANALYSIS FOR SELF-MOVEMENT PERCEPTION

Charles J. Duffy

Departments of Neurology, Neurobiology and Anatomy,
Ophthalmology, and Brain and Cognitive Sciences
and the Center for Visual Science,
University of Rochester, Rochester, New York

I. Introduction

Self-movement perception is central to spatial orientation. The patterned visual motion seen during self-movement constitutes the optic flow field (Gibson, 1966) that provides perceptual cues about the heading of self-movement (Verri *et al.*, 1989; Lappe and Rauschecker, 1993) and the structure of the environment (Werkhoven and Koenderink, 1990a, b; Beusmans, 1993). These cues must be integrated with nonvisual cues about self-movement to optimize adaptation to changing behavioral settings. The work described in this chapter deals with the potential contributions of the medial superior temporal (MST) area of monkey extrastriate visual cortex to the analysis and integration of sensory cues about self-movement.

This chapter describes results obtained from adult Rhesus monkeys studied at the University of Rochester Medical Center, Rochester, NY. Sections II and III include data derived from studies conducted in the laboratories of Dr. R. H. Wurtz at the Laboratory of Sensorimotor Research of the National Eye Institute, Bethesda, MD. All protocols were approved by the Institutional Committee on Animal Research and complied with Public Health Service and Society for Neuroscience policy on the humane care and use of laboratory animals.

0074-7742/00 $30.00

Three conclusions about MST neurons are suggested by our findings. One, they possess a structured sensitivity to optic flow patterns that impart information about the heading of self-movement relative to the direction of gaze. Two, MST neurons show response sensitivity to the speed gradients in optic flow that reflect the structural layout of the environment. Three, MST neurons are influenced by vestibular signals about self-movement such that they might differentiate between optic flow created by self-movement and that which might be created by the movement of large objects in the visual surround. These findings provide convergent evidence of an important role for area MST in self-movement perception and visuospatial orientation.

II. MST Sensitivity to Heading Direction

For simulated heading direction in optic flow, MST neuronal response selectivity was suggested in much of the early work on this area (Tanaka *et al.,* 1986; Saito *et al.,* 1986; Duffy and Wurtz, 1991; Orban *et al.,* 1992; Graziano *et al.,* 1994). Specific response preferences for specific simulated heading directions were demonstrated by Duffy and Wurtz (1995) with a range of heading selectivities across the visual field. Those studies showed that heading directions near the direction of gaze were represented by the greatest number of MST neurons. In addition, neurons that preferred heading directions near the direction of gaze were also the most selective with only much smaller responses to other heading directions. Thus, MST neurons seemed specialized for detecting a centered focus of expansion (FOE) that represents a match between the simulated direction of self-movement and the direction of gaze.

In extending those studies, Duffy and Wurtz (1995) used a stimulus set that concentrated 33 FOEs within the central 40° of the visual field, with stimuli spaced at 10° intervals along eight axes through the fixation point and distributed in 45° increments around the fixation point (Fig. 1). These stimuli simulated self-movement in 33 different directions at and around the fixation point. This dense array of FOE stimuli was designed to test the precision of MST response selectivity within the range of FOE sensitivity that might be required in routine daily behavior.

The results of such studies on three neurons are shown in Figs. 2A–C. For each neuron, the responses to each of the 33 stimuli are illustrated by the oval in the position on the plot that corresponds to the position of the evoking stimulus in Fig. 1. The amplitude of the average response to each stimulus is represented by the length of the oval. The

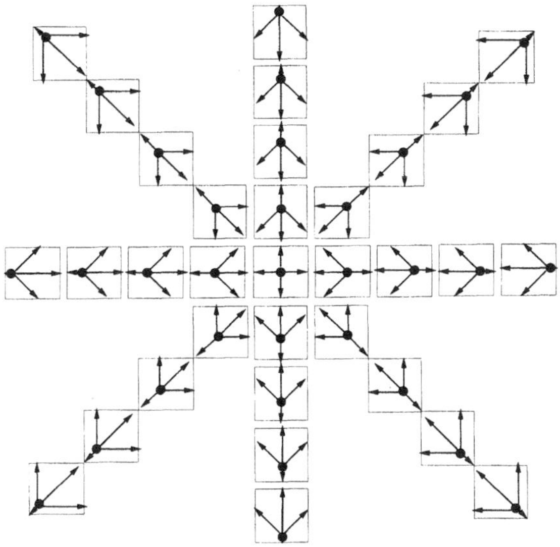

FIG. 1. Schematic representation of the 33 optic flow stimuli used to test the structure of optic flow response fields. Each square frame represents the 90 × 90° stimulus screen. The arrows illustrate the pattern of local directions of dot motion on various parts of the screen in each stimulus. The filled circles show the location of the FOE in each stimulus. In these stimuli, the FOEs were located at 10° intervals along eight axes of displacement, simulating the visual motion seen during observer self-movement in 33 directions distributed around the direction of gaze.

responses in Fig. 2A show a preference for stimuli with FOEs located to the far left of the direction of gaze. The strongest response is to the far left and down FOE stimulus with a gradual decline of response amplitude with increasing distance from that point. The responses in Fig. 2B show a preference for the center FOE stimulus with a gradual decline of responses with increasing stimulus eccentricty, about evenly in all directions around 360°. The responses in Fig. 2C show a preference for the FOEs around the center FOE but a dramatic decline in responsiveness when the FOE is directly over the fixation point.

We studied 12 MST neurons with these stimuli finding an equal number of each of the three types of neurons illustrated in Fig. 2. These findings demonstrate that some MST neurons possess a response organization of potential behavioral utility. Figure 3 illustrates how these neurons might each play different roles in representing self-movement through the visual environment. Figure 3A illustrates how neurons preferring eccentric stimuli might provide a directionally selective indication of a substantial difference between the gaze and heading directions;

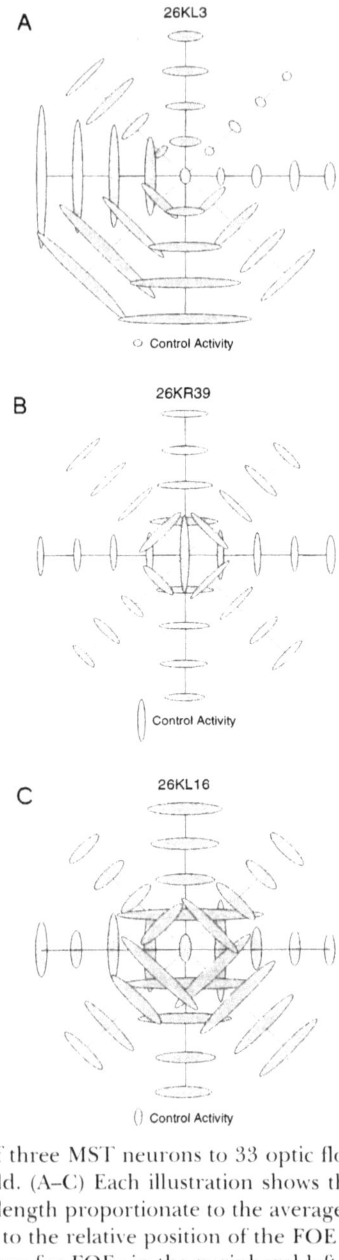

FIG. 2. The response of three MST neurons to 33 optic flow stimuli with FOEs in the central 40° of the visual field. (A–C) Each illustration shows the responses to each of the 33 stimuli as an oval with a length proportionate to the average evoked discharge rate and at a location corresponding to the relative position of the FOE (as in Fig. 1). (A) This neuron showed a clear preference for FOEs in the peripheral left-lower quadrant of the central visual field with smaller responses to FOEs at other locations. (B) This neuron shows its strongest responses to the FOE located directly over the fixation point showing smaller responses with increasing FOE eccentricity in any direction. (C) This neuron shows strong responses to FOEs located around the fixation point with smaller responses when the FOE is in the periphery or located directly over the fixation point.

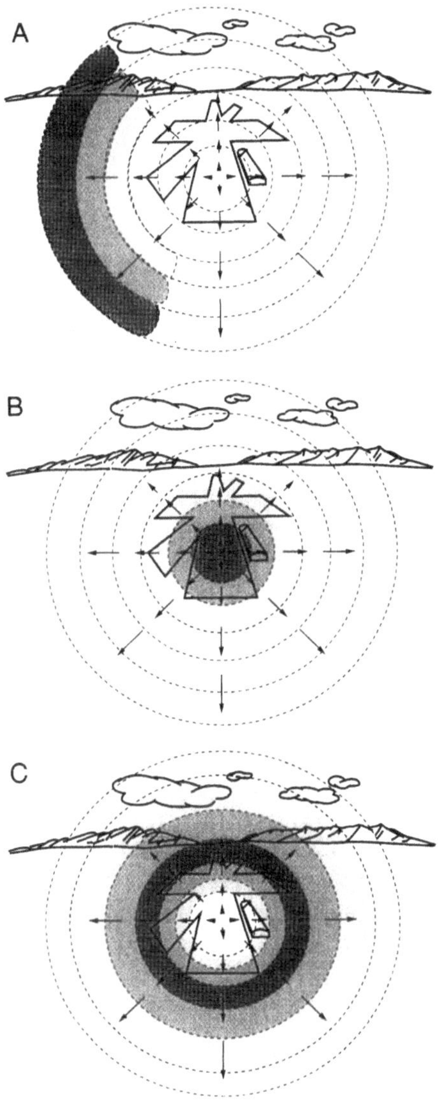

FIG. 3. Diagrams illustrating the potential behavioral utility of optic flow responsive neurons preferring FOEs with various distributions around the visual field. (A–C) Drawings of a pilot's view during approach for a landing at an airport with arrows showing the pattern of visual motion in the optic flow field when the pilot is on course toward the center runway (after Gibson, 1966). The concentric rings illustrate FOE eccentricity when the pilot's heading deviates from the runway. The denser shading represents higher firing rates of neurons that prefer FOEs located in that region. (A) The firing pattern of a neuron that prefers FOEs in the periphery indicating substantial heading deviation to the left. (B) The firing pattern of a neuron that prefers FOEs at the fixation point indicating a heading that matches the direction of gaze. (C) The firing pattern of a neuron that prefers FOEs around the fixation point indicating subtle deviation of heading in any direction away from the direction of gaze.

a difference between where the observer is going and where the observer is looking. Figure 3B illustrates how neurons preferring the center FOE might provide an indicator of matching heading and gaze directions; neurons that are most active when the observer is looking in the direction of self-movement. Figure 3C illustrates how neurons that respond to stimuli surrounding the direction of gaze might function as directionally nonselective indicators of subtle, but potentially important, deviations of the direction of heading from the direction of gaze.

III. MST Sensitivity to the Structure of the Environment

The gradients of speed in optic flow stimuli reflect the structural layout of the visual environment. In studies of MST neurons, Duffy and Wurtz (1997) tested whether graded manipulations of the speed gradients in optic flow stimuli altered the responses to the preferred pattern of visual motion. Stimuli simulating self-movement toward a single surface at a uniform distance were used as the baseline stimulus in which the speed of each dot was specified by a sine * cosine function of its distance from the FOE. We modified the speed gradients in the stimuli by multiplicatively increasing the speed–distance function to create exaggerated speed gradients (Fig. 4A). Such positive speed gradients simulated movement through an environment in which objects near the center of the visual field were much farther from the observer than were objects in the peripheral visual field (Fig. 4C). We used negative multipliers to decrease the speed–distance function and create reversed speed gradients (Fig. 4B). Such negative speed gradients simulated movement through an environment in which objects near the center of the visual field were much closer to the observer than were objects in the peripheral visual field (Fig. 4D).

Most MST neurons show clear response selectivity for speed gradients in optic flow. Figures 4E and 4F show the average amplitude of responses (ordinate) of two MST neurons to the negative (abscissa, left) and positive (abscissa, right) speed gradients. The neuronal responses in Fig. 4E show a clear preference for positive speed gradients; the neuronal responses in Fig. 4F show a clear preference for negative speed gradients. In both neurons, the responses to the simulated flat surface were robust, but the responses to the preferred speed gradients were substantially larger. In addition, the responses to the nonpreferred speed gradients approached the control activity level but did not fall below that level to clearly suggest the involvement of active inhibitory mechanisms.

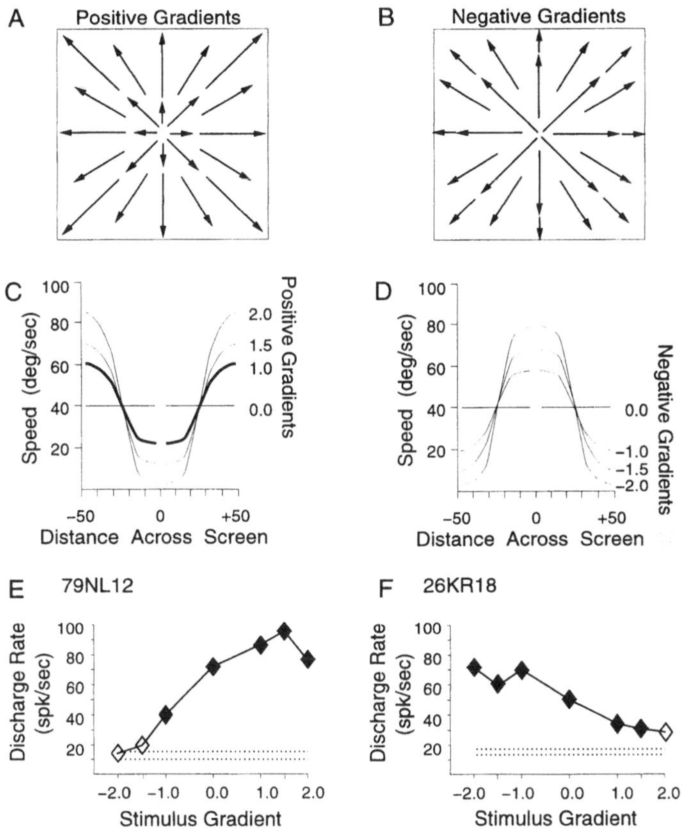

FIG. 4. Speed gradients in optic flow reflect the three-dimensional layout of the visual environment. Optic flow responsive neurons in area MST are sensitive to those gradients. (A–B) Speed gradients in outward radial optic flow are created by changing dot speed as a function of dot eccentricity, here indicated by longer arrows representing faster dot motion. (A) Positive speed gradients have increasing dot speed as a sine * cosine function of distance from the FOE. (B) Negative speed gradients have decreasing dot speed as a sine * cosine function of distance from the FOE. (C–D) Plots of dot speed (left ordinate) versus dot eccentricity (abscissa) for the speed gradients used in these studies. (C) Positive speed gradients were multiplicatively exaggerated (multiplier indicated on right ordinate) to simulate movement through environments in which the central visual field was relatively distant. (D) Negative speed gradients were multiplicatively inverted (multiplier indicated on right ordinate) to simulate movement through environments in which the central visual field was relatively near. (E–F) Graphs of average response amplitude versus stimulus gradients in optic flow for two MST neurons. (E) This neuron showed a strong preference for positive speed gradients. (F) This neuron showed a strong preference for negative speed gradients.

These findings suggest that MST neurons can mediate neural specificity for the visual motion cues that reflect aspects of the three-dimensional structure of the environment. Some neurons would provide greater responses during observer self-movement through an environment in which the central visual field is occupied by relatively distant objects and the peripheral visual field is occupied by relatively near objects (Fig. 5A). Other neurons would provide greater responses during observer self-movement through an environment in which the central visual field is occupied by relatively near objects and the peripheral visual field is occupied by relatively distant objects (Fig. 5B). This assortment of neuronal response sensitivities might contribute to the robust differences in the perceptual impressions created by differences in speed gradients in optic flow.

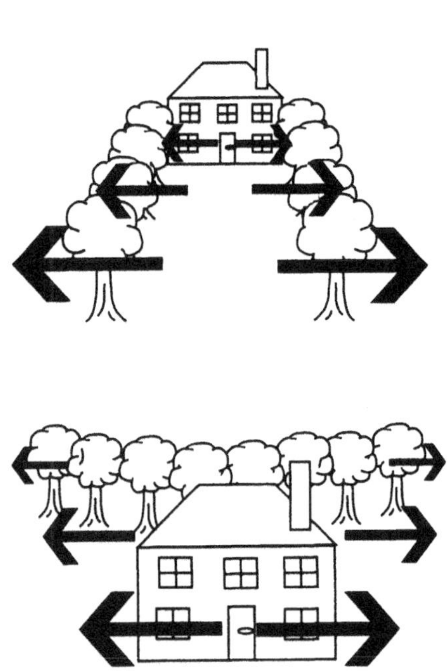

FIG. 5. Diagrams illustrating the environmental structures associated with positive and negative speed gradients in optic flow, as indicated by the length of the arrows in each figure, that evoke different responses from MST neurons. (A) Positive speed gradients have slow movement in the central visual field and fast movement in the periphery simulating observer self-movement through an environment in which objects in the central field are relatively distant. (B) Negative speed gradients have fast movement in the central visual field and slow movement in the periphery simulating observer self-movement through an environment in which objects in the central field are relatively near.

IV. MST Responses to Real Translational Self-Movement

Recently, we (Duffy, 1998) and others (Bremmer *et al.*, 1999) have demonstrated that some MST neurons respond to optic flow stimuli simulating the self-movement scene and also to nonvisual signals about self-movement. In these experiments, the visual neurophysiology laboratory was mounted on a room-sized, two-dimensional sled translator. We used optic flow stimuli simulating the visual scene during observer self-movement in eight directions in the horizontal plane—forward, backward, to the left or right, and the four intermediate oblique directions. The sled moved the monkey, with the visual stimulation equipment and the neuron recording system, using a trapezoidal speed profile. The sled accelerated at 30 cm/s^2 for 1 s, maintained a steady speed of 30 cm/s for 2–3 s, and decelerated at 30 cm/s^2 for 1 s. (Fig. 6).

The optic flow stimuli consisted of 500 white dots on a dark background with dot movement according to the algorithm for the selected optic flow pattern. All dots had a limited screen lifetime randomly assigned from 16 to 2000 ms, with a constant, uniform dot density maintained across the stimulus. The optic flow stimuli were identical under stationary and moving observer conditions with the projector and screen mounted on the sled so that they moved with the animal and covered the central 90o of the visual field.

Translational movements were made with the monkey chair mounted on top of two double-rail drive systems with the upper system making mediolateral movements and the lower system making anteroposterior movements (Fig. 6A). The movement platform carries the neurophysiologic recording system with a primate chair, electronics for single neuron recording, the visual stimulation system, and a two-dimensional magnetic search coil for monitoring eye position. The frame of the search coil is covered by black, opaque plastic that encloses the animal and limits its field of view to the projection screen.

Neuronal discharge records were plotted as spike rasters that were viewed during experiments and replotted off-line as permanent data records. We also used spike density histograms in which discharges were replaced by 20 ms wide Gaussian pulses centered at the discharge time. We measured response amplitude as the mean rate of neuronal discharge during the stimulus period of six to twelve repetitions of each stimulus. Control activity was measured as the discharge rate during stimulus periods in which the animal was required to fixate the red dot on the screen without optic flow or translational movement stimuli.

Figure 7 shows the results of a study of visual and real movement responses recorded from a single MST neuron. The responses to optic

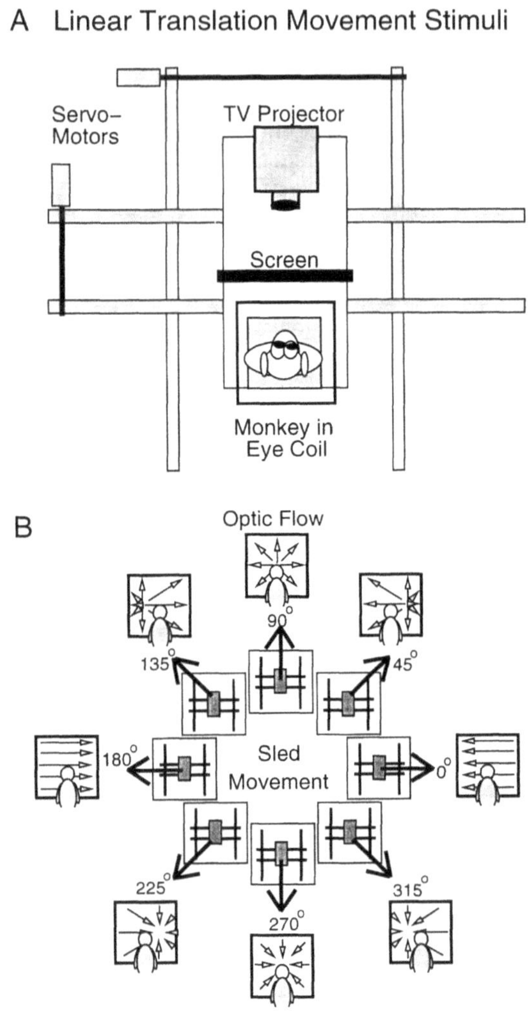

FIG. 6. Real passive translational movement was presented with optic flow stimuli by mounting the visual neurophysiology laboratory on a room-sized sled translator. (A) The monkey and visual projection system were mounted on a moving platform which was attached to a two-dimensional, computer-controlled sled translator. The sled moved the platform in any directional in the horizontal plane. The monkey faced a projection screen that covered the central $90 \times 90°$ of the visual field while the remainder of the visual field was occluded. (B) The sled moved in eight directions distributed around the animal at 45° intervals. Optic flow stimuli were presented separately or in the steady speed period of sled movement with the simulated direction of self-movement in the optic flow always matching the direction of a real movement.

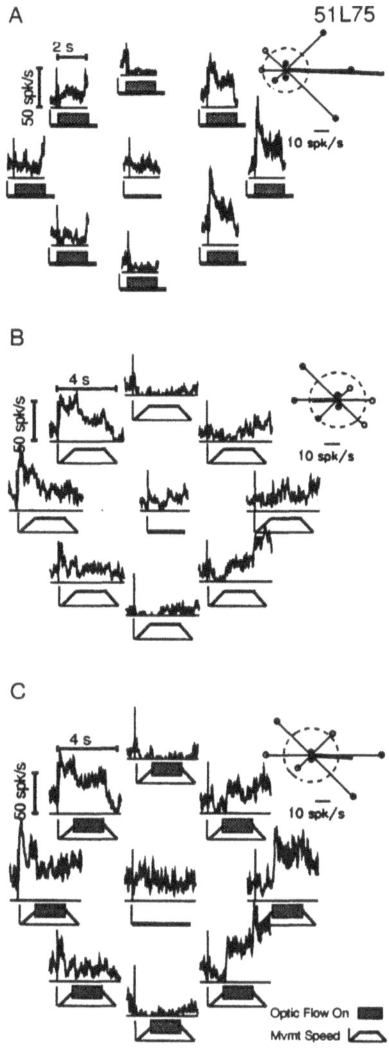

FIG. 7. Responses of an MST neuron to optic flow and real movement stimuli showing the combination of those responses evoked by the naturalistic combination of those stimuli. (A–C) Histograms averaging six stimulus presentations for the eight stimulus directions shown in Fig. 6 with three stimulus conditions—optic flow alone (A), translational movement alone (B), and combined optic flow and translational movement (C). Stimulus onset and the 50 spikes/s activity level are marked by the vertical bar in each histogram; time in the stimulus is marked by the icon under each histogram. (A) Responses to a 2-s optic flow stimulus simulating observer movement show strong preference for simulated rightward observer movement. (B) Responses to real translational movement in eight directions with a trapezoidal velocity waveform showing a preference for leftward and left-forward movement. (C) Responses to combined real translational movement and optic flow with the visual stimulus presented in the steady-speed period of the real movement and showing a combination of response to rightward and leftward stimulus directions.

flow stimuli alone, presented while the sled is stationary, are shown in Fig. 7A. This neuron shows a strong preference for the optic flow stimulus that simulates movement to the right with smaller responses to the other directions and inhibition in the orthogonal directions. The relative amplitude of those responses are shown in the polar plot in the upper right of Fig. 7A. The polar plots show the net vector of each response profile with the mean direction and the resultant length of the net vector indicating the overall direction preference and the strength of that preference. The responses to real translational self-movement, presented while the animal is in complete darkness except for the centered fixation point, are shown in Fig. 7B. This neuron shows a preference for leftward and left-forward self-movement in the acceleration and steady-speed phases with smaller responses to other directions. It also shows responses to deceleration in the opposite direction, a common observation that may be attributable to the fact that acceleration in one direction creates the same forces as deceleration in the opposite direction.

Figure 7C shows the responses of this same neuron to combined optic flow and real translational self-movement. The visual stimuli were presented in the steady-speed phase of the trapezoidal sled movement profile. The visual and movement stimuli were always presented with matching directions in the visual simulation and sled movement (e.g., the centered FOE expanding optic flow stimulus was presented with real translational movement straight ahead). This neuron shows clear evidence of combining the responses to both the optic flow and the movement. There were larger responses evoked by the rightward direction but also clear responses to the left-forward direction.

V. Interactions between Optic Flow and Translational Self-Movement

The responses of this neuron demonstrate that MST can integrate visual and nonvisual cues about self-movement. In studies of 131 neurons, we found a variety of response interactions with combined optic flow and real translational movement stimuli. We have used a two-way analysis of variance to characterize the effects of adding translational movement to the optic flow stimuli. This analysis quantifies the effects of stimulus direction (eight directions at 45° intervals around 360° in the ground plane) and the effects of stimulus modality (visual stimulation alone or visual stimulation with translational movement). The bar graphs in Fig. 8 illustrate the results of this analysis. Each graph shows the percentage of neurons (ordinate) that yielded the *F*-values within the upper limit indicated (abscissa).

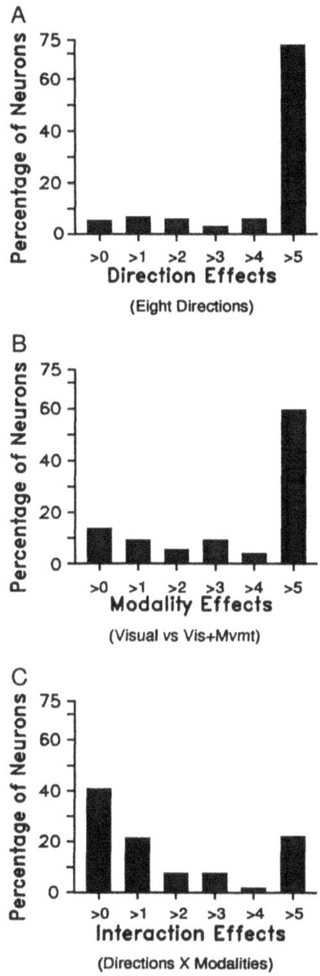

FIG. 8. The results of a two-way analysis of variance with main effects of stimulus direction (eight directions around 360°) and stimulus modality (optic flow alone or optic flow with translational movement). Each bar graph shows the percentage of neurons (ordinate) that yielded the F-values indicated (abscissa). The top graph shows stimulus direction F-values with 82% of the neurons showing significant effects (F-values > 3.0). The middle graph shows stimulus modality F-values with 72% of the neurons showing significant effects. The bottom graph shows direction × modality interaction effects with 31% of the neurons showing significant interactions. Thus, stimulus direction and modality affect the vast majority of neurons, and almost a third show strong interactions between direction and modality effects. (N = 131.)

The graph on the top shows that 82% of the neurons had strong stimulus direction effects (significant F-values > 3.0). The graph in the middle shows that 72% of the neurons had strong stimulus modality effects. The graph on the bottom shows that 31% of the neurons had strong interaction effects between stimulus direction and modality. Thus, we see that stimulus direction affects the vast majority of neurons. However, we also see that the addition of translational movement affects the vast majority of neurons, and that direction-by-modality interaction effects are evident in almost a third.

These findings suggest that there are substantial effects of translational movement in most MST neurons. This is supported by the results of a multiple linear regression analysis in which we tested the model that responses to combined stimulation could be predicted from responses to optic flow alone and real translational movement alone. Figure 9 plots the beta coefficients for translational movement (abscissa) versus optic flow (ordinate) to illustrate the number of neurons that showed various degrees of optic flow and translational movement effects. The largest single group is composed of neurons dominated by optic flow effects with 35% (46/131) of the neurons clustered at the top left of

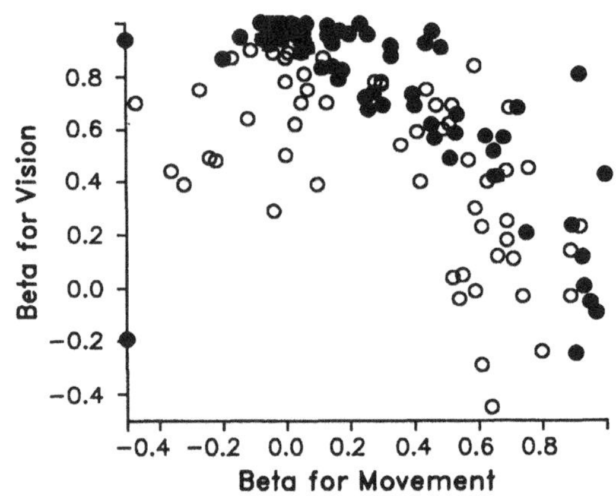

FIG. 9. Results of a multiple linear regression analysis of the influence of optic flow and translational movement on responses to combined stimuli. Each point represents the values obtained for one of the neurons in the sample. The beta coefficients for translational movement are shown on the x-axis and the beta coefficients for optic flow visual stimuli are shown on the y-axis. This analysis reveals a group of neurons that are greatly dominated by visual effects, here represented at the top left of the graph. In addition, there is a large group of neurons with a mixture of visual and movement effects, arcing down and to the right on this graph, with some that are dominated by movement effects, on the bottom right. Good fits to the data were achieved with this simple linear model, the filled dots representing the 58% of the neurons which yielded $R^2 > 0.8$. (N = 131.)

the graph and having betas for optic flow greater than 0.6 with betas for translational movement less than 0.2. Only 13% (17/131) of the neurons are in the corresponding group dominated by translational movement, with the rest of the neurons showing a mixture of optic flow and translational movement effects.

Thus, there is a wide range of translational movement effects in these neurons that support extensive interactions with optic flow responses to produce a continuum of neuronal responses combining visual and vestibular cues about self-movement. This analysis reveals a group of neurons that are greatly dominated by visual effects, here represented at the top left of the graph. In addition, there is a large group of neurons with a mixture of visual and movement effects, arcing down and to the right on this graph, with some that are dominated by movement effects, on the bottom right. Good fits to the data were achieved with this simple linear model, the filled dots representing the 58% of the neurons which yielded Rs^2 greater than 0.80.

By itself, optic flow only encodes the relative movement of the observer and the visual surround. Thus, optic flow is ambiguous as to whether it results from observer self-movement or the movement of a large object in the visual environment of a stationary observer. Neurons that combine optic flow and real translational movement responses may disambiguate optic flow. We have found one subpopulation of neurons that show smaller responses to combined visual and real movement stimuli that would respond more to optic flow induced by object movement than that induced by observer self-movement. Another subpopulation of MST neurons shows stronger responses to combined visual and translational movement stimuli that would respond more to optic flow induced by observer self-movement than that induced by object movement. Thus, subpopulations in MST might disambiguate optic flow by responding differently to optic flow created by object motion and optic flow created by observer self-movement (Duffy, 1998).

VI. MST's Role in Self-Movement Perception

These findings present the consistent impression of important potential contributions of MST neurons to self-movement perception. We have shown response selectivity for optic flow stimuli simulating different directions of observer self-movement as specified by the location of the focus of expansion relative to the direction of gaze. Furthermore, we have found that the response fields of these neurons are organized to provide behaviorally useful signals about heading toward the direction of gaze versus deviation from that heading.

The MST neurons also show response sensitivity to visual motion cues about the three-dimensional structure of the environment as imbedded in the speed gradient of the optic flow field. These neurons either prefer speed gradients that reflect self-movement during approach directly toward a relatively near object with a remote background or through a surround of relatively near objects toward a remote goal.

The MST neurons are also responsive to nonvisual cues about self-movement. These non-visual responses are evoked during passive translational self-movement and are likely to be mediated by the vestibular otoliths. Interactions between optic flow responses and real movement responses create two neuronal subpopulations in MST. One subpopulation would be more active during optic flow from observer self-movement through a field of stationary objects. The other subpopulation would be more active during optic flow from the movement of objects in the visual environment of a stationary observer (Duffy, 1998).

As a result of these response properties, MST neurons might specifically encode relative movement of the observer and objects in the environment. Figure 10 illustrates the fundamental ambiguity in optic flow that might be resolved by visual-vestibular interactions in MST. Here, the centered outward radial flow field is created by a large object in the visual field—the back of a truck as viewed by the driver of a car behind the truck (Fig. 10A). This scene can be created either by backward movement of the truck (Fig. 10B) or forward movement of the car (Fig. 10C). The MST subpopulations influenced by visual and nonvisual self-movement cues could resolve this ambiguity. Some MST neurons are tuned to respond when optic flow is seen in the absence of translational self-movement (when the truck moves backward, as in Fig. 10B), whereas other MST neurons are tuned to respond when optic flow is combined with translational self-movement (when the car moves forward, as in Fig. 10C).

VII. A Distributed Network for Self-Movement Perception

The contribution of MST to perceiving self-movement must be considered in the context of the potential contributions of other cortical areas that respond to optic flow (see Bremmer *et al.*, this volume), especially the ventral intraparietal area (Schaafsma and Duysens, 1996) and area 7A (Siegel and Read, 1997). In addition, cortical areas responsive to vestibular input might also make substantial contributions to self-

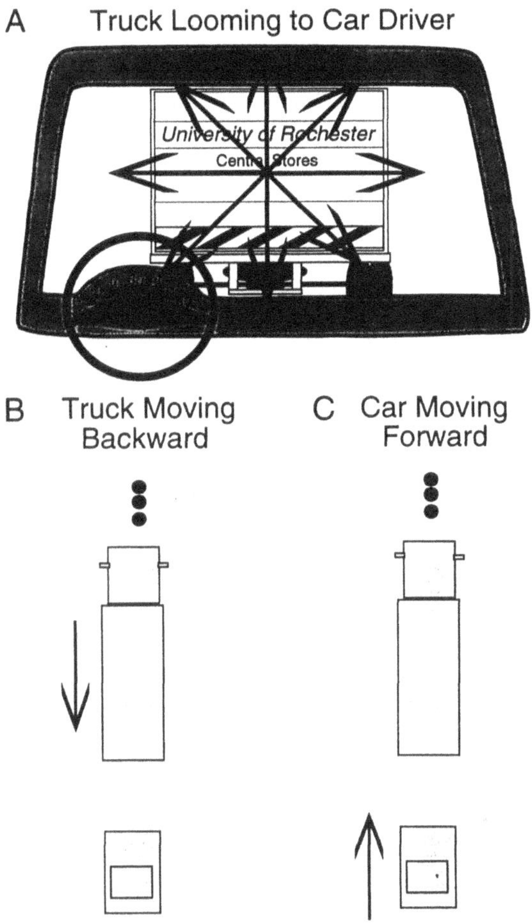

FIG. 10. Schematic illustration of the circumstances in which an observer is confronted by an ambiguous scene (A) with respect to whether the visual motion is from the movement of a large object (truck) in the environment (B) or from observer self-movement (C). (A) The driver of a car stopped behind a large truck sees an outward radial optic flow field created by the looming image of the truck. (B) The outward radial optic flow field (A) might be produced by backward movement of the truck while the observer in the car is stationary. (C) Alternatively, the outward radial optic flow field (A) might be produced by forward self-movement of the observer in the car while the truck is stationary. The MST neurons appear to be influenced by both visual and vestibular signals such that they might contribute to resolving this ambiguity. However, the fact that the same population of neurons are activated by purely visual stimuli creates the potential for visually induced illusions of self-movement.

movement perception, including the parietoinsular vestibular cortex (Guldin *et al.*, 1992; Akbarian *et al.*, 1994), area 2V (Fredrickson *et al.*, 1966; Büttner and Büttner, 1978), and area 7 (Ventre and Faugier-Grimaud, 1988; Faugier-Grimaud and Ventre, 1989). Subcortical centers that respond to visual and vestibular activation might also play an important role in these settings, including the vestibular nuclei (Henn *et al.*, 1974) and the dorsolateral pontine nuclei (Kawano *et al.*, 1992).

A cortical network for self-movement perception must also accommodate the distorting effects of pursuit eye movements on the retinal image of optic flow to establish a veridical match between multisensory self-movement cues. MST neurons are known to show robust pursuit responses (Komatsu and Wurtz, 1988) and it has been suggested that they play an important role in stabilizing perception during pursuit (Wurtz *et al.*, 1990). There is now growing evidence that MST can integrate visual signals from optic flow with pursuit signals to compensate for the distorting effects of pursuit on the retinal image of optic flow (Bradley *et al.*, 1996). This compensatory mechanism may act at the population level in MST, as evidenced by preserved population heading selectivity during pursuit (Page and Duffy, 1999).

All these observations are consistent with the predictions of a neural network model developed to integrate our understanding of sensory and motor signals relevant to self-movement perception in MST (Lappe, 1998). Interactions between the neuronal populations that contribute to the distributed network for self-movement perception must be characterized to describe the underlying mechanisms fully.

Acknowledgments

This work was supported by National Eye Institute grant R01-10287, the Human Frontier Science Program grant RG71/96, and a grant to the University of Rochester Department of Ophthalmology from Research to Prevent Blindness.

References

Akbarian, S., Grusser, O. J., and Guldin, W. O. (1994). Corticofugal connections between the cerebral cortex and brainstem vestibular nuclei in the macaque monkey. *J. Comp. Neurol.* **339,** 421–437.

Beusmans, J. M. (1993). Computing the direction of heading from affine image flow. *Biol. Cybern.* **70,** 123–136.

Bradley, D. C., Maxwell, M., Andersen, R. A., Banks, M. S., and Shenoy, K. V. (1996). Mechanisms of heading perception in primate visual cortex. *Science* **273,** 1544–1547.

Bremmer, F., Kubischik, M., Pekel, M., Lappe, M., and Hoffman, K.-P. (1999). Linear vestibular self-motion signals in monkey medial superior temporal area. *In* "Otholith Function in Spatial Orientation and Movement" (B. Cohen and B. J. M. Hess, Eds.). *Ann. NY Acad. Sci.* **871,** 272–281.

Büttner, U., and Büttner, U. W. (1978). Parietal cortex (2v) neuronal activity in the alert monkey during natural vestibular and optokinetic stimulation. *Brain Res.* **153,** 392–397.

Duffy, C. J. (1998). MST neurons respond to optic flow and translational movement. *J. Neurophysiol.* **80,** 1816–1827.

Duffy, C. J., and Wurtz, R. H. (1991). Sensitivity of MST neurons to optic flow stimuli. II. Mechanisms of response selectivity revealed by small-field stimuli. *J. Neurophysiol.* **65,** 1346–1359.

Duffy, C. J., and Wurtz, R. H. (1995). Response of monkey MST neurons to optic flow stimuli with shifted centers of motion. *J. Neurosci.* **15,** 5192–5208.

Duffy, C. J., and Wurtz, R. H. (1997). Multiple components of MST responses to optic flow. *Exp. Brain Res.* **114,** 472–482.

Faugier-Grimaud, S. and Ventre, J. (1989). Anatomic connections of inferior parietal cortex (area 7) with subcortical structures related to vestibulo-ocular function in a monkey (*Macaca fascicularis*). *J. Comp. Neurol.* **280,** 1–14.

Fredrickson, J. M., Figge, U., Scheid, P., and Kornhuber, H. H. (1966). Vestibular nerve projection to the cerebral cortex of the rhesus monkey. *Exp. Brain Res.* **2,** 318–327.

Gibson, J. J. (1966). "The Senses Considered as Perceptual Systems." Houghton Mifflin, Boston.

Graziano, M. S. A., Andersen, R. A., and Snowden, R. J. (1994). Tuning of MST neurons to spiral motion. *J. Neurosci.* **14,** 54–67.

Guldin, W. O, Akbarian, S., and Grüsser, O. J. (1992). Cortico-cortical connections and cytoarchitectonics of the primate vestibular cortex: A study in squirrel monkeys (*Saimiri sciureus*). *J. Comp. Neurol.* **326,** 375–401.

Henn, V., Young, L. R., and Finley, C. (1974). Vestibular nucleus units in alert monkeys are also influenced by moving visual fields. *Brain Res.* **71,** 144–149.

Kawano, K., Shidara, M., and Yamane, S. (1992). Neural activity in dorsolateral pontine nucleus of alert monkey during ocular following responses. *J. Neurophys.* **67,** 680–703.

Komatsu, H. and Wurtz, R. H. (1988). Relation of cortical areas MT and MST to pursuit eye movements. III. Interaction with full-field visual stimulation. *J. Neurophysiol.* **60,** 621–644.

Lappe, M. (1998). A model of the combination of optic flow and extraretinal eye movement signals in primate extrastriate visual cortex: neural model of self-motion from optic flow and extraretinal cues. *Neural Networks* **11,** 397–414.

Lappe, M., and Rauschecker, J. P. (1993). A neural network for the processing of optic flow from ego-motion in higher mammals. *Neural Computation* **5,** 374–391.

Orban, G. A., Lagae, L., Verri, A., Raiguel, S., Xiao, D., Maes, H., and Torre, V. (1992). First-order analysis of optical flow in monkey brain. *Proc. Nat. Acad. Sci. USA* **89,** 2595–2599.

Page, W. K., and Duffy, C. J. (1999). MST neuronal responses to heading direction during pursuit eye movements. *J. Neurophys.* **81,** 596–610.

Saito, H., Yukie, M., Tanaka, K., Hikosaka, K., Fukada, Y., and Iwai, E. (1986). Integration of direction signals of image motion in the superior temporal sulcus of the macaque monkey. *J. Neurosci.* **6,** 145–157.

Schaafsma, S. J. and Duysens, J. (1996). Neurons in the ventral intraparietal area of awake macaque monkey closely resemble neurons in the dorsal part of the medial superior temporal area in their responses to optic flow. *J. Neurophysiol.* **76**(6), 4056–4068.

Siegel, R. M., and Read, H. L. (1997). Analysis of optic flow in the monkey parietal area 7a. *Cereb. Cortex* **7,** 327–346.

Tanaka, K., Hikosaka, K., Saito, H., Yukie, M., Fukada, Y., and Iwai, E. (1986). Analysis of local and wide-field movements in the superior temporal visual areas of the macaque monkey. *J. Neurosci.* **6,** 134–144.

Ventre, J., and Faugier-Grimaud, S. (1988). Projections of the temporo-parietal cortex on vestibular complex in the macaque monkey (*Macaca fascicularis*). *Exp. Brain Res.* **72,** 653–658. 1988.

Verri, A., Girosi, F., and Torre, V. (1989). Mathematical properties of the two-dimensional motion field: From singular points to motion parameters. *J. Opt. Soc. Am. A* **6**(5), 698–712.

Werkhoven, P., and Koenderink, J. J. (1990a). Extraction of motion parallax structure in the visual system I. *Biol.Cybern.* **63,** 185–191.

Werkhoven, P., and Koenderink, J. J. (1990b). Extraction of motion parallax structure in the visual system II. *Biol. Cybern.* **63,** 193–199.

Wurtz, R. H., Komatsu, H., Dursteler, M. R., and Yamasaki, D. S. (1990). Motion to movement: cerebral cortical visual processing for pursuit eye movements. *In:* "Signal and Sense: Local and Global Order in Perceptual Maps" (G. M. Edelman, W. E. Gall, and W. M. Cowan, Eds.). p. 233–260. Wiley, New York.

NEURAL MECHANISMS FOR SELF-MOTION PERCEPTION IN AREA MST

Richard A. Andersen, Krishna V. Shenoy, James A. Crowell, and David C. Bradley

Division of Biology, California Institute of Technology, Pasadena, California

I. Introduction

Research on the neural circuitry responsible for perception of self-motion has focused on the medial superior temporal area, particularly the dorsal division (MSTd). Cells in this area are selective for the location of the focus of expansion and to pursuit eye movements, two signals necessary for recovering the direction of self-motion (Gibson, 1950). Research reviewed here shows many interesting correlates between the perception of self-motion and the activity of MST neurons. In particular, the focus tuning curves of these cells adjust to take into account motions during eye movements using extra-retinal signals, similar to the results of human perceptional experiments. Eye rotations due to head movements are also compensated for perceptually, and the focus tuning of MST neurons are also compensated for during head-generated eye rotations. Finally, the focus tuning curves compensate for both the direction and the speed of eye rotations, similar to that found in psychophysical studies. However, there are also several aspects of MSTd activity that do not completely mesh with the perception of self-motion; these differences suggest that area MSTd is not the final stage or the only locus of brain activity which accounts for this percept. Finally we offer a "gain field" model, which explains how area MSTd neurons can compensate for gaze rotations.

0074-7742/00 $30.00

II. Area MST—Optic Flow Selectivity

In the middle 1980s, two groups discovered cells in the medial superior temporal (MST) area that were sensitive to complex visual-motion patterns similar to those encountered during self-motion, termed optic flow (Sakata *et al.*, 1985, 1986, 1994; Saito *et al.*, 1986; Tanaka *et al.*, 1986, 1989; Tanaka and Saito, 1989; Gibson, 1950). A number of types of motion patterns were used in these studies (e.g., expansion/contraction, rotation, and laminar motion); cells tended to respond selectively to particular types of motion patterns. Duffy and Wurtz (1991a, b) made the important observation that even though some MSTd neurons were selective for a single type of pattern (single-component cells in their terminology), many others had sensitivity for two (double-component) or even three (triple-component) types of motion patterns. For instance, a triple-component cell might respond to expansion, clockwise rotation, and leftward laminar motion.

A. Spiral Space

One powerful class of computational models for recovering the direction of self-motion is based on linear analyses of local regions of the flow field (Longuet-Higgins and Prazdny, 1980; Koenderink and Van Doorn, 1981; Rieger and Lawton, 1985; Hildreth, 1992). These techniques are used to recover the expansion component of flow, due to observer translation, from the complex visual motions produced by eye rotations. The finding of neurons in MST that were sensitive for expansion/contraction, curl, or laminar motion led to the idea that the brain might in fact be analyzing complex flow fields in terms of these particular components. In other words, the brain might use these simple types of motion patterns as a basis set for describing the more complex patterns observed during self-motion.

The general idea that the brain represents features of the environment using basis sets has been quite fruitful; for instance, color vision is based on three types of photoreceptors with different spectral sensitivities whose relative activations can represent many colors. However, many central locations in the brain tend to use a continuum of selectivities rather than a few basis functions. For instance, the direction of motion in the frontal parallel plane could, in principle, be represented by the relative activities of up/down and left/right detectors (a basis-set description); however, we know that neurons in central motion pathways have cells tuned to a continuous range of motions that completely tile the set of possible motion directions. In fact, the finding by Duffy and Wurtz of double- and triple-component cells suggested that the brain was not using a limited set of basis functions to represent optic flow. However, it was still quite possible that the set of basis functions might also include those sensitive to combinations of the three basic types.

In order to test this idea of basis versus continuous representation of optic flow, Graziano *et al.* (1994) examined the tuning of MSTd neurons in a spiral space. Figure 1A depicts this space; expansion/contraction is represented on the vertical axis, and clockwise/counterclockwise rotation on the horizontal axis. Noncardinal directions in this space represent spirals, and the distance from the origin indicates the magnitude of neural activity. The decomposition hypothesis would predict that MSTd neurons sensitive to rotations or expansions/contractions should have tuning curves which peak along the cardinal axes in this space (e.g., cells sensitive to expansion should also respond selectively to expansion). However, if MSTd is representing optic flow with a continuous array of detectors, then one would predict that there would be many cells which preferred different types of spirals. The data in Fig. 1 shows one example of a cell that is not tuned along one of the cardinal axes, and this result was quite common indicating that the continuous-representation hypothesis is correct. Not only were cells selective for expansions and rotations, but a large number of cells also preferred clockwise and counterclockwise expanding spirals and clockwise and counterclockwise contracting spirals.

In another study, Orban and colleagues (1992) examined the responses of expansion cells when rotary or curl motions were added to the expansion stimulus. They found that the addition of curl reduced neural activity when compared with the response to the expansion pattern alone. This result indicates again that the MSTd cells are not extracting the expansion component from these compound stimuli; if they were, the response of the cells should not have been affected by the introduction of additional motions.

B. Position and Form/Cue Invariance

The MSTd neurons prefer the same patterns of motion when the stimuli are displaced, even if the same part of the retina receives different motion directions depending on the position of the pattern within the receptive field (Graziano *et al.,* 1994). This finding indicates that MSTd neurons extract the overall pattern of motion, even though locally the motion stimulus may be completely different. The cells also demonstrated scale invariance, giving similar responses for large or small flow fields of the same type. The preceding study explored invariance locally and was designed to examine the effect of opposing motions at the same retinal location. Thus typically the invariance tests were made over a 20° diameter area of the MSTd receptive fields. Although the tuning was

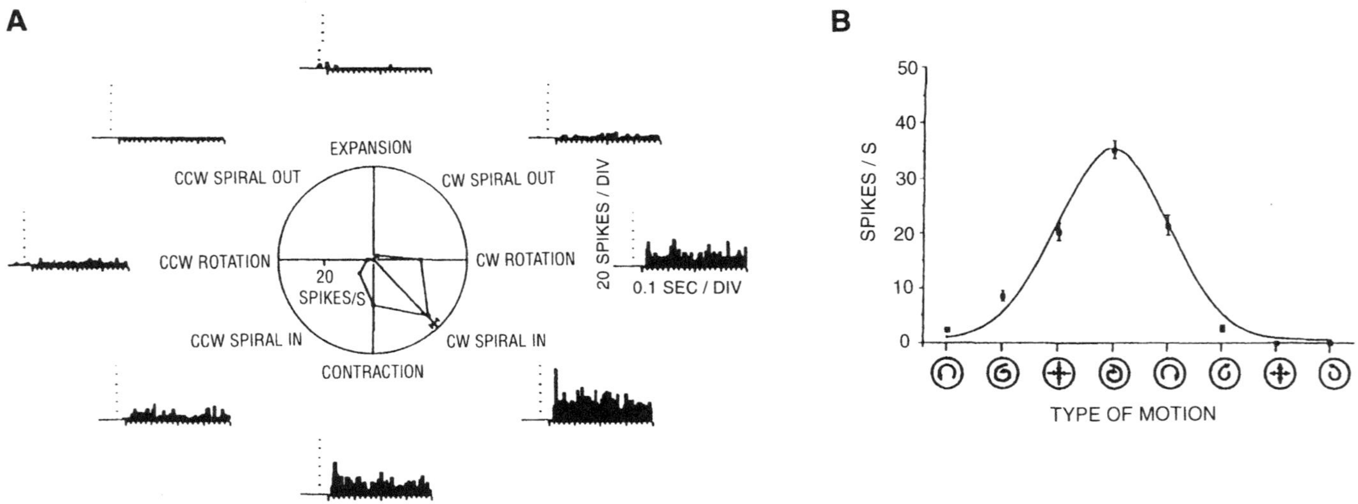

FIG. 1. A cell tuned to clockwise contracting spiral. (A) In this polar plot, the angle represents the type of stimulus, and the radius represents the response magnitude. The line directed at −47° indicates the cell's best direction as determined by a Gaussian curve fit. The error bar shows a standard error of 2.5° for determining the best direction. The response is also shown in histograms at the perimeter, summed over 10 trials. CW, clockwise; CCW, counterclockwise. (B) The tuning curve from (A) plotted in Cartesian coordinates and fitted with a Gaussian function. The icons on the *x*-axis indicate the type of stimulus. The error bars show standard error across ten trials. Reprinted with permission from Graziano *et. al.* (1994), *J. Neurosci.* **14,** 54-67.

amazingly constant in terms of pattern selectivity (i.e., tuning in spiral space), the magnitude of the response sometimes decreased as the stimuli were placed farther away from the center of the receptive field. This effect may explain why some studies reported less invariance, since they examined the magnitude of response to a single pattern of motion as the criterion for position invariance, rather than the invariance in selectivity for different patterns of motion (Duffy and Wurtz, 1995). In a few cells, Geesaman and Andersen (1996) did examine spiral tuning over a 50° diameter area and still found remarkably good invariance. However, since models for self-motion perception which predict a lack of invariance for large displacements in the receptive field have been proposed, it would be important to study more thoroughly pattern selectivity over the entire receptive field of MSTd neurons (Lappe *et al.*, 1996).

If area MST is important for processing optic flow for navigation, it is crucial that the cells exhibit form/cue invariance (i.e., that the exact features providing the motion signals are not important to the overall selectivity of the cells). Geesaman and Andersen (1996) found that MSTd neurons do, in fact, exhibit form/cue invariance; regardless of whether the motion pattern was provided by a single object, a more typical pattern of independent features (dots), or even by non-Fourier motion cues, the cells exhibited the same pattern selectivity. This finding also implies that MSTd cells may have a dual role, processing not only the perception of self motion but also the perception of object motion.

C. Anatomical Organization

Several groups have noted that there appears to be a local clustering of cells with similar preferred stimulus pattern selectivity within MSTd. The topography of MST was examined directly by Geesaman *et al.* (1997). They used a double-label deoxyglucose technique in which the animals viewed one motion pattern in the presence of one metabolic tracer and a second pattern during the administration of a second tracer. Different patterns of patchy cortical columns were found within the superior temporal sulcus, including MSTd, for the different stimulus patterns. Interestingly, when expansion and contraction columns were labeled with the two tracers, they were found to be more widely separated than when expansion and rotation columns were labeled; in other words, cells selective for patterns that were more widely separated in spiral space tended to be farther apart in cortex. Britten (1998) has recently performed electrophysiological mapping experiments and found a columnar organization for different motion patterns within MST.

III. Area MST—Shifting Receptive Fields

As mentioned earlier, one daunting problem for navigation using optic flow is to separate the translational from the rotational component of optic flow (see also van den Berg, this volume, and Lappe and Hoffman, this volume). With the fall of models based on local linear operators, part of the remaining models use templates. According to this class of model, MST neurons are proposed to contain templates for a variety of optic flow conditions, including those which take into account different rates of translation and eye rotation (Perrone and Stone, 1994). The one drawback of this class of model is that it requires a large number of templates, possibly many more than there are neurons in MSTd. An attractive means for reducing the required number of templates is to adjust them dynamically using an extraretinal signal. Psychophysical experiments have shown that an extraretinal signal of eye pursuit speed and direction is used to compensate perceptually for eye rotation (Royden *et al.*, 1992, 1994). Thus the possibility exists that a smaller set of templates could be dynamically shifted to account for eye rotation, rather than a larger number of templates being used to represent all possible eye rotation speeds and directions (Andersen *et al.*, 1996).

This possibility was directly tested in experiments by Bradley *et al.* (1996). It has been known for some years that cells in MSTd are not only active for visual motion stimuli, but also for the direction and speed of eye pursuits (Newsome *et al.*, 1988; Kawano *et al.*, 1984, 1994). This "pure" pursuit neural activity is weaker than is commonly seen for moving visual stimuli. Recently Duffy and Wurtz (1995) showed that MSTd neurons are spatially tuned for the focus of an expansion stimulus. Bradley and colleagues reasoned that this pursuit-related signal may be used to adjust the focus tuning of MST neurons during eye rotation.

In these experiments, the spatial focus tuning of each MSTd neuron was determined with the eyes stationary, as shown in Fig. 2. Next the tuning curve was remapped but with the animal pursuing in the preferred or opposite pursuit direction of the neuron (also determined in earlier tests). The activity of the MSTd neuron was measured when the eye was in approximately the same orbital position as in the previous mapping test, the only difference being that in one case the eye is moving and in the other it is not. The laminar motion caused by the eye rotation combines with the expansion and shifts the focus on the retina in the direction of the eye movement (see Fig. 1, van den Berg, this volume). It was found that the focus tuning curves of many MSTd cells shifted during pursuit, and if they shifted, they were much more likely to shift in the di-

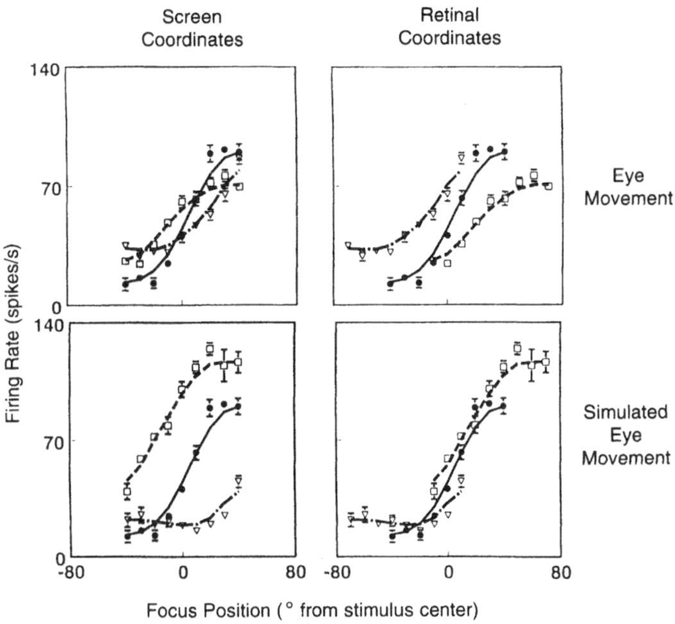

FIG. 2. An MSTd heading cell. In all panels, the solid lines and solid circles represent fixed-eye focus tuning (identical in all four graphs), the dashed lines and open squares are preferred-direction eye movements (real or simulated), and the dot-and-dashed-lines and open triangles are antipreferred-direction eye movements (real or simulated). Data in the left and right columns are identical, except the pursuit curves in the right column were shifted by 30° relative to screen coordinates (thus giving retinal coordinates). The moving-eye focus tuning curves align in the screen coordinates (top left panel) and thus encode the direction of heading. However, for simulated eye movements, the fields align in retinal coordinates. Smooth curves are five-point moving averages of the data. Data points are shown as the mean ± SEM for four replicates, where each replicate is the mean firing during the middle 500 ms of the stimulus-presentation interval. Reprinted with permission from Bradley, D. C., Maxwell, M. A., Andersen, R. A., Banks, M. S., and Shenoy, K. V. (1996). Neural mechanisms for heading perception in primate visual cortex. *Science* **273,** 1544–1547. Copyright 1996 American Association for the Advancement of Science.

rection of the eye movement. These shifts produced activity that was constant with respect to the location of the focus in the world. In other words, these cells code the same focus location regardless of whether the eyes are stationary or moving. In a final experiment, we had the animal hold the eye stationary and simulated the same retinal image as was created when the eyes were moving. This was achieved by moving the display on the screen in the opposite direction to that in which the eye moved in the pursuit condition. We routinely found that the focus tuning curves did not compensate under these conditions, indicating that the compensation observed during real pursuit must be due to an extraretinal pursuit signal. In these experiments not only expansion-selective cells but also contraction- and rotation-selective neurons compensated for pursuit. When a rotary motion is combined with an eye movement, the laminar motion caused by the eye movement shifts the retinal focus of rotation orthogonally to the direction of the eye movement. An eye movement combined with contraction will produce a shift in the focus in the opposite direction of eye pursuit. Interestingly, the focus tuning of curl- and contraction-selective cells compensated in the correct direction—orthogonal for rotation and opposite for contraction. These results suggest that all templates in MST compensate, not just those for expansion. This general compensation can be useful for self-motion perception. For instance, pursing an object on a ground plane to the side while moving forward will produce a spiral or curl-like motion pattern on the retina (see Fig. 5e, Lappe and Hoffman, this volume). Moving backward in the world produces contraction. Thus, templates tuned to these flow patterns can correctly compensate for pursuit eye movements. In these cases, however, unlike expansion, the direction of translation cannot be recovered directly from the focus location, and an additional mapping step is required to recover the direction of self-motion. These results raise a possibility similar to that raised by the cue invariance experiments, namely that MSTd has a more general role in motion perception than just the computation of self-motion. For instance, a horizontal pursuit eye movement over a rotating umbrella does not lead to the perception that the center of rotation displaces up or down, even though the focus of the rotation on the retina does move up or down.

Finally, Britten and van Wezel (1998) have found that microstimulation of MST can effect the perception of heading. They found larger effects during eye pursuit, consistent with the idea that area MST is the site for combining extraretinal pursuit and visual motion signals for self-motion perception.

A. Speed Tuning

Compensation for pursuit eye movements must take into account not only the direction of eye movement but also the speed of pursuit. Experiments currently underway in our laboratory indicate that MSTd neurons' focus tuning curves shift during pursuit in a monotonically increasing fashion with increases in pursuit speed (Shenoy *et al.*, 1998). Thus, the direction and speed of pursuit are both taken into account. Other important variables are the speed of observer translation and distance of objects in the environment; these two variables determine the rate of expansion on the retina during forward self-motion. Duffy and Wurtz (1997) have recently reported that most MSTd neurons' responses are modulated by the expansion rate.

B. Gaze Rotations with Head Movements

Large gaze rotations involve head as well as eye movements and are quite common during locomotion. If area MSTd is responsible for self-motion perception, then it would be expected that its neurons also show compensation during pursuit movements of the head. We have recently tested this idea by requiring monkeys to suppress their VOR to maintain fixation on a target during whole body rotations (Shenoy *et al.*, 1996, 1999). In this paradigm, the head and body are rotated together, and the animal follows a spot of light that moves with them. Compensatory focus tuning shifts were found that were very similar to the compensation observed during pursuit eye movements. In fact, MSTd neurons that compensated during pursuit eye movements generally also compensated during VOR cancellation (Shenoy *et al.*, 1997, 1999). The source of this compensation may be vestibular canal signals. When monkeys are rotated in the dark, many MST neurons are modulated by vestibular stimulation (Thier and Ericksen, 1992). This modulation is larger if the animal suppresses its VOR by tracking a fixation point, suggesting the possibility that an eye pursuit signal directing the eye in the direction of gaze rotation, and thereby canceling the VOR, may also be a factor.

C. Gain Field Model for Compensation

Many, but not all, MSTd neurons show compensatory changes in their focus tuning during the pursuit conditions described earlier. These changes are not necessarily smooth translations of the focus tuning curves on the retina, and many appear as distortions in which one part of the focus tuning curve is suppressed or enhanced. A different, but also very major effect of pursuit on MSTd neurons is to modulate the magnitude of the response (Bradley *et al.*, 1996; Shenoy *et al.*, 1999). Generally the gain increases when the pursuit is in the neuron's preferred pursuit direction.

Figure 3A shows an example of an MSTd neuron which does not show shift compensation, but does exhibit gain modulation by pursuit. We have proposed a simple three-neuron model to explain how the "shifted" tuning curves are produced (Bradley *et al.*, 1996). Figure 3B shows two model input cells that have different focus tuning and are differentially gain-modulated by pursuit. The outputs of these cells are the inputs to a compensating neuron that sums their responses. When

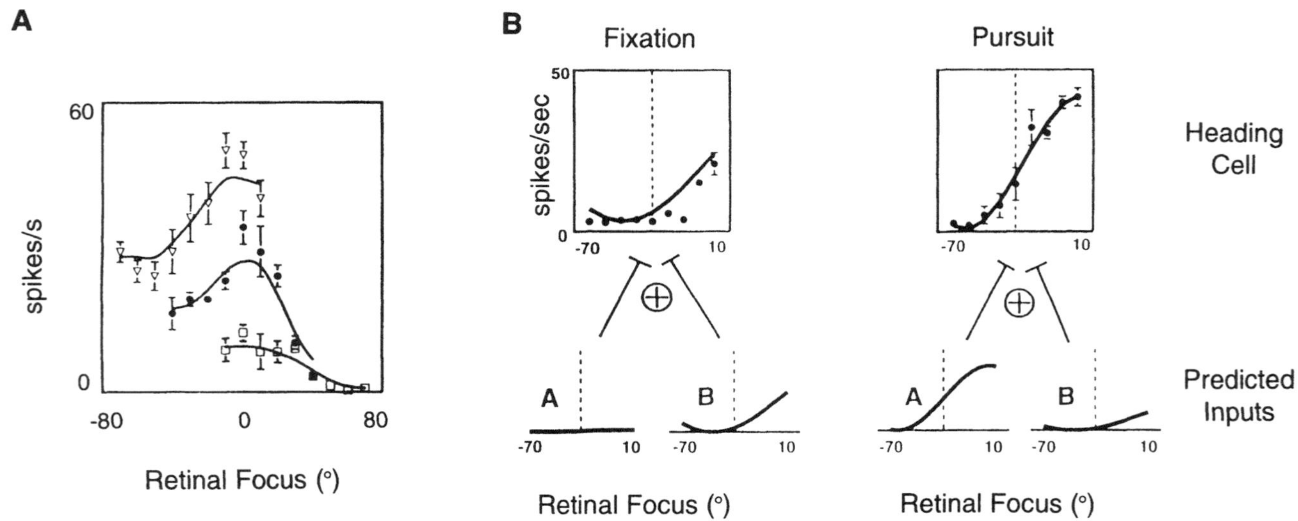

FIG. 3. (A) Example of a retinal (nonshifting) neuron in which the response amplitude is modulated by pursuit. The three curves, corresponding to stationary fixation and opposite directions of pursuit, all peak at ~0°, but their amplitudes vary substantially. This type of cell could serve as input to the heading (shifting) cells. (B) Example of a heading neuron and how it might be constructed from retinal neurons. The upper graphs show the neuron's measured focus tuning during fixation (left) and preferred-direction pursuit (right). The neuron shifted its retinal focus tuning during pursuit in such a way as to compensate for the retinal focus shift induced by that pursuit. Circles, mean response; curves, model fit. (Lower panels) Predicted input functions. Each function is characterized by three sine-wave parameters and multiplied by a gain; two functions are summed to make the focus tuning curve of a heading cell. All parameters were adjusted by nonlinear regression to fit the data (upper panels). The focus tuning shift during pursuit was achieved by increasing the gain on function A while decreasing the gain on function B. The sine-wave parameters (other than gain) were identical for the fixation and pursuit conditions; only the gains were adjusted to simulate the focus tuning shift. The gain-modulated sine functions resemble neurons in the sample that have gain-modulated (nonshifting) focus tuning. Reprinted with permission from Bradley D. C., Maxwell, M. A., Andersen, R. A., Banks, M. S., and Shenoy, K. V. (1996). Neural mechanisms for heading perception in primate visual cortex. *Science* **273,** 1544–1547. Copyright 1996 American Association for the Advancement of Science.

the eyes pursue in one direction, one input cell is modulated upward and the other downward. This produces a compensatory shift in the tuning curve of the output cell. The change in gain modulation of the input cells reverses for pursuit in the opposite direction, producing a compensatory shift in the opposite direction in the output cell. Using this very simple model with few parameters, we were able to recreate accurately the focus tuning compensation found in MSTd neurons (Bradley *et al.*, 1996).

Similar gain modulation effects were found for gaze rotations during VOR cancellation (Shenoy *et al.*, 1999). Thus, this gain model could account for compensation for gaze rotations due to eye or head movements. van den Berg and Beintema (1997, 1998) have recently proposed a model for self-motion perception which uses a similar gain modulation via an extraretinal pursuit signal. Lappe (1998) has recently proposed a model which uses the distribution of compensatory shifts to arrive at self-motion perception performance similar to that found in humans.

D. Psychophysics

As mentioned earlier, we have found that MST neurons compensate for eye rotations due to passive head movements. However, it has never been determined if humans correctly perceive the direction of self-motion during head-generated gaze rotations. We have recently examined this issue and have found that subjects have the same complete compensation for gaze rotation with head movements as is found with eye movements (Crowell *et al.*, 1997, 1998a). It is difficult to isolate experimentally the source of the extraretinal signal for pursuit eye movement compensation because it could be due to efference copy or muscle and orbital proprioceptive signals. Such experiments would require, among other things, moving the eyes passively. On the other hand, the head is much more accessible for examining the source of compensation signals. There are three obvious sources for head movement compensation—vestibular, neck-proprioceptive, and efference copy signals.

We examined the relative importance of these three extraretinal signals for self-motion perception by having observers make judgments about displays simulating linear forward motion across a ground plane. In conditions in which compensation is incomplete or nonexistent, our subjects inaccurately perceived motion along a path that curved in the direction of the gaze movement. We found that self-motion perception is most accurate when all three extraretinal signals are present, less accurate when only two (efference copy and neck-proprioception or vestibular and neck-proprioception) are present, and very inaccurate when only one signal (vestibular or neck-proprioception) is present (Crowell *et al.*, 1997, 1998a). This above finding is very interesting in the context of our physiological findings. When vestibular stimulation was presented alone in the psychophysical experiments, the human subjects performed the same VOR cancellation task as the monkeys in the physiology experiments. However, in humans there was no perceptual compensation. This perceptual finding makes good ecological sense. There are an infinite number of self-motion paths that all create the same retinal velocity field; for example, a linear translation with a pursuit gaze shift and a curvilinear translation can give rise to the same instantaneous retinal velocity pattern, although over time the two would differ. Thus, it is plausible that the visual system uses extraretinal signals to distinguish between these possibilities. However, a vestibular canal signal considered in isolation is consistent either with a head rotation during linear self-motion or with self-motion moving on a curved path. In fact, because head turns also generate efference copy and neck-proprioceptive signals, a canal signal alone is actually more consistent with curvilinear self-motion. The fact that MST neurons demonstrate compensation that is not observed perceptually during VOR cancellation suggests that area MSTd may not be the final stage in the computation of self-motion. In MSTd, there are populations both of compensating and non-compensating cells (Shenoy *et al.*, 1997, 1999); it is thus possible that different populations of MSTd neurons are read out at a downstream sight depending on the presence or absence of other cues such as neck-proprioception and efference copy.

As mentioned earlier, not only do MSTd expansion neurons compensate during pursuit, but so do curl-selective cells. We have recently asked whether humans compensate equally well for the shift of the focus of rotation of curl stimuli and the focus of expansion of expanding stimuli during pursuit (Maxwell *et al.*, 1997; Crowell *et al.*, 1998b). We find that human subjects do compensate with these types of stimuli, but only partially—approximately 45% of the amount required using expansions and only 25% using curls. This is another inconsistency between psychophysics and physiology, since MST curl cells compensate by approximately the same amount as expansion cells. However, there are many fewer curl than expansion cells in MSTd, and this difference may contribute to the smaller degree of compensation. Consistent with this population code idea, Geesaman and Qian (1996) have found that dots moving at the same velocity are perceived as moving much faster if they are part of a global expansion compared to a global curl. They also propose that the predominance of expansion- over curl-selective cells may contribute to this perceptual difference.

To summarize, comparison of perceptual and physiological responses to simulated self-motion has uncovered many parallels and some inconsistencies. The many parallels suggest that cortical area MSTd plays an important role in self-motion perception; the inconsistencies suggest that it is not the only or final cortical site involved.

IV. Conclusion

Physiological experiments which demonstrate a possible neurophysiological foundation for self-motion perception are described. The particularly challenging problem of correctly estimating observer translation during eye rotations appears to be accomplished by a convergence and interaction of pursuit and visual signals in area MSTd. Specifically, the focus tuning of MSTd neurons is shifted by the extraretinal signal to compensate for eye rotation. This shift appears to be accomplished by a gain modulation mechanism. Such a gain mechanism has previously been shown to provide a possible basis for spatial constancy (Andersen and Mountcastle 1983; Andersen 1997). Eye position and vestibular signals have been found to gain-modulate retinal visual signals in the posterior parietal cortex (Brotchie *et al.,* 1995). These modulations allow representation of objects in head, body, or world coordinates. The results from MSTd could be considered a velocity analog of this mechanism, in which eye velocity signals gain-modulate visual motion signals. These results suggest that the gain modulation mechanism is a very general method for performing computations in the brain, especially those computations related to spatial constancy and spatial perception.

References

Andersen, R. A. (1997). Multimodal integration for the representation of space in the posterior parietal cortex. *Phil. Trans. Roy. Soc. Lond. B* **352,** 1421–1428.

Andersen, R. A., Bradley, D. C., and Shenoy, K. V. (1996). Neural mechanisms for heading and structure-from-motion perception. Cold Spring Harbor Symposia on Quantitative Biology, Cold Spring Harbor Lab. Press **LXI,** 15–25.

Andersen, R. A., and Mountcastle, V. B. (1983). The influence of the angle of gaze upon the excitability of the light sensitive neurons of the posterior parietal cortex. *J Neurosci.* **3,** 532–548.

Beintema, J. A., and van den Berg, A. V. (1998). Heading detection using motion templates and eye velocity gain fields. *Vision Res.* **38,** 2155–2179.

Bradley, D. C., Maxwell, M. A., Andersen, R. A., Banks, M. S., and Shenoy, K. V. (1996). Neural mechanisms for heading perception in primate visual cortex. *Science* **273,** 1544–1547.

Britten, K. H. (1998). Clustering of response selectivity in the medial superior temporal area of extrastriate cortex in the macaque monkey. *Vis. Neurosci.* **15,** 553–558.

Britten, K. H., amd van Wezel, J. A. (1998). Electrical microstimulation of cortical area MST biases heading perception in monkeys. *Nat. Neurosci.* **1,** 59–63.

Brotchie, P. R., Andersen, R. A., Snyder, L. H., and Goodman, S. J. (1995). Head position signals used by parietal neurons to encode locations of visual stimuli. *Nature* **375,** 232–235.

Crowell, J. A., Banks, M. S., Shenoy, K. V., and Andersen, R. A. (1997). Self-motion path perception during head and body rotations. *Invest. Opthalmol. Vis. Sci. Abstr.* **38,** 481.

Crowell, J. A., Banks, M. S., Shenoy, K. V., and Andersen, R. A. (1998a). Visual self-motion perception during head turns. *Nat. Neurosci.* **1,** 732–737.

Crowell, J. A., Maxwell, M. A., Shenoy, K. V., and Andersen, R. A. (1998b). Retinal and extra-retinal motion signals both affect the extent of gaze-shift compensation. *Invest. Opthalmol. Vis. Sci. Abstr.* **39,** 1093.

Duffy, C. J., and Wurtz, R. H. (1991a). Sensitivity of MST neurons to optic flow stimuli. I. A continuum of response selectivity to large-field stimuli. *J. Neurophysiol.* **65,** 1329–1345.

Duffy, C. J., and Wurtz, R. H. (1991b). Sensitivity of MST neurons to optic flow stimuli. II. Mechanisms of response selectivity revealed by small-field stimuli. *J. Neurophysiol.* **65,** 1346–1359.

Duffy, C. J., and Wurtz, R. H. (1995). Response of monkey MST neurons to optic flow stimuli with shifted centers of motion. *J. Neurosci.* **15,** 5192–5208.

Duffy, C. J., and Wurtz, R. H. (1997). Medial superior temporal area neurons respond to speed patterns in optic flow. *J. Neurosci.* **17,** 2839–2851.

Geesaman, B. J., and Andersen, R. A. (1996). The analysis of complex motion patterns by form/cue invariant MSTd neurons. *J. Neurosci.* **16,** 4716–4732.

Geesaman, B. J., and Qian, N. (1996). A novel speed illusion involving expansion and rotation patterns. *Vision Res.* **36,** 3281–3292.

Geesaman, B. J., Born, R. T., Andersen, R. A., and Tootell, R. B. H. (1997). Maps of complex motion selectivity in the superior temporal cortex of alert macaque monkey: A double-label 2-deoxyglucose study. *Cereb. Cortex* **7,** 749–757.

Gibson, J. J. (1950). "The Perception of the Visual World." Houghton Mifflin, Boston.

Graziano, M. S. A., Andersen, R. A., and Snowden, R. J. (1994). Tuning of MST neurons to spiral motions. *J. Neurosci.* **14,** 54–67.

Hildreth, E. C. (1992). Recovering heading for visually-guided navigation. *Vision Res.* **32,** 1177–1192.

Kawano, K., Sasaki, M., and Yamashita, M. (1984). Response properties of neurons in posterior parietal cortex of monkey during visual-vestibular stimulation. I. Visual tracking neurons. *J. Neurophysiol.* **51,** 340–351.

Kawano, K., Shidara, M., Watanabe, Y., and Yamane, S. (1994). Neural activity in cortical area MST of alert monket during ocular following responses. *J. Neurophysiol.* **71,** 2305–2324.

Koenderink, J. J., and Van Doorn, A. J. (1981). Exterospecific component of the motion parallax field. *J. Opt. Soc. Am.* **71,** 953–957.

Lappe, M. (1998). A model for the combination of optic flow and extraretinal eye movement signals in primate extrastriate visual cortex. *Neural Networks* **77,** 397–474.

Lappe, M., Bremmer, F., Pekel, M., Thiele, A., and Hoffmann, K-P (1996). Optic flow processing in monkey STS: A theoretical and experimental approach. *J. Neurosci.* **16,** 6265–6285.

Longuet-Higgins, H. C., and Prazdny, K. (1980). The interpretation of a moving retinal image. *Proc. Roy. Soc. Lond. B Biol. Sci.* **211,** 151.

Maxwell, M. A., Crowell, J. A., Bradley, D. C., and Andersen, R. A. (1997). Comparison of eye pursuit effects on expanding and rotating motion patterns. *Soc. Neurosci. Abstr.* **23,** 173.

Newsome, W. T., Wurtz, R. H., and Komatsu, H. (1988). Relation of cortical areas MT and MST to pursuit eye-movements. II. Differentiation of retinal from extraretinal inputs. *J. Neurophysiol.* **60,** 604–620.

Perrone, J. A., and Stone, L. S. (1994). A model of self-motion estimation within primate extrastriate visual-cortex. *Vision Res.* **34,** 2917–2938.

Orban, G. A., Lagae, L., Verri, A., Raiguel, S., Xiao, D., Maes, H., and Torre, V. (1992). First-order analysis of optical flow in monkey brain. *Proc. Nat. Acad. Sci. USA* **89,** 2595–2599.

Rieger, J. H., and Lawton, D. T. (1985). Processing differential image motion. *J. Opt. Soc. Am. A* **2,** 354–360.

Royden, C. S., Banks, M. S., and Crowell, J. A. (1992). The perception of heading during eye movements. *Nature* **360,** 583–585.

Royden, C. S., Crowell, J. A., and Banks, M. S. (1994). Estimating heading during eye movements. *Vision Res.* **34,** 3197–3214.

Sakata, H., Shibutani, H., Kawano, K., and Harrington, T. (1985). Neural mechanisms of space vision in the parietal association cortex of the monkey. *Vision Res.* **25,** 453–463.

Sakata, H., Shibutani, H., Ito, Y., and Tsurugai, K. (1986). Parietal cortical neurons responding to rotary movement of visual stimulus in space. *Exp. Brian Res.* **61,** 658–663.

Sakata, H., Shibutani, H., Ito, Y., Tsurugai, K., Mine, S., and Kusunoki, M. (1994). Functional properties of rotation-sensitive neurons in the posterior parietal association cortex of the monkey. *Exp. Brain Res.* **101,** 183–202.

Saito, H., Yukie, M., Tanaka, K., Hikosaka, K., Fukada, Y., and Iwai, E. (1986). Integration of direction signals of image motion in the superior temporal sulcus of the macaque monkey. *J. Neurosci.* **6,** 145–157.

Shenoy, K. V., Bradley, D. C., and Andersen, R. A. (1996). Heading computation during head movements in macaque cortical area MSTd. *Soc. Neurosci. Abstr.* **22,** 1692.

Shenoy, K. V., Crowell, J. A., Bradley, D. C., and Andersen, R. A. (1997). Perception and neural representation of heading during gaze-rotation. *Soc. Neurosci. Abstr.* **23,** 15.

Shenoy, K. V., Crowell, J. A., and Andersen, R. A. (1998). The influence of pursuit speed upon the representation of heading in Macaque cortical area MSTd. *Soc. Neurosci. Abstr.* **24,** 1746.

Shenoy, K. V., Bradley, D. C., and Andersen, R. A. (1999). Influence of gaze rotation on the visual response of primate MSTd neurons. *J. Neurophysiol.* **81,** 2764–2786.

Tanaka, K., Hikosaka, K., Saito, H., Yukie, M., Fukada, Y., and Iwai, E. (1986). Analysis of local and wide-field movements in the superior temporal visual areas of the macaque monkey. *J. Neurosci.* **6,** 134–144.

Tanaka, K., and Saito, H. A. (1989). Analysis of motion of the visual field by direction, expansion/contraction, and rotation cells clustered in the dorsal part of the medial superior temporal area of the macaque monkey. *J. Neurophysiol.* **62,** 626–641.

Tanaka, K., Fukada, Y., and Saito, H. A. (1989). Underlying mechanisms of the response specificity of expansion/contraction and rotation cells in the dorsal part of the medial superior temporal area of the macaque monkey. *J. Neurophysiol.* **62,** 642–656.

Thier, P., and Erickson, R. C. (1992). Responses of visual tracking neurons from cortical area MST-l to visual, eye and head motion. *Eur. J. Neurosci.* **4,** 539–553.

van den Berg, A. V., and Beintema, J. A. (1997). Motion templates with eye velocity gain fields for transformation of retinal to head centric flow. *NeuroReport.* **8,** 835–840.

COMPUTATIONAL MECHANISMS FOR OPTIC FLOW ANALYSIS IN PRIMATE CORTEX

Markus Lappe

Department of Zoology and Neurobiology, Ruhr University Bochum, Bochum, Germany

I. Introduction

In the visual cortex of the primate, the information-processing steps necessary to analyze optic flow occur in a hierarchical system of specialized motion-sensitive areas. Computational models of optic flow processing that employ neural network techniques are useful to interpret the neuronal data obtained from these areas. This is important because behaviorally relevant parameters are not encoded in the single neuron activity but rather distributed across a neuronal population. The goal of such models is twofold. On the one hand, they have to reproduce physiological data from single unit studies, and ultimately strive to explain the mechanisms underlying the neuronal properties. On the other hand, they have to concern themselves with the generation of behavior and show how the properties of individual neurons in turn relate to psychophysical measurements at the system level.

0074-7742/00 $30.00

During the last ten years, psychophysical studies have described how humans perceive their direction of heading from the patterns of optic flow and how the various visual and nonvisual cues are functionally combined to solve this task (see e.g. the chapter by van den Berg, this volume). At the same time, experimental physiological studies have provided an account of optic flow processing in the primate visual cortex (see chapters by Bremmer *et al.*, Andersen *et al.*, and Duffy, this volume). It is well documented that visual motion information proceeds from the primary visual cortex (V1) to the middle temporal area (MT) and then to the medial superior temporal (MST) area and other areas in the parietal cortex. Area MT contains a preprocessed representation of the optic flow field that is well suited to serve as a basis for flow-field analysis. Area MST subsequently analyzes the flow field to estimate self-motion. In the course of this transformation from local image motion to global self-motion, additional signals that support self-motion estimation are combined with the optic flow. These are oculomotor signals, retinal disparity, and vestibular signals. To understand the complex information processing that occurs along the pathway of motion analysis in the primate cortex, it is useful to complement single-unit neurophysiology and behavioral/psychophysical observations with theoretical and computational considerations. This requires the formulation and evaluation of biologically plausible models.

II. Foundations and Goals of Modeling

The primary goal of the modeling approaches described in this chapter is to understand at a computational level how neurons and neuronal populations analyze optic flow and contribute to the control and perception of self-motion. By mathematically devising biologically plausible models, one can summarize and formalize experimental findings, provide ways to test hypotheses about the function of neurons and cortical areas quantitatively, and develop unifying concepts of how the brain solves complex computational problems.

Computational models are important for interpreting neuronal data and formulating testable predictions. Hence they must directly interact with physiological experiments. Empirical findings are the basis on which models are built and constrain the elements that can be used in the construction of a model. A valid neurobiological model must capture as much of the physiological and anatomical properties of the structure it wants to model. To demonstrate that this is the case, it is necessary to compare the model to experimental findings. The comparison of model behavior with physiological data will therefore be a central part of this chapter.

However, a useful model must not only reproduce and predict neuronal properties but also show how these properties contribute to a behavioral function or task. The models that will be discussed here address the task of heading estimation from retinal flow. Heading estimation is an important part of successful goal-directed movement and is involved in many daily activities. Visual computation of heading from retinal flow is a difficult task. During straightforward self-movement, heading is indicated by the focus of expansion. In general, however, heading detection is more complicated than a simple search for the focus of expansion. This is because we perform smooth pursuit eye movements to track a visual target during self-motion. Such tracking eye movements induce retinal image slip. When combined with the optic flow during self-translation, image slip induced by eye rotation transforms the structure of the retinal flow field and obscures the focus of expansion (see Lappe and Hoffmann, this volume, for more information). The primate visual system is thus often confronted with a very complex flow pattern. It needs mechanisms to analyze efficiently retinal flow fields that are perturbed by eye movements.

Mathematically, the task of heading estimation can be formulated as follows. The visual motion seen by the eye of a moving observer can—like any rigid body motion—be described by its translational and rotational components. Each of these has in principle three degrees of freedom. Heading detection requires the determination of the direction of translation. This is a problem with many unknown parameters. These are the six degrees of freedom in the self-motion plus the distances of all elements of the flow field from the eye. The latter are involved because the visual motion of an element of the flow field depends on the parameters of the self-motion and on the distance of the element from the eye. For translational self-movements, the visual speed of each element scales with inverse distance. This is known as motion parallax. Motion parallax is an important cue to segregate translational from rotational motion because rotational motion induces equal angular speed in all image points, independent of distance. Accurate measurement of the retinal flow provides information to solve the heading task, namely the direction and speed of every moving point. This allows us to decompose the flow mathematically into translational and rotational components and to determine the direction of heading if more than six moving points are registered (Longuet-Higgins and Prazdny, 1980). Many computational algorithms have been developed for the computational of self-motion from optic flow (see Heeger and Jepson, 1992b, for an overview; see Sinclair *et al.,* 1994; Fermüller and Aloimonos, 1995, for more recent work). Several models that compute heading from optic flow using neuronal elements have been proposed. The next section provides an overview of different classes of models. The subsequent sections will compare these models to experimental findings from monkey neurophysiology.

III. Models of Optic Flow Processing in Primates

The models that are discussed in this paper consist of neuronlike elements that respond to optic flow and that can be compared to optic flow processing neurons in primate visual motion pathway. The typical model layout consists of two layers of neurons, representing areas MT and MST, respectively. The properties of the neurons in the second (MST) layer mainly depend on their synaptic connections with the first (MT) layer neurons. Different models can be distinguished by the way in which these neuronal elements and their connections are constructed.

A. Models Based on Learning Rules

One class of models uses learning rules that originate from artificial neural networks theory to specify synaptic connections. These are back-propagation networks (Hatsopoulos and Warren, 1991; Beardsley and Vaina, 1998) or unsupervised learning mechanisms (Zhang *et al.*, 1993; Wang, 1995; Zemel and Sejnowski, 1998). The synaptic connections between the neuronal elements are generated by repetitively presenting a learning set of optic flow stimuli as input to the model and each time adjusting the synaptic connections according to the learning rule. The properties of the second-layer neurons then depend on the choice of input flow fields and learning rule. For instance, the basic response properties of MST neurons to expansion, rotation, and translation of large-field random-dot patterns can directly be generated by presenting various combinations of such flow patterns using unsupervised learning techniques (Zhang *et al.*, 1993; Wang, 1995; Zemel and Sejnowski, 1998). However, the neurons in this case only learn to form associations between input patterns but not between the input and a function or behavior, as would be required for the determination of self-motion. Therefore, such models do not directly address the issue of how optic flow is analyzed nor in which way the neurons contribute to the processing of optic flow. Zemel and Sejnowski (1998) used a learning procedure to generate a sparse encoding of typical flow fields obtained from moving scenes. They then demonstrated that heading can be estimated from this encoding. This required a further computational step, however, which was not part of the original learning procedure. Hatsopoulos and Warren (1991) trained a back-propagation network to determine the location of the focus of expansion. This covers the case of heading direction during simple translation. However, this approach did not generalize to the case of combined observer translation and eye rotation. The models that are presented next address the task of heading estimation from retinal flow in the general case.

B. Template Matching Models

Template matching models attempt to solve the task of heading estimation by constructing a priori (i.e., without learning) neurons that are tuned to individual optic flow patterns (Perrone, 1992; Perrone and Stone, 1994; Warren and Saunders, 1995). In these models, each neuron forms a template for a specific flow pattern; hence, these models are called template models. The response of an individual neuron in a template model depends on the match between the input flow field and the template of that neuron. Sensitivity for the direction of heading is obtained by building templates for all flow fields that could possibly occur for any given direction of heading. This immediately results in a major problem because infinitely many flow patterns could arise from a single direction of heading. This is because eye rotations and the structure of the visual environment modify the pattern of flow on the retina but leave the direction of heading unchanged. The original template model of Perrone (1992) hence suffered from the fact that an unrealistically large number of templates would be required. Later work has attempted to cut down the number of templates. Perrone and Stone (1994) opted to do this by constraining the eye movements. Their model considered only those eye movements which stabilize gaze on an environmental target. This is the most prominent natural oculomotor behavior. However, complete reliance on this constraint is not consistent with human psychophysical data (Crowell, 1997).

A different approach toward fewer templates was taken by van den Berg and Beintema (1997, 1998). Instead of constructing individual templates for any combination of observer translation and eye rotation, they proposed to approximate such templates by the combination of only two first-order templates. The first template would be tuned to pure observer translation. The second template represents the derivative of the first template with respect to the amount of rotation in the flow. Formally, this is equivalent to approximating a mathematical function by the first two terms of its Taylor series. The activity of the derivative template is used to compensate changes in the activity of the pure translation template when the eye rotates. Such a combination of templates is tuned to heading because it always prefers the same observer translation irrespective of eye rotation. The benefit of this approach clearly is that fewer templates are needed. However, the approximation induces systematic errors for high rotation rates, because the compensation is done to the first order only. Yet, a dependence of the error on rotation rate is also often seen in human psychophysical data (van den Berg, this volume). The error can be overcome by the inclusion of an extraretinal eye movement signal. This signal modulates the activity of the derivative template and extends the effective range of rotations for which compensation is successful.

C. Differential Motion Parallax

Other models draw on research in computer vision algorithms for the recovery of camera motion. These models implement computer vision algorithms with physiologically plausible neural processing elements. Two computational procedures have received particular attention. They are presented in this and the next section. The first algorithm originated from work of Rieger and Lawton (1985) who determined heading from differential motion parallax. The differential-motion-parallax algorithm uses not the individual motion vectors in the optic flow but rather the difference vectors between adjacent flow vectors. During a combination of translation and rotation, these difference vectors always point toward the direction of heading, much like the optic flow vectors during pure translation point toward the focus of expansion. This procedure has been used to model human psychophysical data (Hildreth, 1992a, b). Recently it has also been put into a neurobiological framework (Royden, 1997). This framework proposes that MT neurons compute the motion parallax field by center-surround mechanisms. This appears possible because the center and the surround of the receptive field of MT neurons indeed show opposite motion sensitivities (Allman *et al.*, 1985; Raiguel *et al.*, 1995).

D. Optimal Approximation: The Population Heading Map Model

Another technique often used in computer vision procedures employs optimization methods. In the case of self-motion, this means finding a set of motion parameters (translation and rotation) that optimally predict a measured flow field. Mathematically, this is achieved by minimizing the mean squared difference between the measured flow field and all flow fields constructed from any possible combination of observer translation and rotation (Bruss and Horn, 1983; Koenderink and van Doorn, 1987). Finding the self-motion parameters that minimize this difference is equivalent to finding the actual self-motion. However, because of motion parallax, any candidate flow field also depends on the 3-D

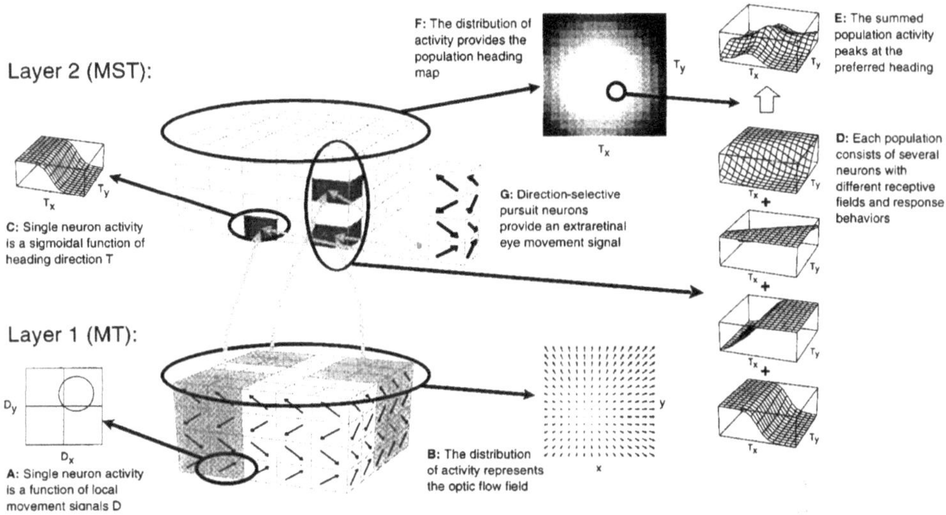

FIG. 1. The population heading map model (Lappe and Rauschecker, 1993a,b; Lappe *et al.*, 1996) consists of two layers of neurons. They correspond to areas MT and MST in monkey cortex. The first (MT) layer contains neurons that are selective to local speed and direction of motion (A). Their receptive fields are arranged in a retinotopic map. Each map position consists of a hypercolumn containing neurons with many different selectivities to local visual motion. Together, the neurons of one hypercolumn encode the visual motion at the corresponding part of the visual field. The distribution of the activities of all hypercolumns in the first layer represents the optic flow field (B).

The second (MST) layer contains neurons that analyze the optic flow and determine self-motion. This layer also contains a topographic map. But unlike the first layer, this is a map of heading directions. The map is constructed in two steps. First, the response of individual neurons to optic flow fields depends on the direction of heading. The dependency has a sigmoidal shape (C). Second, groups of neurons are collected into columns that each represent a specific heading. Each column of neurons receives input from different parts of the visual field (shaded areas in the first layer) and contains cells with different optic flow response properties (D). The activities of all neurons within one such column are summed into a population activity (E). The population activity is maximal when the preferred heading of the population and the true heading of the optic flow are the same. The distribution of the activities of all populations in the second layer provides a computational map of heading (F). The activity peak in this map signals the true heading of the observer.

Recently, this model has been modified to include extraretinal eye movement signals (Lappe, 1998). Pursuit neurons, which form a separate population in the MST layer (G), are active during pursuit eye movements and are selective for the direction of the eye movement. Their activity reflects an extraretinal (i.e., nonvisual) input such as an efference copy. This signal is fed into the optic flow processing neurons in the MST layer. The optic flow processing neurons hence can use both visual and extraretinal signals to compensate for eye movements in heading estimation.

structure of the visual scene. Therefore, the minimization must include not only all possible self-motion parameters but also all possible 3-D scene layouts. As with the template-matching method discussed earlier, this amounts to a very large number of possibilities—six degrees of freedom of the observer's motion plus one for each visible point in the scene. Fortunately, if one is only interested in determining the translational heading then the number of parameters can be dramatically reduced. Heeger and Jepson (1992b) have presented a modified version of the least-squares optimization method that is much more economical and can be easily implemented in parallel processing elements.

Based on this method, Lappe and Rauschecker (1993a, b) have developed a population-based neural network model of optic flow processing in areas MT and MST (Fig. 1). In this model, populations of optic flow processing neurons compute the mean-squared differences for many individualized headings in parallel. Each heading is represented by a small population of neurons. The activity of this population defines the momentary likelihood of this specific heading. This results in a heading map in which each position in the map contains a set of neurons that act together as a population, representing a predefined heading direction. No single neuron of a population can determine heading alone. Only the combination of individual neuron responses into the population gives the appropriate signal. Any individual neuron in a single population contributes only to part of the calculation. Therefore, the properties of a neuronal population in the map and of its constituent neurons are not directly equal. This is an distinctive feature of this model which will become important in Section VII.

IV. Comparisons with Physiology: Optic Flow Representation in Area MT

Area MT is the first area in the primate visual pathway that is dedicated specifically to the processing of motion. Most models assume that area MT forms the cortical representation of the optic flow field. Area MT contains a retinotopic map of the contralateral visual field (Albright and Desimone, 1987). Most neurons in area MT are direction- and speed-selective (Maunsell and Van Essen, 1983). Cells in area MT are organized in direction columns similar to the orientation columns in V1 (Albright *et al.*, 1984). The MT analogon of a V1-hypercolumn could provide all essential information for encoding local motion at a single visual field location. The combination of many such hypercolumns would encode the full optic flow field. Typically, models employ such

MT hypercolumns as the starting point of the representation of the flow field.

Modeling studies have shown how some aspects of the global organization of MT can be used for optic flow processing. These include properties of the retinotopic mapping in MT, the antagonistic center/surround organization of the receptive fields, and the disparity sensitivity of MT neurons.

A relation to optic flow is already apparent in the retinotopic mapping in area MT. Preferred speeds of MT neurons increase with the eccentricity of their receptive field (Maunsell and Van Essen, 1983) similarly to the way optic flow speeds naturally do. The number of direction-sensitive neurons preferring motion away from the fovea is significantly higher than the number of neurons preferring motion toward the fovea (Albright, 1989). This property is well adapted to the centrifugal structure of the flow field under natural self-motion conditions (Lappe and Rauschecker, 1994, 1995b).

The receptive fields of many MT neurons consist of a direction-selective central region complemented by an antagonistic surround (Allman *et al.*, 1985; Raiguel *et al.*, 1995). Surround motion in the same direction as in the center of the receptive field inhibits the neuron. Such an arrangement of selectivities could yield detectors that compute the local motion parallax field. From such detectors, heading could be estimated using the differential-motion-parallax algorithm (Royden, 1997) or the optimization algorithm of Heeger and Jepson (1992a), which forms the computational basis of the population heading map model.

A detailed model of the representation of visual motion in area MT showed that two further receptive field properties of MT neurons benefit the representation of optic flow (Lappe, 1996). First, receptive field sizes in area MT grow with eccentricity of the receptive field center (Albright and Desimone, 1987). Second, MT neurons respond to motion signals from within their receptive field in a disparity-selective way (Bradley *et al.*, 1995). The combination of these two factors provides an effective way to enhance the representation of the flow field in the presence of noise (Fig. 2). The extended receptive fields of MT neurons provide a spatial smoothing of the flow field which reduces motion noise. Because the structure of the flow field is very fine in the center but much courser and more uniform in the periphery, it makes sense to vary the scale of the smoothing (the sizes of the receptive fields) with eccentricity. However, extensive smoothing of the flow field might remove signals that are necessary for heading direction. Especially important in this regard is motion parallax, i.e., the difference in speed of objects at different depths. Motion parallax carries important information to separate

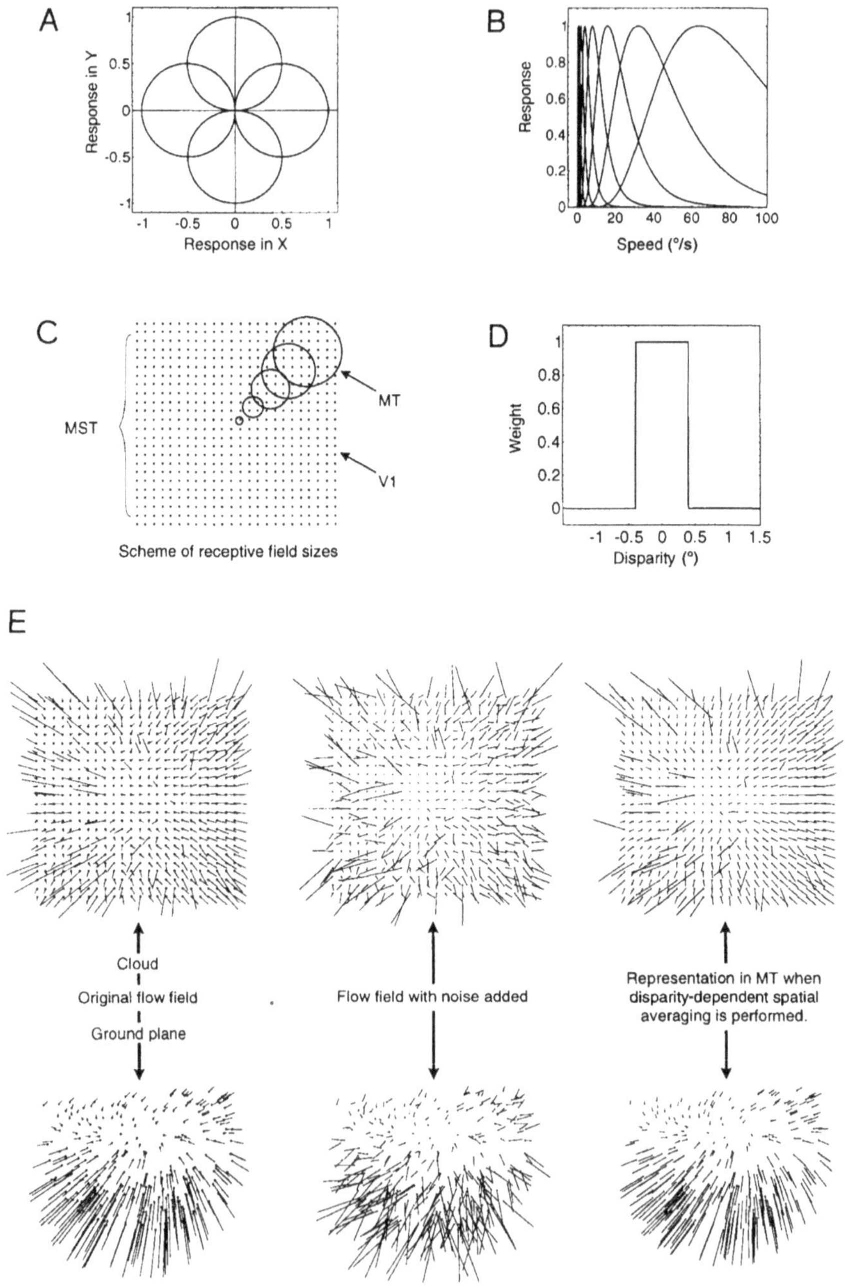
A
Response in Y
Response in X
1
0.5
0
-0.5
-1
B
Response
Speed (°/s)
0.8
0.6
0.4
0.2
20
40
60
80
100
C
MST
MT
V1
Scheme of receptive field sizes
D
Weight
Disparity (°)
1.5
E
Cloud
Original flow field
Ground plane
Flow field with noise added
Representation in MT when disparity-dependent spatial averaging is performed.

translational and rotational flow components. Such a separation is required for heading detection during eye movements. A loss of motion parallax information because of extensive smoothing is therefore not desirable. Figure 2 shows how this unwanted effect is overcome when the spatial smoothing is made disparity-dependent, as it is in MT neurons. Spatial smoothing is performed only within depth planes. In this way, noise is reduced, while motion parallax information is retained. This model of disparity-dependent spatial averaging of the flow field can explain the enhanced robustness of heading perception from stereoscopic flow stimuli observed in humans (Lappe, 1996; van den Berg and Brenner, 1994).

V. Comparisons with Physiology: Optic Flow Selectivity in Area MST

Area MST receives major input from area MT and is thought to analyze optic flow and determine self-motion (Bremmer *et al.*, this volume; Anderson *et al.*, this volume; Duffy, this volume). Most effort in modeling optic flow processing has focussed on area MST and on comparing the properties of model neurons with those of neurons recorded from this area. In comparing the behavior of model neurons with their MST counterparts, several key properties of MST neurons must be consid-

---

FIG. 2. Model of the representation of the optic flow field in MT (Lappe, 1996). For each position in a retinotopic map, the model assumes a set of direction-selective neurons with different tuning properties. These contain four cosine direction tunings along the cardinal axes (A) and eight Gaussian speed tunings on a logarithmic scale between 0.5 and 64° (B). Direction and speed of local motion are determined from the activities of these neurons by a population code.

The size of receptive fields in area MT varies with eccentricity (Albright and Desimone, 1987). The model assumes that the response of a neuron is determined by the spatial average of the visual motion inside its receptive field (C). Consistent with electrophysiological findings (Bradley *et al.*, 1995), the response also depends on the disparity of motion signals. The contribution of each motion signal to the spatial summation in the receptive field is weighted according to its disparity (D). Only motion signals with a disparity close to the preferred disparity of a neuron contribute.

Such a representation is very robust against noise in the flow field (E). On the left, two flow fields which are experienced during a combination of observer translation and eye rotation are shown. The upper flow field depicts motion in a random 3-D environment (a cloud of random dots). The lower flow field depicts motion across a horizontal ground plane. The central column presents noisy versions of the same flow fields. When these noisy flow fields are processed by the MT model (right column), much of the noise is removed. The representation in MT is very close to the original flow patterns.

ered: the selectivity for several types of flow patterns such as expansion, contraction, rotation, and unidirectional motion (Duffy and Wurtz, 1991a) or spirals (Graziano *et al.*, 1994), the dependence of the response on the location of the focus of expansion (Duffy and Wurtz, 1995; Lappe *et al.*, 1996), the position invariance of the selectivity (Graziano *et al.*, 1994; Lagae *et al.*, 1994), and the combination with extraretinal eye movement signals (Erickson and Thier, 1991; Bradley *et al.*, 1996; Page and Duffy, 1999).

Several of these characteristics are also found in area VIP (Schaafsma and Duysens, 1996; Bremmer *et al.*, 1997, this volume). At present, it is not clear how these two areas differ with respect to optic flow processing. Here we will focus on comparisons between model and MST neurons. This is done with the understanding that similar comparisons can be made to VIP neurons. As more experimental findings will become available, differences between the two areas might become apparent.

A. Selectivity for Multiple Optic Flow Patterns

When Tanaka and Saito (1989a, b) first systematically investigated the optic flow response properties of MST neurons, they used a set of "elementary" optic flow stimuli. This stimulus set consisted of unidirectional motion, rotation, expansion, and contraction. For each of these elementary flow fields, they found neurons that responded selectively to only a single one. Tanaka and Saito required a genuine optic-flow neuron to respond to expansion, contraction, or rotation but not to unidirectional motion. Later studies found that the majority of MST neurons responded to several different flow stimuli (Duffy and Wurtz, 1991a, b). Duffy and Wurtz proposed a classification into triple, double, and single component neurons, with the explicit understanding that this reflects a continuum of selectivities in MST. Triple-component neurons respond selectively to one direction of unidirectional motion, one sense of rotation, and either expansion or contraction. Double-component neurons respond to unidirectional motion and to either rotation or expansion/contraction. Very few neurons in Duffy and Wurtz's study responded to rotation and expansion/contraction but lacked direction selectively. The most selective, but least populous group consists of single-component neurons that responded only to one of the stimuli: 9% of all neurons respond to expansion or contraction; 4% respond to rotation. The predominance of triple- and double-component neurons and the relative scarcity of single-component neurons has since been confirmed in two subsequent studies (Duffy and Wurtz, 1995; Lappe *et al.*, 1996).

In models, such a variety of selectivities can arise in two conceptually different ways. Either the model begins with highly selective neurons that for instance respond only to pure expansions, and then adds mechanisms that induce sensitivity also for other flow patterns. This is the approach proposed by the template models (Perrone, 1992; Perrone and Stone, 1994). Or alternatively, the model starts with broadly selective neurons such as the triple-component neurons in MST and refines its selectivity successively until more selective response properties are reached. This is the approach of the optimization and learning models (Lappe and Rauschecker, 1993a, b; Lappe *et al.*, 1996; Zemel and Sejnowski, 1998). Probably in favor of the latter is the observation that many more MST neurons respond to multiple flow patterns than to only a single one. This would seem to suggest that these single-component neurons represent a higher degree of abstraction, which might be achieved by internal convergent connections in MST. Lappe *et al.* (1996) have suggested such a two-step convergence. A large number of broadly selective neurons are combined in populations in the heading map. Higher, more selective neurons could read-out the population heading map. This arrangement would have some analogy to the way complex cells in V1 are constructed from converging inputs of simple cells. It would result in a substantially smaller number of highly selective (single-component) neurons than broadly selective (triple- and double-component) neurons.

In the heading map model, different degrees of component selectivity can be obtained by assuming different selectivities for combinations of self-motion and eye movements (Lappe and Rauschecker, 1993a; Lappe *et al.*, 1996). This applies to triple- and double-component neurons. Single-component radial (expansion/contraction) neurons can be generated at a higher level by convergence. The template model of Perrone and Stone contains single-component radial and unidirectional neurons as well as one type of double-component selectivity (radial + unidirectional). But it fails to predict the other types of selectivity found in MST (Perrone and Stone, 1998), which is actually the majority of neurons. Both models, however, do not generate true single-component rotation selectivity.

Selective responses to many different flow patterns can be observed in MST. Yet, some flow patterns clearly appear more related to self-motion than others. It is easy to see why expansion selectivity could be required for heading detection. It corresponds to forward movement. This is much less clear for contraction selectivity (backward movement) or even rotation selectivity. Full field rotation would require to rotate the head around the axis of gaze which is unlikely to occur frequently in the natural behavior of ground living animals. These neurons might be involved in tasks other than self-motion estimation (e.g., the perception of the motion of objects) (Graziano *et al.*, 1994; Zemel and Sejnowski, 1998). On the other hand, very many MST neurons combine responses to expansion and rotation (and to unidirectional motion), suggesting that such a combination might be useful also for self-motion processing. Some MST cells even prefer vectorial combinations of rotation, and expansion/contraction over the two separate patterns, and display a selectivity for spiral motion (Graziano *et al.*, 1994). The question arises as to why such combinations might be useful for optic flow processing.

Theoretical considerations and models have provided a possible answer to this question. It is based on the structure of naturally occurring flow patterns and the benefits of broad selectivity in a population code. Spiraling retinal motion patterns are quite common when the motion of the observer consists of combined self-movement and eye movement (Lappe *et al.*, 1998; Warren and Hannon, 1990; Lappe and Rauschecker, 1994). Thus, responses to spiraling flow patterns might be expected from neurons that process self-motion in the presence of eye movements. Therefore, most models of heading detection contain neurons that respond to spiraling flow patterns. When the selectivities of such neurons are broad, they also include selective responses for pure rotations as do the triple-component neurons in MST. However, these model neurons are not designed to respond selectively to rotational flow patterns. Rather the rotational or spiral selectivity is a consequence of their selectivity for heading in the presence of eye movements.

B. Selectivity for the Location of the Focus of Expansion

Another important question is the dependence of optic flow responses on spatial parameters of the stimulus. This concerns especially the position invariance of the selectivity to optic flow components and the sensitivity for the location of the focus of expansion. Position invariance describes the observation that many MST neurons preserve their selectivity for a small optic flow pattern even when this pattern is moved to another location within the receptive field of the neuron (Duffy and Wurtz, 1991a; Graziano *et al.*, 1994; Lagae *et al.*, 1994). This is a very illustrative feature of the complexity of the response behavior of MST neurons. It gives a clear demonstration that these neurons truly respond to entire optic flow patterns and not just to part of the local motions within the patterns. However, it has been troublesome for models of heading detection from optic flow. Shifting a radial motion pattern to a different location also shifts the focus of expansion along with it. Yet, the response of any heading detection neuron should vary with the location of the focus of expansion. Models deal with this problem in two ways. The first is to observe that retaining a preference for expansion versus contraction at different positions in the receptive field does not imply that the response strength to expansion has to be equal at these positions. Hence a variation of response strength with the location of the focus of expansion can occur despite a conserved preference of expansion over contraction. Secondly, position invariance occurs more frequently for restricted variations of spatial location of small stimuli (Lagae *et al.*, 1994; Graziano *et al.*, 1994) and less often for larger stimuli and displacements (Duffy and Wurtz, 1991b; Lappe *et al.*, 1996). Positional invariance within restricted regions of the visual field is often found in models (Lappe *et al.*, 1996; Perrone and Stone, 1998; Zemel and Sejnowski, 1998).

The neural specificity for the location of the focus on expansion in full-field optic flow stimuli is more directly related to self-motion processing. Lappe *et al.* (1996) tested predictions of the population heading map model. They investigated neuronal sensitivity for the focus of expansion in area MST and compared the results to computer simulations of model neurons. In these experiments, large field (90° by 90°) computer-generated optic flow stimuli were presented. They simulated forward (expansion) and backward (contraction) self-motion in different directions with respect to a random cloud of dots in three-dimensional space. The dependence of neuronal responses on the focus of expansion were determined for up to 17 different locations. The response profile of an example MST neuron is shown in Fig. 3A. The neuronal responses to expansion and contraction vary smoothly with the position of the singular point, saturating as the focus of expansion is moved into the visual periphery. Responses to both expansion and contraction can be elicited by proper placement of the singular point. These experimental findings are predicted by computer simulations of a model neuron (Fig. 3B).

Different models predict different shapes of the response profile for variations of the focus of expansion. In the population heading map, sigmoidal response profiles like the ones depicted in Fig. 3 are used for the constituent neurons of the heading populations (Fig. 1D). They form the majority of cells in the model. Neurons that read out the population activities—necessarily a minority of model neurons—retain a different selectivity profile and show a peaked tuning curve for the focus of expansion (Fig. 1E). Indeed, some MST neurons have a peak-shaped response profile rather than a sigmoidal one (Duffy and Wurtz, 1995; Lappe *et al.*, 1996). The proportion of these neurons varies somewhat between studies. Lappe *et al.* (1996) reported very few peak-shaped responses,

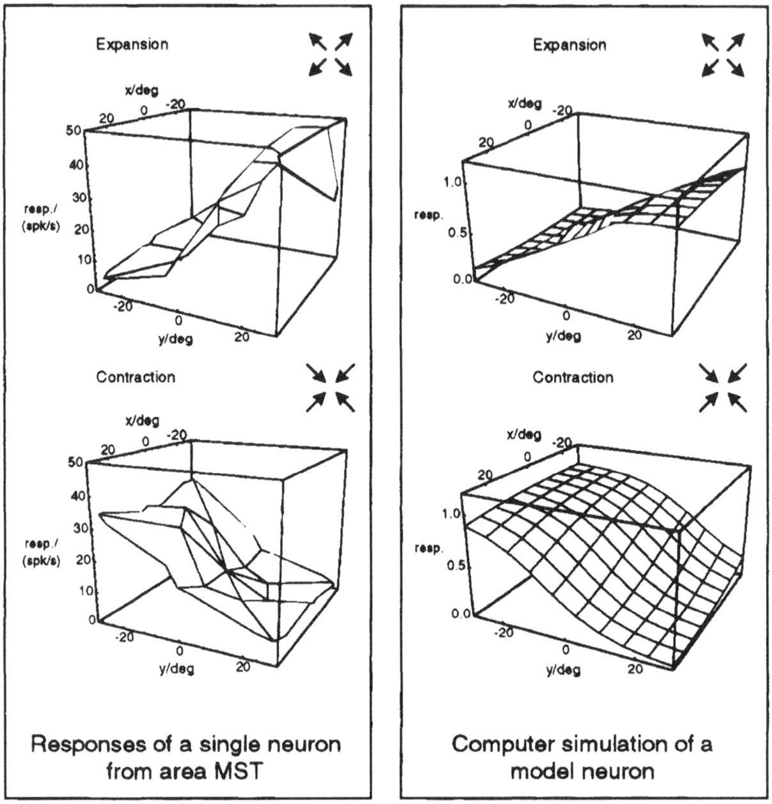

FIG. 3. Electrophysiological experiments in area MST (left) and computer simulations of a single neuron from the population heading map model (right) show very similar results (Lappe *et al.*, 1996). In both cases, the stimuli consisted of full field expanding or contracting optic flow patterns in which the retinal position of the focus was systematically varied. The modulations of response strength with the retinal position of the focus are displayed by 3-D surface plots. The (x, y)-plane represents the positions of the focus, the z-axis neuronal activity.

whereas the data of Duffy and Wurtz (1995) indicates a higher proportion. But in both studies such neurons are a minority. This is consistent with the predictions of the population model. The template model of Perrone and Stone (Perrone, 1992; Perrone and Stone, 1994) exclusively uses Gaussian (peak)-shaped tuning curves. This results in difficulties to account for the sigmoid tuning properties often observed in MST. Gaussian-tuned model neurons with centers very far in the periphery might resemble the sigmoidal tuning curves in MST (Perrone and Stone, 1998). However, in the model they would not be expected to make up a substantial proportion of cells, as in fact they do in MST. The template model of van den Berg and Beintema (1997, 1998) arrives at one class of neurons with a tuning curve very similar to the sigmoidal tuned neurons described by Lappe *et al.* (1996). In their model, these neurons are required for the estimation of eye rotation and the compensation of rotational motion for heading detection.

C. Optic Flow Selectivity during Eye Movements: Integration of Visual and Extraretinal Signals

For a complete and unambiguous representation of self-motion, optic flow needs to be combined with other sensory signals (van den Berg, this volume). The combination with stereoscopic disparity (Lappe, 1996) has already been mentioned. Even more important is the combination of optic flow with signals related to eye movements. Eye movements add visual motion to the optic flow field on the retina and severely complicate its analysis. Extraretinal eye movement signals can be used to overcome this problem. Interestingly, areas MT and MST have been shown to play a major role not only in optic flow analysis but also in the control of eye movements. Areas MT and MST are involved in the generation of smooth pursuit movements (Dürsteler and Wurtz, 1988; Komatsu and Wurtz, 1989), and in optokinetic (Hoffmann *et al.*, 1992; Ilg, 1997) and ocular following (Kawano *et al.*, 1994; Inoue *et al.*, 1998) reflexes. Eye-movement-related signals in area MST contain extraretinal components. Visual tracking neuron continue to fire during ongoing pursuit when the tracked target is briefly extinguished (Sakata *et al.*, 1983; Newsome *et al.*, 1988). They even fire during pursuit of a figure that never directly stimulates their visual receptive field (Ilg and Thier, 1997). Other MST neurons show a contrary behavior: they respond to motion of a visual stimulus during fixation but do not respond when the same retinal image motion is induced by an eye movement (Erickson and Thier, 1991). In these neurons, extraretinal eye movement information acts to differentiate true motion in the world from self-induced motion.

Could these extraretinal signals be used by optic flow processing neurons? Lappe *et al.* (1994, 1998) have argued that pursuit neurons in MST could provide a signal of current eye velocity that might allow optic flow processing neurons to compensate to some degree for the visual disturbance by the eye movement. In this model, the pursuit signal is used to substract the eye-movement-induced visual motion component from the total retinal flow. However, this process is likely to be incomplete. For instance, the speed of eye movement seems to be less well represented by the pursuit neurons than its direction. Precise heading detection would still have to rely on the visual signal but would be supplemented by extraretinal compensation. Such a hybrid model can account for the most prevalent conditions in which human heading detection has been shown to rely on extraretinal input (Lappe *et al.*, 1994). Moreover it directly leads to neurons that implement the extraretinal influences observed in MST (Erickson and Thier, 1991) with unidirectional motion patterns (Lappe, 1998).

The use of pursuit signals can also be seen in actual recordings of MST neuronal responses to optic flow stimuli during ongoing smooth pursuit. Bradley *et al.* (1996) recorded selectivity of single MST cells to the focus of expansion while the monkey performed a smooth pursuit eye movement. These experiments compared the neuronal responses in two situations. In one condition, an expanding optic flow pattern was presented while the monkey tracked with his eyes a small dot that moved in the preferred or antipreferred pursuit direction of the neuron. In the second condition, the monkey was required to keep its eyes stationary and the optic flow pattern now included not only the expansion but also an overall translation. This combination of expansion and translation visually simulated the effects of the previously performed eye movement. Therefore, the two conditions presented identical visual but different extraretinal input. Many optic flow selective MST neurons responded differently in the two conditions. In the condition that included the active eye movement, the neurons responded to the focus of expansion of the stationary optic flow pattern, even though the eyes moved. Thus, an extraretinal signal in MST can compensate for the disturbances of the optic flow field due to eye movements. However, this compensation is far from complete for most neurons. On average, neurons compensated for only about 50% of the eye movement. Individual neurons strongly undercompensated or even in some cases overcompensated the eye-movement induced retinal motion.

The results of Bradley *et al.* have been reproduced in two models that are concerned with the combination of visual and extraretinal signals in MST (Lappe, 1998; Beintema and van den Berg, 1998). Figure 4 shows results from the model of Lappe (1998). The eye-movement-induced retinal image motion is compensated by extraretinal input. Then heading is determined from the compensated flow field. Different levels of compensation in individual neurons are modeled by varying the strength of the extraretinal signal. The model produces accurate heading judgments even with the incomplete compensation observed in MST. This is possible because the model uses a visual backup mechanism parallel to the extraretinal compensation. The model of van den Berg and Beintema

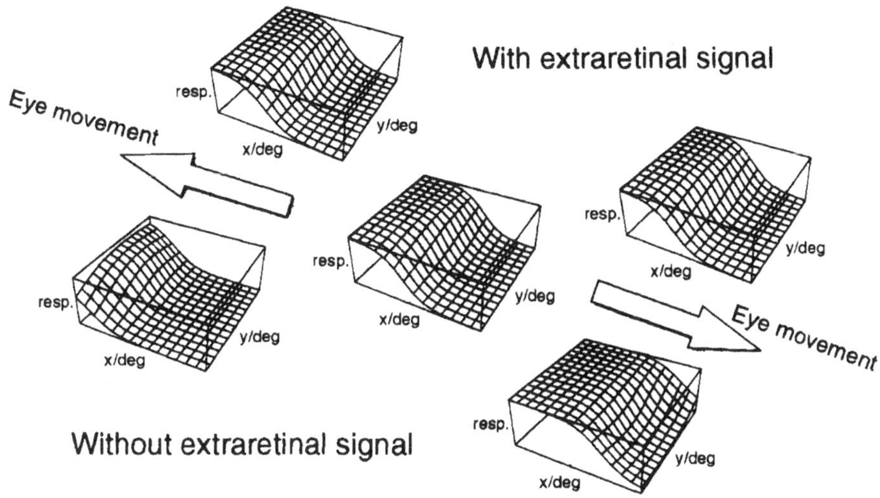

FIG. 4. A simulation of the experiment of Bradley *et al.* (1996). The plots show responses of an optic flow selective neuron to combined optic flow and simulated eye movement stimuli (after Lappe, 1998). The central plot shows the responses to pure expansional flow stimuli. It is dependent on the location (x, y) of the focus of expansion. The outer plots show the responses during eye movements. In the absence of an extraretinal signal, the response curve shifts along the direction of the eye movement. In contrast, when extraretinal input is available, the response curve stays fixed, independent of the eye movement. The response is the same as to an undisturbed pure expansion.

(1997, 1998) uses two types of templates. Retinal motion templates respond to the retinal position of the focus of expansion and only partially compensate for eye movements using the properties of the flow field. Heading templates receive an additional extraretinal eye movement signal and respond to the direction of heading thus compensating for eye movements. This model can also account for the broad range of extraretinal compensation strength in area MST.

The experiments of Bradley *et al.* (1996) tested the selectivity of MST neurons to heading along one axis of pursuit. Complete mappings of two-dimensional heading selectivity during pursuit in different directions were obtained by Page and Duffy (1999). Most MST neurons in their experiments showed significant pursuit influences on the tuning for heading. Many neurons responded to different headings during pursuit rather than during fixation. Page and Duffy concluded that individual MST neurons do not accurately compensate for the influences of pursuit and cannot directly signal heading. But correct heading estimates could be obtained from a population analysis. This is consistent with models that propose a population code for heading in area MST (Lappe and

Rauschecker, 1993b; Lappe *et al.*, 1996; Zemel and Sejnowski, 1998) but not with models that use single neurons as heading detectors (Perrone, 1992; Perrone and Stone, 1994; Beintema and van den Berg, 1998).

VI. Receptive Fields of Optic Flow Processing Neurons

Optic flow response properties of MST neurons must be generated at least in part from the combination of selectivities for local 2-D motion signals from area MT. The question is how the local 2-D motion sensitivities are organized within the receptive field of optic flow processing neurons. This is an important issue that has so far escaped a satisfactory answer. An analysis of the receptive field structure of neurons in models of optic flow processing might provide some help in understanding this relationship and guiding further experiments. It turns out that several models independently arrive at similar predictions for the structure of optic flow processing receptive fields.

Let us first consider the simple idea that the arrangement of local motion selectivities is directly equal to the preferred optic flow pattern. This is the basic assumption of early template models (Perrone, 1992; Perrone and Stone, 1994, 1998). They construct, for instance, expansion-selective cells by arranging 2-D motion sensitivities in a radial expanding pattern inside the receptive field. Experimental studies in MST have clearly disproved such a simple arrangement. Duffy and Wurtz termed it the "direction mosaic" hypothesis. They tested it directly by comparing optic flow selectivity to the selectivity to small 2-D motion stimuli in different parts of the receptive field (Duffy and Wurtz, 1991b). Lagae *et al.* (1994) performed a similar test. Clearly, for true expansion-selective cells, the 2-D motion selectivities in subparts of the receptive field did not match the hypothesis of a radial arrangement as proposed by the template models. According to the template model, single-component expansion cells possess the most basic and most simple tuning. In contrast, in the experimental data, the tuning of single-component neurons to optic flow is most difficult to explain directly from 2-D motion inputs in the subfields (Lagae *et al.*, 1994; Duffy and Wurtz, 1991b). It is somewhat easier for the triple-component neurons. These findings are again consistent with the view that single-component neurons form the highest level of optic flow analysis in MST, as proposed by the population model (Lappe and Rauschecker, 1993b; Lappe *et al.*, 1996).

Several other models make different predictions for the structure of the receptive fields of optic flow processing neurons. These models employ differences between flow vectors as the main computational element. Differences between flow vectors are useful because of the properties of motion parallax. For translational self-movement, points at different distances move at different visual speeds. For rotational (eye) movements, all image points move at the same angular speed, independent of their distance from the observer. Motion parallax is therefore a major cue to separate rotational and translational components in the flow field. Because the rotational component of all flow vectors is identical, the difference between any two flow vectors depends only on the translational component. For instance, differences between neighboring flow vectors can be used to compute a rotation-independent local-motion-parallax field which reconstructs the focus of expansion (Rieger and Lawton, 1985). This procedure is used in the models of Hildreth (1992a,b) and Royden (1997). It could be implemented by motion-selective neurons with center-surround opponent-motion selectivity which are found in area MT. Sensitivity for local opponent motions is also an important mechanism in the model of Beintema and van den Berg. The neurons that represent the derivative templates (i.e., the neurons that subserve the decomposition of translation and rotation) actually compute local motion parallax (Beintema and van den Berg, 1998).

Finally, the population heading map model, too, uses differences between flow vectors as its basic computations. The construction of the synaptic connections consists of comparisons between small groups of two to five flow vectors (Lappe and Rauschecker, 1995b). If these flow vectors are near each other, then the comparison is equivalent to an opponent motion detector (Heeger and Jepson, 1992a).

However, the usefulness of differences between flow vectors is not restricted to local motion parallax. One might argue that parallax information from widely separated regions of the visual field could often even be more useful because most local image regions contain only small depth variations (i.e., limited motion parallax). In the population heading map model, the comparisons need not be local. Comparisons might be performed between any two motion vectors. If these are far apart in the visual field, the computation becomes more complex, however, requiring the comparison of both the speed and direction of motion. Local opponent motion can therefore be regarded as a special case of a more global flow analysis in this model.

In summary, several current models contain similar operators despite the fact that they start out from very different computational approaches. This suggests that these operators reflect a common principle of optic flow processing. As of yet, the structure of the receptive field of flow field processing neurons in MST is unknown. It will be interesting to see whether the prediction derived from these models can be found in the properties of real optic flow processing neurons.

VII. The Population Heading Map

In many areas of the brain, behaviorally relevant parameters are represented in the form of a topographic map. By analogy, such a map has been hypothesized also for the representation of heading (Lappe and Rauschecker, 1993b; Perrone and Stone, 1994; Duffy and Wurtz, 1995). Questions then arise about the structure of this map and the properties of its constituents. Area MST has only a very crude receptive field retinotopy (Desimone and Ungerleider, 1986). Instead, there is some indication that neurons with similar optic flow sensitivities are clustered together (Duffy and Wurtz, 1991a; Lagae *et al.*, 1994; Geesaman *et al.*, 1997; Britten, 1998). The model of Lappe and Rauschecker (1993b) was the first to propose a computational map of heading direction as the functional organization of flow-field-sensitive neurons in area MST. This map is based on a distributed representation of heading by populations of neurons. The population heading map has a number of features that will be discussed in this section.

A. Properties of the Population Heading Map

The population heading map proposes a retinotopic encoding (or map) of heading. This is different from the typical retinotopy of the receptive fields in other visual areas. This retinotopic encoding associates each heading direction in visual field coordinates with a corresponding map location. Heading leftward of the direction of gaze is represented on the left and rightward to the direction of gaze, on the right. The preferred heading of an individual neuron (i.e., its place in the heading map) does not necessarily coincide with the spatial location of its receptive field. The receptive field consists of the manifold of all inputs to the neuron. The preferred heading, on the other hand, is the neuron's contribution to the functional organization of the map. The two must not be confused. A neuron-encoding leftward heading can have a receptive field in the right field of view. Yet, its place in the heading map is in the left hemifield. This dissociation is very important from the computational perspective because information about any given heading can be gathered from any part of the visual field. This is necessary to ensure that all visual input contributes to the determination of heading.

However, the retinotopic representation of a higher motion parameter such as heading precludes the existence of a retinotopic visual field representation in the same neural structure. This is consistent with the poor retinotopy of the receptive fields in MST (Desimone and Ungerleider, 1986) and the clustering of optic flow selectivities instead (Duffy and Wurtz, 1991a; Lagae *et al.*, 1994; Geesaman *et al.*, 1997).

The population heading map is composed not of individual neurons but of populations of neurons that jointly represent a specific heading. Thus, there is an additional computational step between the individual neuron's sensitivity to optic flow and the representation of the direction of heading. In this intermediate step, the responses of several neurons are combined into a population response. These neurons share the preference for the same direction of heading and occupy the same location in the population heading map. The sum of their respective responses, i.e., the population response, is proportional to the current likelihood of the preferred heading of this population. However, the optic flow sensitivities of the constituent neurons need not be all alike. For instance, different neurons receive input from different parts of the visual field (i.e., possess different receptive fields). More importantly, they also possess different response selectivities to optic-flow stimuli. This is evident from Fig. 1D. The response curves for heading are quite different for the four neurons. Yet, they share one heading to which all neurons of the population respond. This is the heading associated with the population and with the map location. The aggregation of the four activities into the population response yields a peak-tuning for the preferred heading.

There are two important consequences of the two-step process by which the map is constructed. First, it is the flanks of the response profiles of the individual neurons and not their respective best response that contribute to the heading map. Figure 1E shows that peak activity of the population coincides with the intersection of the response flanks of the individual neurons in Fig. 1D. Second, the behavior of individual neurons can be quite different from the behavior of the population. Both properties are important for relating experimental findings to model predictions and for an interpretation of the contribution of MST neuronal activity to the task of heading estimation.

B. Analysis of Population Data from Area MST

We want to compare the features of the population heading map with the properties of optic flow processing in area MST. The first question is whether the MST population activity can estimate the direction of heading. To demonstrate this, Lappe *et al.* (1996) computed maps of the location of the focus of expansion from the activity of a set of recorded neuronal responses. They used a least-mean-square minimization scheme to derive the position of the focus of expansion from the actual recording data. The procedure is similar in spirit to the population heading map. The grayscale maps of Fig. 5 present the mean-squared heading error for nine focus positions within the central 15° of the visual field. From it, the location of the focus can be retrieved with an average precision of 4.3° of visual angle. Clearly, the MST population provides enough information to locate the focus of expansion.

Recent results by Page and Duffy (1999) show that the same conclusion holds for heading estimation during eye movements. The responses of single neurons are not invariant against eye movements. But the population provides sufficient information to determine heading.

Further experimental support for a population heading map in area MST is provided by the analysis of a perceptual illusion in which the focus of expansion appears to shift in the field of view. This illusion occurs when a radially expanding random-dot pattern is transparently overlapped by unidirectional motion of a second pattern of random dots (Duffy and Wurtz, 1993). The focus of expansion is perceived not at its original location but displaced in the direction of the overlapping motion. This illusion is reproduced in computer simulations of the model (Lappe and Rauschecker, 1995a). Recently, Charles Duffy and I have compared the responses of individual MST neurons and of individual model neurons to the illusory stimulus (Lappe and Duffy, 1999). We used a paradigm in which two sets of optic flow stimuli were compared (Fig. 6A). One set contained the transparent overlap stimuli which give rise to the illusory shift of the focus of expansion. The other set contained stimuli in which the two motion patterns were vector summed. Vector summation yields a pure expansion pattern with a true eccentric focus. Hence in both cases the focus positions (true or illusory) are eccentric. However, these eccentric positions are 180° opposite for each pair from the two stimulus sets!

Different models predict different response behavior of individual neurons in this case. Template matching predicts that transparently overlapping unidirectional motion should cause neurons encoding the true center of expansion to stop firing, whereas neurons encoding the illusory center of expansion should start firing. The population model predicts that transparently overlapping unidirectional motion should cause graded changes in the responses of all neurons to alter the aggregate response in a manner that shifts the net center of motion. In the

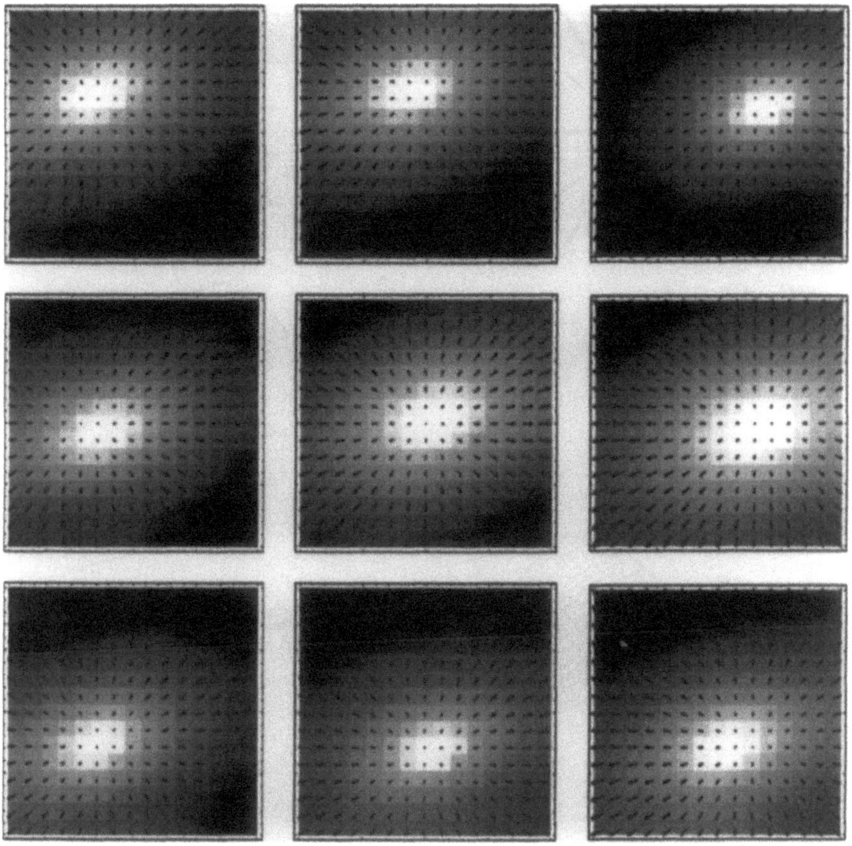

FIG. 5. Computational heading maps obtained from the recorded neuronal activities of 31 MST neurons (Lappe *et al.*, 1996). Each panel shows a retinotopic heading map. The brightness gives the likelihoods of each specific heading. For comparison with the true location of the focus of expansion, the optic flow stimuli are plotted on top of the grayscale maps.

population heading map, the behavior of single-model neurons is actually very different from the behavior seen in the human perception and at the population level in the model. Rather than shift or reverse their response behavior between the two types of stimuli, individual neurons rotate their response gradients with variable amounts of rotation (Figs. 6B and 6C). In the model, such a rotation of the response profiles of single neurons is sufficient to result in the observed shift of the focus of expansion at the population level. The population activity is derived from

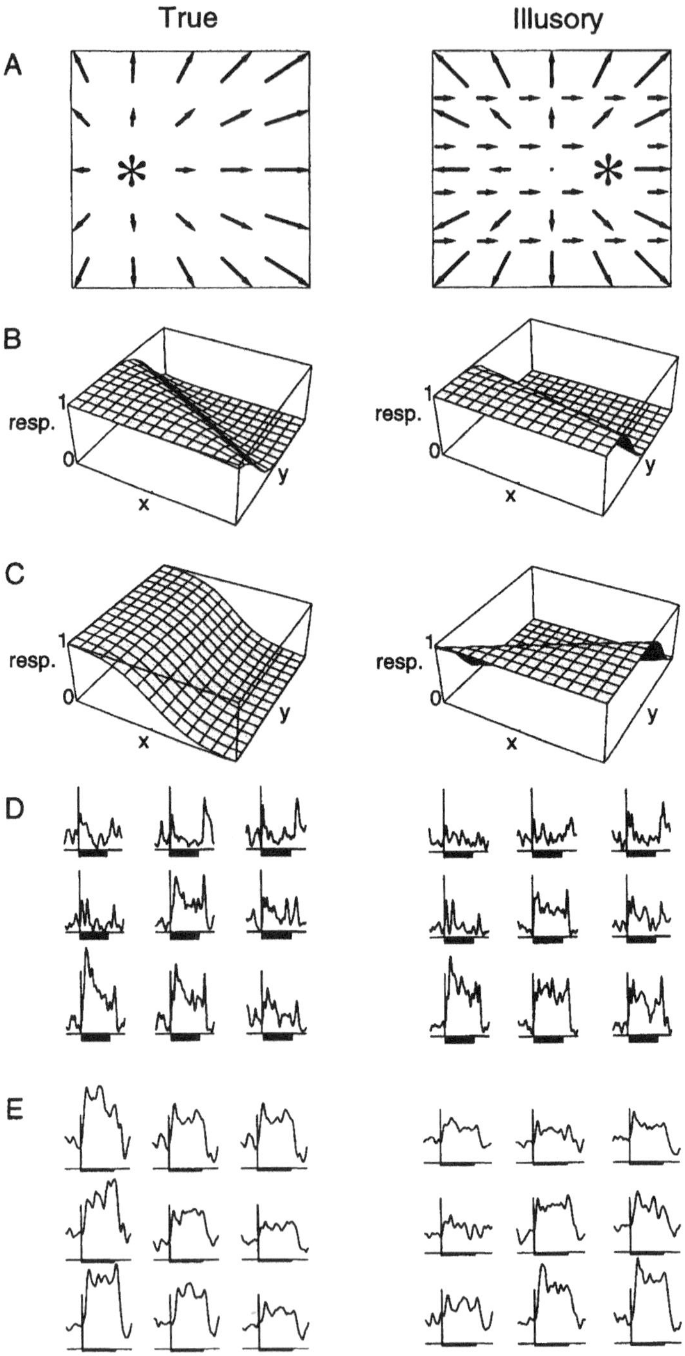
True
Illusory
A
B
resp.
1
0
x
y
C
D
E

the overlap of the response profiles of individual neurons. The shift of the population response is therefore the result of the combination of many individual response profiles (Fig. 7). Model neurons are grouped in populations that encode different directions of heading, i.e., different locations of the focus of expansion. For the vector-summed stimuli, which contain a true eccentric focus of expansion, the response profiles of all neurons in one such population are arranged such that the neuronal responses cohere maximally when the focus of expansion is at that position. In the transparent overlap condition, the individual response profiles are rotated. Maximum coherence now occurs for the opposite location in the visual field (i.e., at the illusory focus position).

We represented the vector sum and transparent overlap stimuli to 228 MST neurons and compared the results to the analysis of 300 model neurons using comparable stimuli (Lappe and Duffy, 1999). The model not only predicted the illusory shift of the center of expansion but also predicted the behavior of individual MST neurons (Figs. 6B–6E). The findings are compatible with a population code for heading but not with the prediction from template matching. For the illusory stimuli, the behavior of single MST neurons is qualitatively different from the perceptual findings. The perceived center of motion in the vector sum and the transparent overlap conditions shift, whereas the response profiles of

---

FIG. 6. Neuronal responses to true and illusory displacements of the focus of expansion (Lappe and Duffy, 1999). (A) The illusion occurs when a radial and a unidirectional motion pattern are presented simultaneously as a transparent overlapping motion (right panel) (Duffy and Wurtz, 1993). The focus of the expansion is perceived to be displaced in the direction of the overlapping unidirectional motion, which is to the right in this example. This asterisk indicates the perceived location of the focus of expansion. The illusory focus position is opposite the focus position obtained when the two motion patterns are simply vector summed (left). In this case, a pure expansion stimulus is generated with a true focus on the left. (B, C) Single-model neurons respond differently in the two conditions but their responses do not exhibit a similar shift. Neuron B responds maximally when the true center of expansion is located in the lower-left hemifield. In the illusory condition, the neuron responds strongest when upward motion is presented transparently overlapping (right). Perceptually, and also at the population level in the model, this motion pattern results in an upward displacement of the center of expansion, opposite from the location of the focus in the pure expansion stimuli. In contrast, the response profile of the neuron is merely slightly rotated between the two conditions. C shows an example of a larger rotation of the response profile but still not a complete reversal. (D, E) Spike density histograms show mean responses over six presentations of each stimulus for two different MST neurons. Neuron D responded similarly to the two stimulus sets, much like the model neuron in B. Neuron E shows different behavior. It responded best to left centers of motion in the pure expansion stimuli and for up- and leftward motion in the transparent overlap condition. The response behaviors of model and MST neurons are very similar.

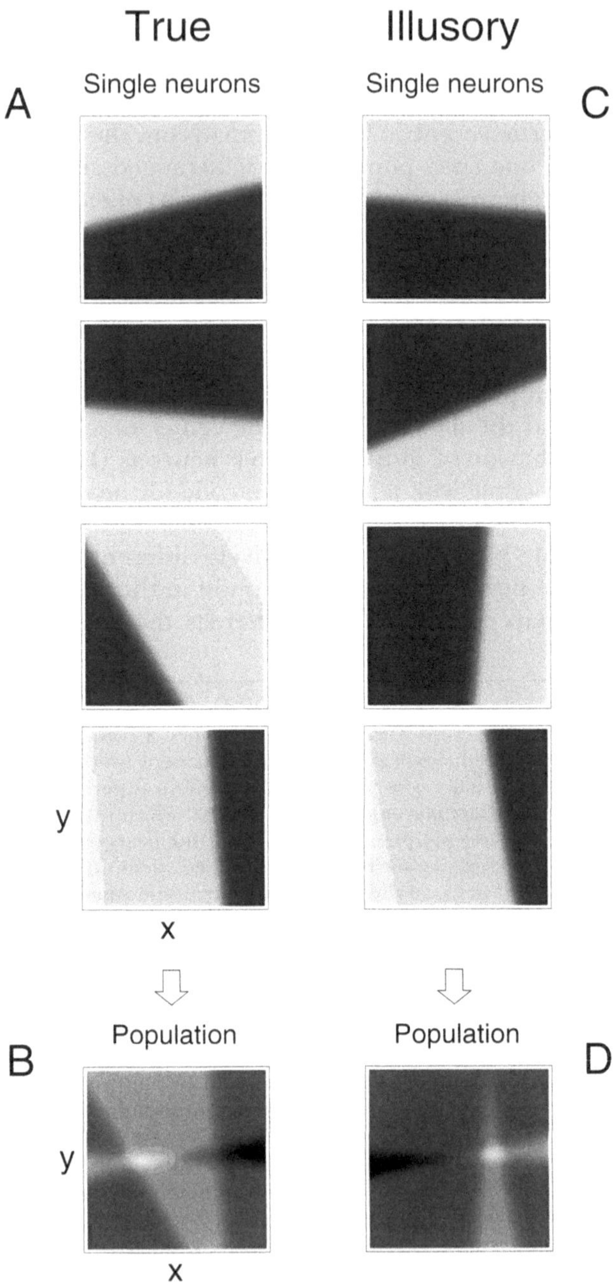
True
Illusory
Single neurons
Single neurons
A
C
y
x
Population
Population
B
D
y
x

single neurons in the two conditions are rotated against each other. The population heading map reconciles this apparent mismatch. It reproduces both the single neuron behavior and the perceptual shift. This shows that the graded response profile rotations observed in MST can provide enough modulation to the distribution of neural activity to induce the illusory shift. The population heading map suggests that the distribution of rotation angles, not the individual rotation of individual neurons, subserves the perceptual effect. We therefore compared the distribution of rotations of the response profiles of single MST neurons to that in the model. The distribution of rotation angles in MST closely matched the model prediction (Lappe and Duffy, 1999).

The illusory stimuli contain large differences in direction and speed of individual dot motions. This is similar to strong motion parallax. The rotation of the response profiles of single model neurons is related to their analysis of motion parallax. A similar response profile rotation in model and MST neurons is also observed with more realistic motion parallax stimuli which simulate self-motion in an depth-rich environment containing (Lappe *et al.*, 1996; Pekel *et al.*, 1996). Motion parallax is an important cue for compensating eye-movement-induced perturbations of the retinal flow. Such perturbations displace the retinal projection of the center of expansion (see van den Berg, this volume, or Lappe and Hoffmann, this volume). Model neurons use motion parallax to shift the perceived center of expansion (the population response) back to the correct position. The magnitude of the illusory shift in humans is consistent with such a visual compensation mechanism (Lappe and Rauschecker, 1995a). The similarity between the behavior of single MST and model neurons could suggest that MST can use motion parallax cues to compensate for eye movements, but this compensation occurs at the population level.

FIG. 7. Schematic illustration of the population encoding of the perceived shift of the center of expansion (Lappe and Duffy, 1999). We consider four individual neurons that form a single population tuned to leftward heading (see also Figs. 1D and 1E). (A) Individual response profiles for true displacements of the focus of expansion. Grayscale maps correspond to viewing 3-D surface plots as in Fig. 6 from above. Brightness represents response activity. (B) Population response profile obtained from summing the responses of the four neurons. The individual response profiles in A are differently oriented such that they maximally overlap at a position left of the center. The population reaches peak activity when the center of motion is at that point (i.e., heading is to the left). (C) Response profiles of the same four neurons for a presentation of the illusory stimuli. Individual response gradients are rotated by different amounts. (D) The summation results in a different population response profile. Maximum activity (i.e., optimum overlap of the rotated profiles) now occurs on the right.

The population heading map might also explain puzzling results of microstimulation in MST (Britten and van Wezel, 1998). A monkey had to discriminate between leftward and rightward heading in optic-flow flow stimuli. Electrical stimulation was applied in order to influence heading judgments and thereby demonstrate an involvement of area MST in this task. Stimulation at sites within area MST that contained many neurons which responded best to the same heading (e.g., leftward) biased the behavioral choices of the monkey toward the predicted (leftward) heading in only 67% of the cases. In the remaining 33%, there was instead a bias in the opposite direction (Britten and van Wezel, 1998). How can the population heading map explain such a result? The prediction of the effect of microstimulation in the experiment was estimated from the preferences of the stimulated neurons. When several neurons within a recording penetration respond strongest to leftward heading, it was assumed that this site only contributed to the percept of leftward heading. However, neuronal populations in the heading map are not formed by neurons with the same best response (the neurons in Fig. 1D clearly have different best responses). Instead they are formed by neurons with optimum overlap of their response profiles. In the population heading map, the effects of microstimulation cannot be predicted from observations of the tuning of individual neurons but only from the populations. Microstimulation in the population heading map could shift the population response in a direction different from the best response of individual neurons. When only two directional choices are available (left/right), a considerable amount of responses in the apparently opposite direction would be expected. Further experiments will establish more precisely this global organization of optic flow processing in population of neurons.

VIII. Conclusion

The neuronal analysis of optic flow is a task which can only be understood by a multidisciplinary approach combining theoretical and experimental work. A number of models of optic flow processing in primate cortex that can replicate the basic response properties of neurons in areas MT and MST have been proposed. Several of these models appear to use similar basic mechanisms for the analysis of motion parallax. They differ, however, in the way by which these basic measurements are combined. Experimental findings together with modeling considerations suggest that heading in area MST is represented by a population map.

Acknowledgments

I am grateful to Jaap Beintema for comments on the manuscript and to the Human Frontier Science Program and the Deutsche Forschungsgemeinschaft for financial support.

References

Albright, T. D. (1989). Centrifugal directionality bias in the middle temporal visual area (MT) of the macaque. *Vis. Neurosci.* **2,** 177–188.

Albright, T. D., and Desimone, R. (1987). Local precision of visuotopic organization in the middle temporal area (MT) of the macaque. *Exp. Brain Res.* **65,** 582–592.

Albright, T. D., Desimone, R., and Gross, C. G. (1984). Columnar organization of directionally selective cells in visual area MT of the macaque. *J. Neurophysiol.* **51,** 16–31.

Allman, J. M., Miezin, F., and McGuinness, E. (1985). Stimulus specific responses from beyond the classical receptive field: Neurophysiological mechanisms for local-global comparisons in visual neurons. *Ann. Rev. Neurosci.* **8,** 407–430.

Beardsley, S. A., and Vaina, L. M. (1998). Computational modelling of optic flow selectivity in MSTd neurons. *Network* **9,** 467–493.

Beintema, J., and van den Berg, A. V. (1998). Heading detection using motion templates and eye velocity gain fields. *Vision Res.* **38,** 2155–2179.

Bradley, D., Maxwell, M., Andersen, R., Banks, M. S., and Shenoy, K. V. (1996). Mechanisms of heading perception in primate visual cortex. *Science* **273,** 1544–1547.

Bradley, D., Qian, N., and Andersen, R. (1995). Integration of motion and stereopsis in middle temporal cortical area of macaques. *Nature* **373,** 609–611.

Bremmer, F., Duhamel, J.-R., Ben Hamed, S., and Graf, W. (1997). The representation of movement in near extrapersonal space in the macaque ventral intraparietal area (VIP). *In:* "Parietal Lobe Contributions to Orientation in 3D-Space," (P. Thier and H.-O. Karnath, Eds.), vol. 25 of Exp. Brain Res. Ser., pp. 619–630. Springer, Heidelberg.

Britten, K. H. (1998). Clustering of response selectivity in the medial superior temporal area of extrastriate cortex in the macaque monkey. *Vis. Neurosci.* **15,** 553–558.

Britten, K. H., and van Wezel, R. J. A. (1998). Electrical microstimulation of cortical area MST biases heading perception in monkeys. *Nat. Neurosci.* **1,** 59–63.

Bruss, A. R., and Horn, B. K. P. (1983). Passive navigation. *Comp. Vis. Graph Image Proc.* **21,** 3–20.

Crowell, J. (1997). Testing the Perrone and Stone (1994) model of heading estimation. *Vision Res.* **37,** 1653–1671.

Desimone, R., and Ungerleider, L. G. (1986). Multiple visual areas in the caudal superior temporal sulcus of the macaque. *J. Comp. Neurol.* **248,** 164–189.

Duffy, C. J., and Wurtz, R. H. (1991a). Sensitivity of MST neurons to optic flow stimuli. I. A continuum of response selectivity to large-field stimuli. *J. Neurophysiol.* **65**, 1329–1345.

Duffy, C. J., and Wurtz, R. H. (1991b). Sensitivity of MST neurons to optic flow stimuli. II. Mechanisms of response selectivity revealed by small-field stimuli. *J. Neurophysiol* **65**, 1346–1359.

Duffy, C. J., and Wurtz, R. H. (1993). An illusory transformations of optic flow fields. *Vision Res.* **33,** 1481–1490.

Duffy, C. J., and Wurtz, R. H. (1995). Response of monkey MST neurons to optic flow stimuli with shifted centers of motion. *J. Neurosci.* **15,** 5192–5208.

Dürsteler, M. R., and Wurtz, R. H. (1988). Pursuit and optokinetic deficits following chemical lesions of cortical areas MT and MST. *J. Neurophysiol.* **60,** 940–965.

Erickson, R. G., and Thier, P. (1991). A neuronal correlate of spatial stability during periods of self-induced visual motion. *Exp. Brain Res.* **86,** 608–616.

Fermüller, C., and Aloimonos, Y. (1995). Direct perception of three-dimensional motion from patterns of visual motion. *Science* **270,** 1973–1976.

Geesaman, B. J., Born, R. T., Andersen, R. A., and Tootell, R. B. H. (1997). Maps of complex motion selectivity in the superior temporal cortex of the alert macaque monkey: A double-label 2-deoxyglucose study. *Cereb. Cortex* **7,** 749–757.

Graziano, M. S. A., Andersen, R. A., and Snowden, R. (1994). Tuning of MST neurons to spiral motions. *J. Neurosci.* **14**(1), 54–67.

Hatsopoulos, N. G., and Warren, Jr., W. H. (1991). Visual navigation with a neural network. *Neural Networks* **4**(3), 303–318.

Heeger, D. J., and Jepson, A. (1992a). Recovering observer translation with center-surround motion-opponent mechanisms. *Invest. Ophthalmol. Vis. Sci. Suppl.* **32,** 823.

Heeger, D. J., and Jepson, A. (1992b). Subspace methods for recovering rigid motion I: Algorithm and implementation. *Int. J. Comput. Vision* **7**, 95–117.

Hildreth, E. C. (1992a). Recovering heading for visually-guided navigation. *Vision Res.* **32,** 1177–1192.

Hildreth, E. C. (1992b). Recovering heading for visually-guided navigation in the presence of self-moving objects. *Philos. Trans. Roy. Soc. Lond. B* **337,** 305–313.

Hoffmann, K.-P., Distler, C., and Ilg, U. (1992). Callosal and superior temporal sulcus contributions to receptive field properties in the macaque monkey's nucleus of the optic tract and dorsal terminal nucleus of the accessory optic tract. *J. Comp. Neurol.* **321,** 150–162.

Ilg, U. J., and Thier, P. (1997). MST neurons are activated by pursuit of imaginary targets. *In:* "Parietal Lobe Contributions to Orientation in 3D-Space," P. Thier, and H.-O. Karnath, Eds.), pp. 173–184. Springer, Berlin, Heidelberg, New York.

Ilg, U. (1997). Responses of primate area MT during the execution of optokinetic nystagmus and afternystagmus. *Exp. Brain Res.* **113,** 361–364.

Inoue, Y., Takemura, A., Kawano, K., Kitama, T., and Miles, F. A. (1998). Dependence of short-latency ocular following and associated activity in the medial superior temporal area (MST) on ocular vergence. *Exp. Brain Res.* **121,** 135–144.

Kawano, K., Shidara, M., Watanabe, Y., and Yamane, S. (1994). Neural activity in cortical area MST of alert monkey during ocular following responses. *J. Neurophysiol.* **71,** 2305–2324.

Koenderink, J. J., and van Doorn, A. J. (1987). Facts on optic flow. *Biol. Cybern.* **56,** 247–254.

Komatsu, H., and Wurtz, R. H. (1989). Modulation of pursuit eye movements by stimulation of cortical areas MT and MST. *J. Neurophysiol* **62,** 31–47.

Lagae, L., Maes, H., Raiguel, S., Xiao, D.-K., and Orban, G. A. (1994). Responses of macaque STS neurons to optic flow components: A comparison of areas MT and MST. *J. Neurophysiol.* **71,** 1597–1626.

Lappe, M. (1996). Functional consequences of an integration of motion and stereopsis in area MT of monkey extrastriate visual cortex. *Neural Comp.* **8,** 1449–1461.

Lappe, M. (1998). A model of the combination of optic flow and extraretinal eye movement signals in primate extrastriate visual cortex. *Neural Networks* **11,** 397–414.

Lappe, M., Bremmer, F., and Hoffmann, K.-P. (1994). How to use non-visual information for optic flow processing in monkey visual cortical area MSTd. *In:* ICANN 94—Proceedings of the International Conference on Artificial Neural Networks, 26–29 May 1994, Sorrento, (M. Marinaro and P. G. Morasso, Eds.), pp. 46–49. Springer, Berlin, Heidelberg, New York.

Lappe, M., Bremmer, F., Pekel, M., Thiele, A., and Hoffmann, K.-P. (1996). Optic flow processing in monkey STS: A theoretical and experimental approach. *J. Neurosci.* **16,** 6265–6285.

Lappe, M., & Duffy, C. (1999). Optic flow illusion and single neuron behavior reconciled by a population model. *Eur. J. Neurosci.* **11,** 2323–2331.

Lappe, M., Pekel, M., and Hoffmann, K.-P. (1988). Optokinetic eye movements elicited by radial optic flow in the macaque monkey. *J. Neurophysiol.* **79,** 1461–1480.

Lappe, M., and Rauschecker, J. P. (1993a). Computation of heading direction from optic flow in visual cortex. *In:* "Advances in Neural Information Processing Systems," vol. 5 (C. L. Giles, S. J. Hanson, and J. D. Cowan, Eds.), pp. 433–440. Morgan Kaufmann, San Mateo, CA.

Lappe, M., and Rauschecker, J. P. (1993b). A neural network for the processing of optic flow from ego—motion in man and higher mammals. *Neural Comp.* **5,** 374–391.

Lappe, M., and Rauschecker, J. P. (1994). Heading detection from optic flow. *Nature* **369,** 712–713.

Lappe, M., and Rauschecker, J. P. (1995a). An illusory transformation in a model of optic flow processing. *Vision Res.* **35,** 1619–1631.

Lappe, M., and Rauschecker, J. P. (1995b). Motion anisotropies and heading detection. *Biol. Cybern.* **72,** 261–277.

Longuet-Higgins, H. C., and Prazdny, K. (1980). The interpretation of a moving retinal image. *Proc. Roy. Soc. Lond. B* **208,** 385–397.

Maunsell, J. H. R., and Van Essen, D. C. (1983). Functional properties of neurons in middle temporal visual area of the macaque monkey. I. Selectivity for stimulus direction, speed, and orientation. *J. Neurophysiol.* **49**(5), 1127–1147.

Newsome, W. T., Wurtz, R. H., and Komatsu, H. (1988). Relation of cortical areas MT and MST to pursuit eye movements. II. Differentiation of retinal from extraretinal inputs. *J. Neurophysiol.* **60**(2), 604–620.

Page, W. K., and Duffy, C. J. (1999). MST neuronal responses to heading direction during pursuit eye movements. *J. Neurophysiol.* **81,** 596–610.

Pekel, M., Lappe, M., Bremmer, F., Thiele, A., and Hoffmann, K.-P. (1996). Neuronal responses in the motion pathway of the macaque to natural optic flow stimuli. *NeuroReport* **7**, 884–888.

Perrone, J. A. (1992). Model for the computation of self-motion in biological systems. *J. Opt. Soc. Am* **9**, 177–194.

Perrone, J. A., and Stone, L. S. (1994). A model of self-motion estimation within primate extrastriate visual cortex. *Vision Res.* **34,** 2917–2938.

Perrone, J. A., and Stone, L. S. (1998). Emulating the visual receptive field properties of MST neurons with a template model of heading estimation. *J. Neurosci.* **18,** 5958–5975.

Raiguel, S., Van Hulle, M., Xiao, D., Marcar, V., and Orban, G. (1995). Shape and spatial distribution of receptive fields and antagonistic motion surrounds in the middle temporal area (V5) of the macaque. *Eur. J. Neurosci.* **7,** 2064–2082.

Rieger, J. H., and Lawton, D. T. (1985). Processing differential image motion. *J. Opt. Soc. Am. A* **2,** 354–360.

Royden, S. C. (1997). Mathematical analysis of motion-opponent mechanisms used in the determination of heading and depth. *J. Opt. Soc. Am. A* **14,** 2128–2143.

Sakata, H., Shibutani, H., and Kawano, K. (1983). Functional properties of visual tracking neurons in posterior parietal association cortex of the monkey. *J. Neurophysiol.* **49,** 1364–1380.

Schaafsma, S., and Duysens, J. (1996). Neurons in the ventral intraparietal area of awake macaque monkey closely resemble neurons in the dorsal part of the medial superior temporal area in their responses to optic flow patterns. *J. Neurophysiol.* **76,** 4056–4068.

Sinclair, D., Blake, A., and Murray, D. (1994). Robust estimating of egomotion from normal flow. *Int. J. Computer Vision* **13,** 57–69.

Tanaka, K., and Saito, H.-A. (1989a). Analysis of motion of the visual field by direction, expansion/contraction, and rotation cells clustered in the dorsal part of the medial superior temporal area of the macaque monkey. *J. Neurophysiol.* **62**, 626–641.

Tanaka, K., and Saito, H.-A. (1989b). Underlying mechanisms of the response specificity of expansion/contraction and rotation cells in the dorsal part of the medial superior temporal area of the macaque monkey. *J. Neurophysiol.* **62**, 642–656.

van den Berg, A. V., and Beintema, J. (1997). Motion templates with eye velocity gain fields for transformation of retinal to head centric flow. *NeuroReport* **8,** 835–840.

van den Berg, A. V., and Brenner, E. (1994). Why two eyes are better than one for judgements of heading. *Nature* **371,** 700–702.

Wang, R. (1995). A simple competitive account of some response properties of visual neurons in area MSTd. *Neural Comp.* **7,** 290–306.

Warren, W. H., and Saunders, J. A. (1995). Perceiving heading in the presence of moving objects. *Perception* **24,** 315–331.

Warren, W. H., and Hannon, D. J. (1990). Eye movements and optic flow. *J. Opt. Soc. Am. A* **7**, 160–169.

Zemel, R. S., and Sejnowski, T. J. (1998). A model for encoding multiple object motions and self-motion in area MST of primate visual cortex. *J. Neurosci.* **18,** 531–547.

Zhang, K., Sereno, M. I., and Sereno, M. E. (1993). Emergence of position-independent detectors of sense of rotation and dilation with hebbian learning: An analysis. *Neural Comp.* **5,** 597–612.

HUMAN CORTICAL AREAS UNDERLYING THE PERCEPTION OF OPTIC FLOW: BRAIN IMAGING STUDIES

Mark W. Greenlee

Department of Neurology, University of Freiburg, Germany

I. Introduction

Visual motion processing is believed to be a function of the dorsal visual pathway, where information from V1 passes through V2 and V3 and onto medial temporal (MT) and medial superior temporal (MST) areas, also referred to as V5/V5a, for motion analysis (Zeki, 1971, 1974, 1978; Van Essen *et al.*, 1981; Albright, 1984; Albright *et al.*, 1984). From there, the motion information passes to cortical regions in the parietal cortex as part of an analysis of spatial relationships between objects in the environment and the viewer (Andersen, 1995, 1997; Colby, 1998). Additional information is passed to the frontal eye fields (FEF) in prefronal cortex (lateral part of area 6) and is used in the preparation of saccadic and smooth pursuit eye movements (Schiller *et al.*, 1979; Bruce *et al.*, 1985; Lynch, 1987).

In this chapter, we review electrophysiological and brain-imaging studies that have investigated the cortical responses to visual motion and optic flow. The aim of this chapter is to determine the extent to which the cortical responses, as indexed by stimulus-evoked changes in blood flow and tissue oxygenation, are specific to optic flow fields. We first review the current literature on electrophysiological recordings in monkey cortex and functional imaging of human cortical responses to visual mo-

0074-7742/00 $30.00

tion and optic flow. A brief review is given of the studies on the effect of focal attention on brain activation during visual motion stimulation. We also describe the method of retinotopic mapping, which has been used in the past to mark border regions between retinotopically organized visual cortex. We next discuss methods used to record eye movements during brain-imaging experiments. In a recent study, we apply these methods to better understand how different visual areas respond to optic flow and the extent to which responses to flow fields can be modulated by gradients in speed and disparity.

A. Electrophysiological Studies

Electrophysiological studies applying single- and multiunit recording techniques have been conducted, before and after cortical lesioning, in behaving monkeys that were trained to perform motion discrimination tasks (Newsome *et al.*, 1985; Newsome and Paré, 1988; Britten *et al.*, 1992). Lesions in the areas MT/MST (V5/V5a) lead to an impairment in the ability of monkeys to discriminate between different directions and speeds of random dot motion stimuli (Pasternak and Merigan, 1994; Orban *et al.*, 1995). The MST neurons appear to be particularly well suited for the task of analyzing optic flow fields with their large receptive fields, which are often greater than 20° and extend into the ipsilateral hemifield. Indeed single-unit responses in MST have been shown to depend on the complex motion gradients present in optic flow fields (Tanaka *et al.*, 1986, 1989; Tanaka and Saito, 1989; Lappe *et al.*, 1996; Duffy and Wurtz, 1997; for a review see Duffy, this volume).

Recent investigations suggest that neurons in area MT are also sensitive to the disparity introduced in motion stimuli (DeAngelis and Newsome, 1999). Electrical stimulation of MT neurons, found to be sensitive to binocular disparity, affects the depth judgments of monkeys performing visual motion discrimination tasks. Binocular disparity could potentially provide an important cue in resolving depth information inherent in dichoptically presented motion sequences. Recent work from Bradley and Andersen (1998) suggest that neurons in area MT can make use of binocular disparity to define motion planes of different depths.

Electrical stimulation of MST neurons shifts the heading judgments made by monkeys while they viewed optic flow fields (Britten and van Wezel, 1998). Possible interactions between heading judgments and pursuit eye movements in optic flow fields on the responses of MST neurons have recently been studied by Page and Duffy (1999). They found a significant effect of pursuit eye movements on the responses of MST neurons (see Section II.B).

B. Brain Imaging Studies of Motion Perception

Various attempts have been made in the last 10 years to characterize motion-sensitive cortical regions in human visual cortex. These methods have been based primarily on EEG/MEG, PET, and fMRI techniques. The most important studies are reviewed in this section.

1. *EEG/MEG*

Earlier visually evoked potential work points to a negative potential with a latency of around 200 ms, which has been related to the motion-onset of grating patterns (Müller *et al.,* 1986; Müller and Göpfert, 1988). Comparisons between pattern-evoked responses and motion-onset VEPs point to a more lateral location in occipitotemporal cortex of the latter (Göpfert *et al.,* 1999), and the amplitude of this component is reduced by prior motion adaptation (Bach and Ullrich, 1994; Müller *et al.,* 1999).

Studies that use multichannel electroencephalography (EEG) and magnetoencephalography (MEG) in subjects who viewed moving gratings or dot motion displays have been conducted. Probst *et al.* (1993) located an electrical dipole in the temporoparietooccipital (TPO) region in the contralateral hemisphere for eccentric hemifield stimulation, which was associated with visual motion. Anderson *et al.* (1996) measured multichannel MEG responses to sinewave grating motion and found dipoles centered in the occipitotemporal cortex. The location varied somewhat across subjects, and response amplitude showed some degree of stimulus selectivity. All these studies point to the occipitotemporal and/or temporoparietooccipital junction region as the site for the human V5/V5a areas. The higher temporal resolution makes the EEG/MEG methods attractive, despite their lower spatial resolution compared to PET and fMRI.

2. *PET*

Zeki *et al.* (1991) used positron emission tomography (PET) with the short-life radioactive tracer $H_2{}^{15}O$ to map cortical responses to random dot motion (1° wide black squares on white background). Their stimuli moved with a speed of 6°/s in one of eight directions. They found significant responses in the human homologue of area V5/V5a. Watson *et al.* (1993) and Dupont *et al.* (1994) could replicate and expand these findings.

De Jong *et al.* (1994) used $H_2{}^{15}O$-PET to explore the hemodynamic correlates of the cortical response to optic flow fields. Six subjects viewed simulated optic flow fields (consisting of small bright dots on a dark background) under binocular viewing conditions. Comparisons were made between displays with 100% coherent motion (radial expansion from a virtual horizon) and 0% coherent motion (same dots and speed gradients, but random direction). The average speed was 7.6°/s (coherent motion) and 17.8°/s (random condition). Anatomical MRI was conducted in four of the six subjects. Overlays of the functional activations were then superimposed onto the anatomical MR images. The reported Talairach coordinates (based on 3-D stereotactic atlas of Talairach and Tournoux, 1988) correspond to the human V5/V5a complex (MT/MST) in the border region between areas 19 and 37, in the inferior cuneus in area 18 (the human homologue of V3), in the insular cortex, and in the lateral extent of the posterior precuneus in occipitoparietal cortex (areas 19/7). Cheng *et al.* (1995) had ten subjects monocularly view an 80° (virtual) field, while luminous dots moved coherently in one of eight directions. The control conditions consisted either of incoherent motion sequences or mere fixation. The authors used electrooculography (EOG) to control for eye movements during the PET scans. The results indicate that several visual areas respond to visual motion stimuli. Some of the occipitotemporal (V5/V5a, BA 19/37) and occipitoparietal (V3A, BA 7) responses were more pronounced during coherent motion perception.

3. *fMRI*

Using functional magnetic resonance imaging (fMRI), Tootell *et al.* (1995a) mapped BOLD responses in striate and extrastriate visual areas to visual motion stimuli (expanding–contracting radial gratings). They found that the human V5/V5a responds well to low-stimulus contrast levels and saturates already at 4% contrast. Prior adaptation to unidirectional motion prolongs the decay of the BOLD response in V5/V5a, which has been related to the perceptual motion aftereffect (Tootell *et al.*, 1995b). In a further study, Reppas *et al.* (1997) found that the human homologue of V3a is more sensitive to motion-defined borders than V5/V5a. Tootell *et al.* (1997) reported that V3a responds well to both high- and low-contrast motion stimuli. It is obvious from these initial studies that several areas in visual extrastriate and associational cortex respond selectively to visual motion.

Orban and collaborators have performed several extensive studies on the effects of direction and speed of frontoplanar dot motion, using PET and fMRI methods (Dupont *et al.*, 1994, 1997; van Oostende *et al.*, 1997; Cornette *et al.*, 1998; Orban *et al.*, 1998). Subjects performed psychophysical tasks of direction and speed discrimination during scanning. Careful documentation of the visual areas responding to various forms of dot motion suggests that several areas beyond V5/V5a, including areas in the lingual gyrus and cuneus, show selective responses to the direction and speed of visual motion. Orban and colleagues identified an area in the ventral portion of extrastriate cortex, which they refer to as the kinetic occipital (KO) cortex (Dupont *et al.,* 1994; van Oostende *et al.,* 1997). KO responds well to motion-defined borders within complex motion displays.

Using fMRI and flat maps of visual cortex, we have compared BOLD responses to first-order (luminance-derived) and second-order (contrast-derived) motion stimuli (Smith *et al.,* 1998). Our results suggest that several areas respond to both types of motion. Area KO, which we refer to as V3b, appears to respond preferentially to certain types of second-order motion. These results are in agreement with results from a study on patients with focal lesions in occipitotemporal and occipitoparietal cortex (Greenlee and Smith, 1997). Impairments in the ability to discriminate the speed of first- and second-order motion stimuli were highly correlated among the patients, suggesting a common pathway for both types of motion.

C. Effects of Attention on rCBF/BOLD Responses to Visual Motion

The effects of selective and divided attention have been studied using both PET and fMRI methods. Corbetta *et al.* (1991) had subjects attend to either the speed, color, or shape of sparse randomly moving blocks. In the selective attention condition, subjects were instructed to attend to one of the three stimulus dimensions and the stimuli differed only along that dimension. In the divided attention condition, the stimuli could differ along any one of the three dimensions and the subjects had to detect whether a change occurred or not. In the selective attention condition, the authors found a shift in activation depending on the stimulus dimension to which the subject attended. When subjects attended to the speed of the moving stimuli, the activation occurred in lateral occipitotemporal cortex (most likely in the V5/V5a region, but also in more anterior regions in BA 21 and BA 22).

The subject's attention level has been shown to affect the BOLD response in fMRI experiments. Beauchamp *et al.* (1997) presented subjects with complex motion displays. Random dot motion was sequentially interleaved with motion displays containing a circular annulus defined by coherently moving dots. The subjects were instructed to attend to the central fixation point and passively view the motion. In two further conditions, subjects were instructed either to attend to both the location and speed of the dots within the annulus or to attend only to the color of the dots within the annulus. The BOLD signal in the human homologue of V5/V5a was highest for the condition with attention to both speed and location, and the response decreased to 60 and 45% when attention was shifted to the dot color or the fixation point.

O'Craven *et al.* (1997) asked subjects to attend either to moving or static random dots (black dots moving among static white dots). This dynamic stimulus remained constant during the entire MR-image acquistion period. During alternating intervals, subjects attended either to the static or the dynamic dots (cued by an instruction). The authors found a modulation in the BOLD signal depending on which instruction set the subjects followed: attention to the dynamic components of the displays led to larger responses in the V5/V5a region. In a similar fashion, Gandhi *et al.* (1999) instructed subjects to attend to a cued moving target (a grating moving within a semicircular window). BOLD signals were largest when subjects attended to the grating in the contralateral visual field. The effect was about 25% of the stimulus-evoked response (driven by left-right alternation of the grating stimulus). Similar effects have been reported for spatial attention to flickering checkerboard patterns (Tootell *et al.*, 1998).

In summary, the effects of attention tend to raise the BOLD signal in association with the visual stimulus. The effects reported so far vary between 25 and 50% of the motion-evoked response. Similar effects have been reported for static patterns (Kastner *et al.*, 1998). Based on the results to the studies reviewed earlier, attention appears to modulate a stimulus-evoked response, but it has not been shown to evoke responses in otherwise silent areas.

II. New Techniques in Brain Imaging

Brain imaging studies are highly reliant on the postprocessing of image data. Several methods for extracting the BOLD signal from T2*-weighted MR-images have been presented (Friston *et al.*, 1995; Sereno *et al.*, 1995; Cox, 1996; Engel *et al.*, 1997; Smith *et al.*, 1998). We cannot go into detail here on these various methods but rather would like to describe briefly two techniques that are relevant to the study of optic flow perception.

The first method concerns the demarcation of retinotopically defined visual areas in striate and extrastriate visual cortex. By applying stimuli located in different parts of the visual field sequentially, a retinotopic map of the left and right visual cortex can be derived. We are able to use this information to compare the responses to visual motion and optic flow stimuli in the first five visual areas (V1–V5) in human cortex.

A further method for monitoring eye movements in the MR-scanner is presented. An important prerequisite for interpreting the results of visual experiments is to know to what extent the subjects moved their eyes during stimulation and rest periods. Using a infrared-reflection technique we can accurately monitor with high temporal and spatial resolution the horizontal components of eye movements. By this means, we can compare fixation and pursuit during visual motion and optic-flow stimulation.

A. Retinotopic Mapping of Visual Areas

The human cerebral cortex is approximately 2 mm in depth. It is bordered by the white matter on one side and the cerebral spinal fluid and supportive brain tissues (pia mater, arachnoidea, dura mater) on the other side. The cortical surface is rarely flat. Rather the cortex is a complexly folded 3-D structure. Several visual areas are located in the posterior portion of the cerebral cortex. Within each early visual area, the visual field is represented according to the meridian angle and eccentricity, which is referred to as retinotopy (Van Essen and Zeki, 1978). To view more readily the results of functional imaging studies, this complex structure can be unfolded into the form of a "flat map." Various techniques have been put forth for the segmentation and unfolding of the cortex (Dale *et al.*, 1999; Fischl *et al.*, 1999; Sereno *et al.*, 1995; Teo *et al.*, 1997). The basic idea is that the gray matter should be segmented first from the other tissues and fluids, and then this segmented cortical sheet should be interconnected with the least amount of spatial distortion. Afterward, this interconnected grid can be flattened to make a flat map of that part of the cortex. Once this flat map has been obtained, the functional results from fMRI experiments can be mapped onto the surface of this grid using the resultant transformation matrix. An example of the segmentation of the left hemisphere of one subject is shown in Fig. 1 (see color insert). We used the program MrGray from the Stanford group (e.g., Teo *et al.*, 1997). Voxel intensities are chosen such as to optimally classify grey and white matter. Some interactive correction can be performed on a slice-by-slice basis to improve the result. This segmented matrix is stored in a file, which is subsequently used in the MatLab program MR-Unfold from Wandell and colleagues (Teo *et al.*, 1997).

The two major techniques of sequential retinopic mapping are based on phase and eccentricity encoding. Examples of each of these methods are given in Fig. 2 (see color insert). One segment of a radial checkerboard is presented on a medium gray background. While the subject views the central fixation dot, the checkerboard segment shifts either in a clockwise or counterclockwise direction. The checkerboard flickers at 8 Hz during its presentation. Each step can be synchronized with the image acquisition, so that the exact timing of each image can be determined and accounted for in the analysis of the fMRI responses. Since the blood oxygen level responses have been shown to be sluggish, with a time constant around 6 s, each revolution of the checkerboard segment is made over 54 s (18 positions, each position shown for 3 s). The segment is rotated a total of four times. This yields a BOLD response with a total of four periods with a period length of 54 s. By extracting the phase of the fundamental response, we can track the phase as it shifts over the cortical surface (see color key in Fig. 2a). For example, if we begin by stimulating the upper vertical meridian of the right visual field (RVF) and then shift the checkerboard wedge in a clockwise direction, we can map the responses of the right upper visual quadrant by plotting the phase of the response at each location within the left visual cortex. Since the upper vertical meridian is known to correspond to the V1-V2 border of the ventral visual cortex and the horizontal meridian to the fundus of the calcarine, we can follow the representation of this visual quadrant in V1 by plotting the temporal phase of the response within the calcarine fissure. The border region between V1 and V2 is demarcated by a change in the phase angle of the fundamental response, such that V1 is characterized by horizontal-to-vertical meridian phase changes, whereas V2 is marked by a vertical-to-horizontal phase transition (Engel *et al.*, 1994, 1997; Sereno *et al.*, 1995). An example for one subject is given in Fig. 2a. The first three visual areas can be clearly demarcated using this method. With this information, the responses of the different visual areas to optic flow, and other forms of visual motion stimulation, can be determined. Figure 2b shows the results of the eccentricity-encoding method.

B. Eye Movement Recording during Brain Imaging

An obvious source of experimental error in imaging studies of visual processing is the extent to which the subjects move their eyes during the scanning period. Although movements of the head are restrained by various methods, and the effects of head motion can be partially eliminated by postacquisition motion correction (Cox, 1996; Woods *et al.*, 1998a, b; Kraemer and Hennig, 1999), the effects of eye movements have been largely ignored in the past. Some form of prescan training has been em-

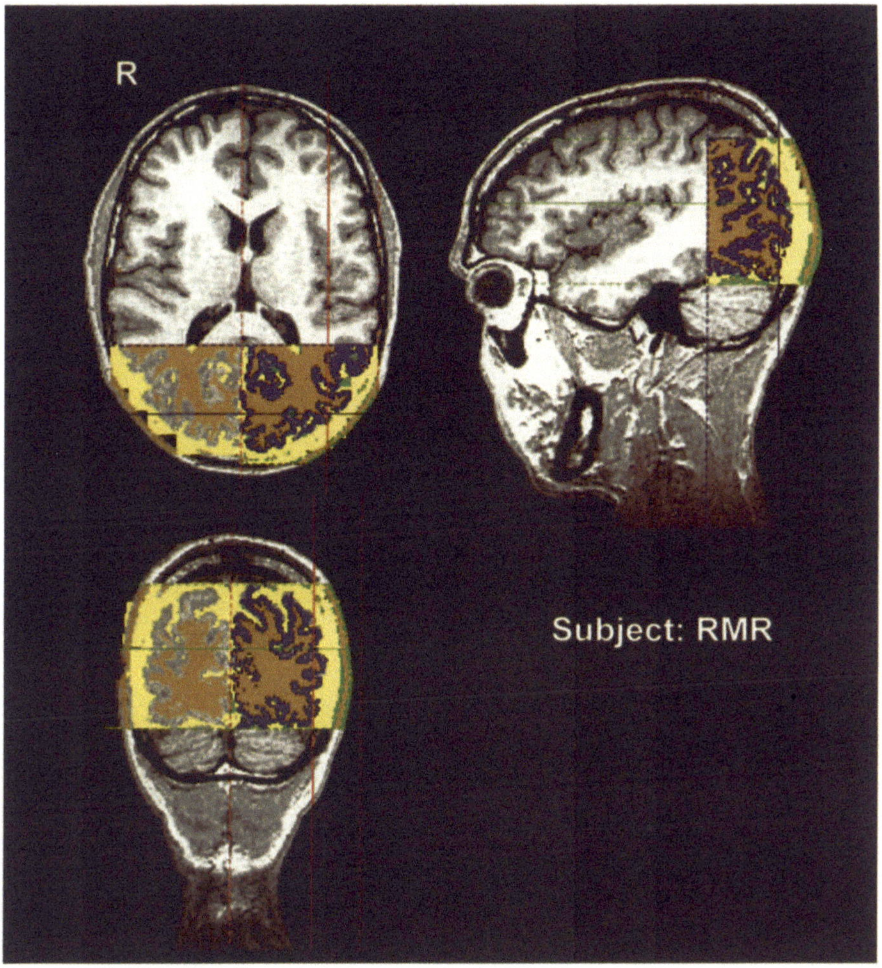

FIG. 1. Example of cortex segmentation program MrGray. The left visual cortex (blue pixels) has been segmented from the white matter (beige pixels) and the supportive tissue and CSF (yellow pixels). The unfolded cortex is shown in the form of a flat map (Fig. 2).

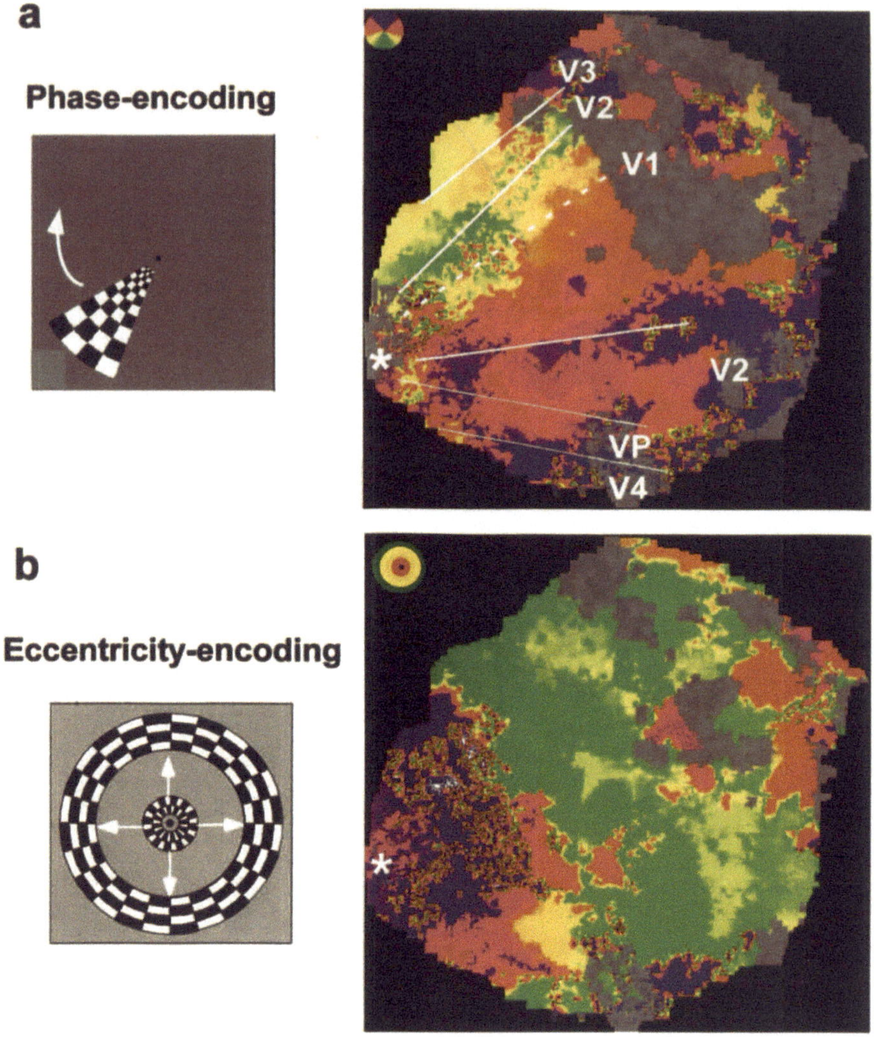

FIG. 2. Examples of cortical flat maps. Two methods of sequential stimulus presentation are depicted. (a) Schematic of the phase-encoding method, in which a flickering checkerboard segment is shifted stepwise over a 54-s period (18 steps, 3 s each). The MR-image aquiisition is synchronized to the stimulus progression. Four revolutions yield a BOLD response with four periods with a period duration of 54 s. The colors give the phase of the fundamental response at that location (see color key). (b) The results for the eccentricity-encoding method (see color key).

ployed in the hope that the subjects conform to the instructions during the entire scan period (mostly around 60 min in duration). Our experience suggests that this is often not the case especially for such long scan periods. The effects of pursuit during motion perception (Barton *et al.*, 1996) and a comparison between saccades and pursuit (Petit *et al.*, 1997) without eye position monitoring have been published. In trained monkeys viewing expanding optic-flow fields, it has been shown that the horizontal eye position and tracking movements of the eyes are influenced by the focus of expansion and flow direction (Lappe *et al.*, 1998). Similar results have been presented for human observers (Niemann *et al.*, 1999).

In an attempt to determine eye position during the MR-scan period, Felbinger *et al.* (1996) modified EOG electrodes and recorded saccades during functional imaging. The eye position trace was noisy due to magnetically induced currents in the recording system, but the authors were able to monitor large (30°) saccades with this method. Freitag *et al.* (1998) monitored EOG during motion perception under fixation and pursuit conditions and found larger BOLD responses in V5/V5a during pursuit. Obviously, the EOG method is too noisy in the MR scanner to provide precise information about the quality of fixation, the presence of small corrective saccades, and the gain of smooth pursuit.

We have designed a fiber-optic device that uses the well-established method of infrared light reflection at the iris-sclera edge to track the horizontal position of the eye (Kimmig *et al.*, 1999). An example of a recording is given in Fig. 3. As can be seen in Fig. 3b, the infrared signal is not affected by the magnetic field or by fast changes in field strength (i.e., gradients) within the head coil because only infrared light enters the scanner via completely nonferrous fiber-optic cables (Fig. 3a). We have successfully recorded BOLD responses during saccade, antisaccade and pursuit tasks (Kimmig *et al.*, 1999). An example of such a recording is given in Fig. 4b. It shows the T2* effect [mean region-of-interest (ROI) activation over the left FEF; Fig. 4a] as acquired with the echo-planar technique as a function of time during both resting and stimulation periods. The integrated saccadic activity, normalized to the mean and standard deviation of saccade frequency and amplitude, is shown in the other trace. There is a close correspondence between the saccadic activity and the amplitude of the BOLD signal.

An example of the BOLD response in area V5/V5a is given in Fig. 5 for saccadic and pursuit tasks. Figure 5b shows the result during a typical saccadic reaction time task, in which the subject made leftward or rightward saccades to a single target. Figure 5c presents the results from the same subject while the subject performed a smooth pursuit task. The stimulus frequency and displacement amplitude were identical in both

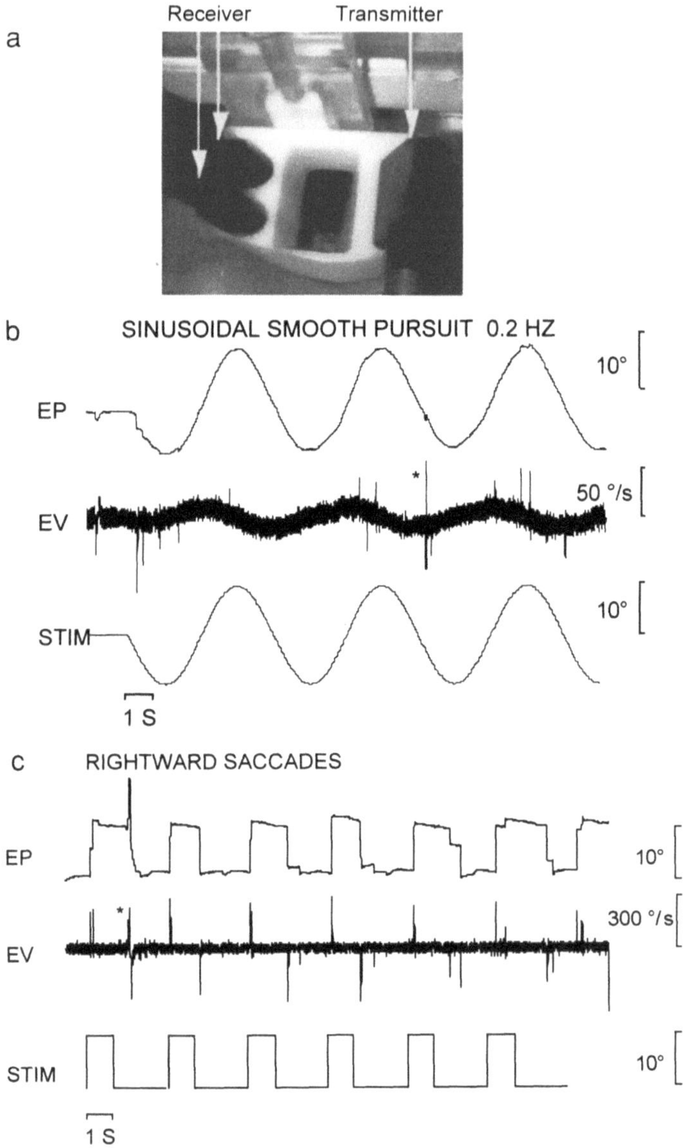

FIG. 3. MR-Eyetracker as described by Kimmig *et al.* (1999). (a) The right eye of a subject in the headcoil. Infrared light is created by photodiodes outside of the scanner and transported via fiber-optics to the eye (Transmitter). The light is reflected by the iris and the sclera, and this reflected light is picked up by the two receiver cables (Receiver). The intensity of this reflected light is then compared to yield a linear estimate of the horizontal eye position. (b) Eye position (EP), eye velocity (EV), and stimulus position (STIM) over time while the subject performed a smooth pursuit task during echo-planar image acquisition. (c) The data collected during a saccadic task while acquiring EPI. Otherwise as in panel (b).

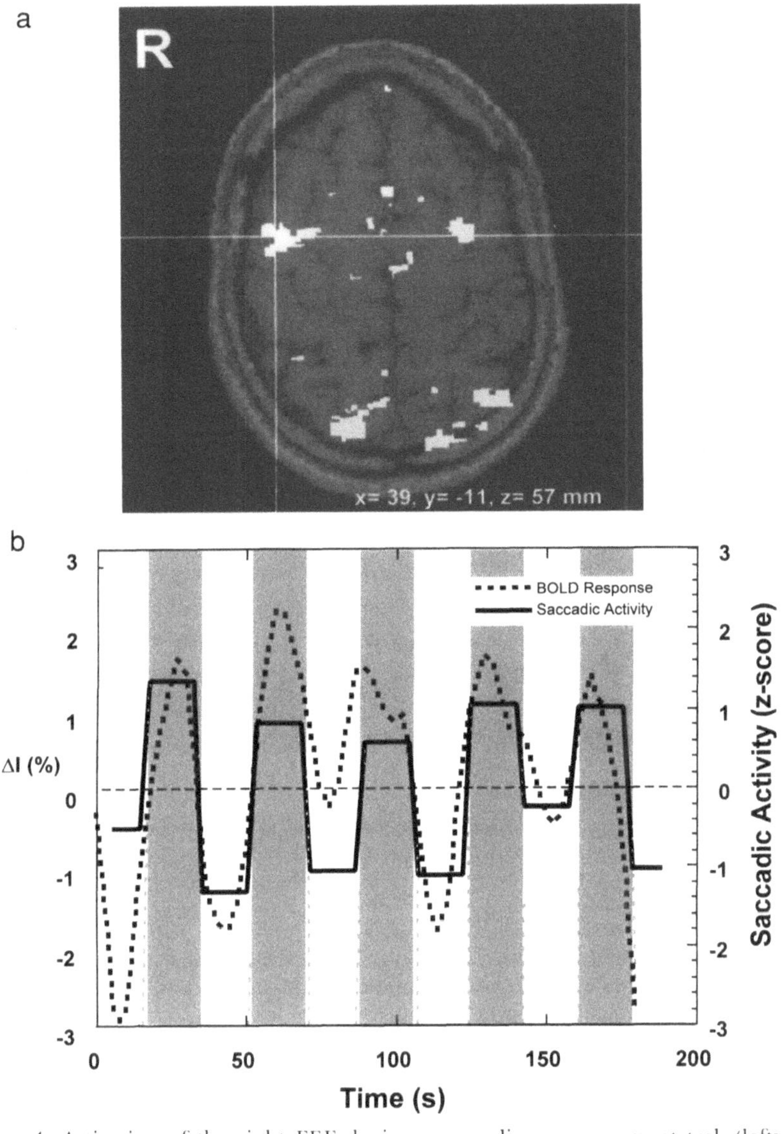

FIG. 4. Activation of the right FEF during a saccadic eye movement task (leftward and rightward saccades during five 30-s periods). (a) The significantly activated voxels (z-scores > 3 shown in white) and the cross-hairs denote the position of the ROI used in this analysis (right FEF). (b) The T2*-signal time course. Gray stripes show activation periods; white stripes indicate rest periods. The dotted trace presents the BOLD response for one subject (taken from the ROI shown in panel a) and the dark trace shows the integrated saccadic activity during stimulation (saccade task) and rest. The BOLD signal is shifted by 3–6 s, but otherwise correlates well with the normalized saccadic activity.

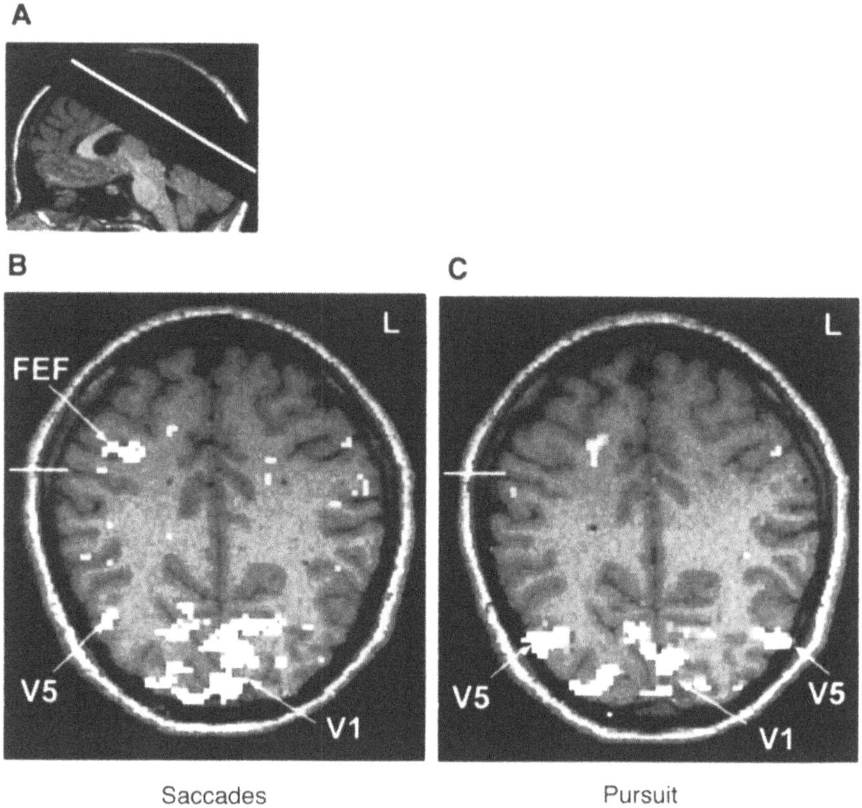

FIG. 5. BOLD responses during saccadic and pursuit eye movement tasks. The subject was positioned in the scanner and was instructed to pursue a red dot on a gray random-dot background. The dot moved 10° left or right of the center. (a) The 12 slices acquired over the scan period (hatched region). The white line shows the acquisition slice presented in panels (b) and (c). (b) Significant activation (positive correlation with shifted stimulus time course > 0.5) for one subject during a prosaccade task. Significant activation was evident in V1, V5, and FEF in this slice. (c) The results for the smooth pursuit task in which the subject tracked a sinusoidally moving target. The white horizontal bar denotes the position of the central sulcus (right hemisphere). Activation level is in white, as in Fig. 4.

tasks. Despite these similiarites, the pursuit task evoked considerably more activity in the V5/V5a complex than did the saccadic task (Kimmig *et al.*, 1999).

The results of these eye movements studies indicate that the human homologue of V5/V5a receives, in addition to the visual stimulation, a pronounced extraretinal input, and this input is reflected by larger BOLD signals. It follows that studies of optic flow should control for the eye movements made by the subjects, since activation in V5/V5a reflects both retinal and extraretinal sources.

C. Optic Flow and Functional Imaging

In our study on the BOLD responses to optic flow (Greenlee *et al.*, 1998), we presented dynamic, random-dot kinematograms dichoptically, one field to the left eye and one field to the right eye. The two fields were presented noninterlaced at a 72-Hz framerate with the help of a liquid crystal display (LCD) projector. We used polarizing optical filters together with adjustable right-angle prisms to superimpose the two fields optically.

Three-speed vectors were employed to create dynamic dot displays. The average speed was 10°/s, and the maximum speed was 17°/s. The different conditions of random-dot kinematograms were defined by the relation of the motion vectors within the flow field (Fig. 6). Aligning the motion vectors in a radial fashion led to the impression of expanding optic flow (Fig. 6a). Assigning the same dots a random direction with a mean free pathlength of 2.4° led to the impression of a random walk (Fig. 6b). Assigning the dots a random direction with a mean free pathlength of 0.2° led to the impression of a random jitter. A further condition of optic flow was studied by adding to the expanding flow field a rotational component. This manipulation led to spiral motion with clockwise or counterclockwise rotation, at approximately 1 rotation every 3 s.

Based on the electrophysiological studies cited earlier, we expect that the human homologue of V5/V5a should respond more to optic flow fields with disparity cues for depth. Thus, we introduced a further condition in which we varied the binocular disparity of the flow fields. Two conditions of binocular disparity were employed to evaluate its effects on responses to the optic flow fields. The dichoptic flow fields were identical to each other except with respect to the spatial displacements required to simulate disparity. In the condition we call "appropriate binocular disparity" the size of the disparity increased in proprotion to the increasing eccentricity from the center of expansion. Dots in the center of expansion had zero disparity, whereas the maximum disparity corresponded to 46 min of arc of uncrossed disparity. In the control condition, two identical flow fields were presented, one in each eye and no disparity was introduced. We measured BOLD responses in the following regions-of-interest: in the striate-extrastriate cortex (V1,V2), in the precuneus (V3,V3a), in the occipitotemporal area (V5/V5a), and in the kinetic occipital region (KO/V3b).

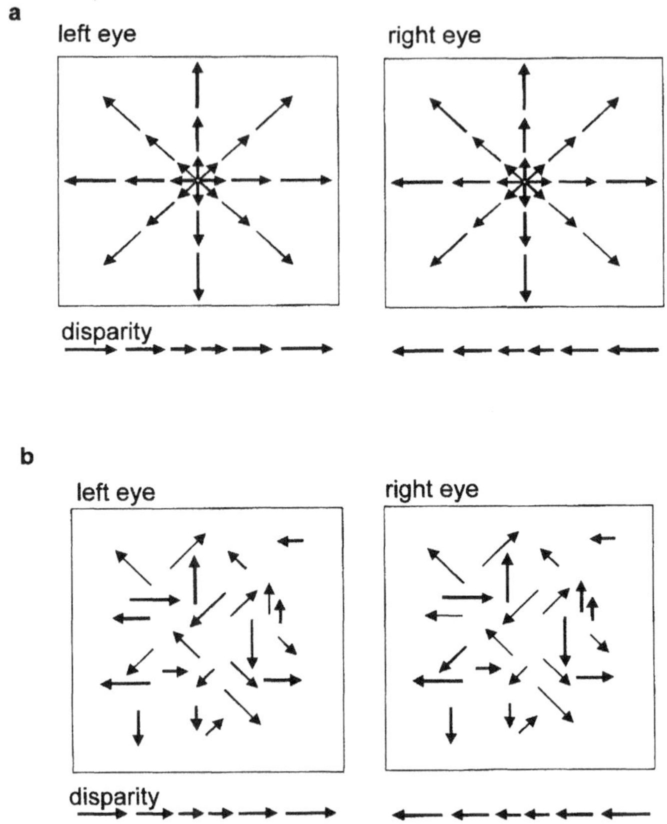

FIG. 6. Schematic representation of dichoptic optic flow fields. (a) The motion vectors in the flow fields presented to the left and right eye. Arrows denote the direction of motion, and arrow length depicts the speed of motion. The disparity of the left and right retinal images are coded by the horizontal arrows along the bottom, with zero disparity at the focus of expansion and 46 arc min of disparity at the perimeter. (b) The random-walk condition, where each dot is assigned a random direction and speed, otherwise as in panel (a).

Figure 7a presents the time course of the experiment. Following an initial rest period, the subject was presented with a 30-s epoch of optic flow, followed by an epoch containing the same number of static dots. The static stimulation was followed by a second rest period in which the stimulation was only a central fixation point. This rest-motion-static sequence was cycled four times. We could thus contrast the effects of motion stimulation to rest, static stimulation to rest and motion to static stimulation. An example from one subject is given in Fig. 7b.

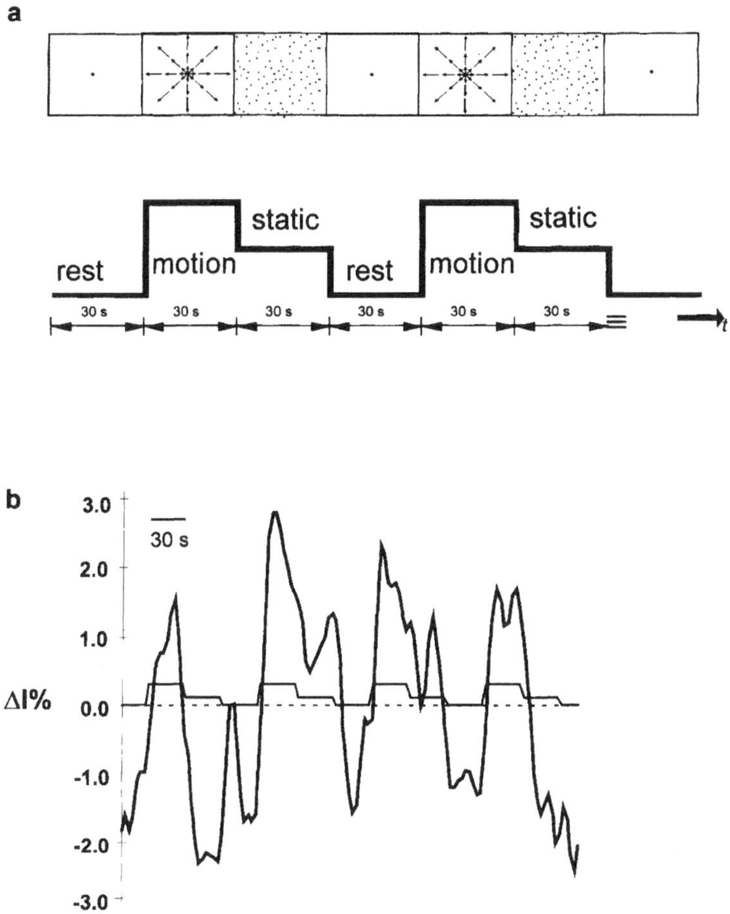

FIG. 7. Schematic illustration of the time course of stimulation during the fMRI experiments with optic flow. (a) The stimulus sequence. Periods of rest (30-s duration) were followed by a period of motion stimulation and a period of static stimulation. Each rest-motion-static sequence was repeated four times. (b) The time course of the BOLD signal over the V5/V5a region for one subject.

Imaging was performed on a 1.5-Tesla Siemens Vision Magnetom equipped with a fast gradient system for echo planar imaging. We used T2*-weighted sequences with a 128 × 128 matrix yielding 2-mm in-plane resolution. Twelve 4-mm planes were sampled every 3 s, and a total of 125 volumes were acquired in runs lasting 375 s. The echo-planar images were first corrected for head motion and then smoothed with a Gaussian filter having a standard deviation (SD) of two voxels. The image series was additionally filtered over time with a Gaussian having an SD of two images. The resulting time series was cross-correlated voxel-by-voxel with a temporally smoothed version of the stimulus boxcar. Pixels that surpass a correlation threshold of 0.5, corresponding to a z-score of 3.0 or greater, are highlighted. Regions of interest were determined in selected areas in occipital, temporal, and parietal cortex. The level of activation within a ROI was assessed by multiplying the mean SD of the time course of each voxel within the ROI with the standardized correlation coefficient (Bandettini *et al.*, 1993). In other words, the overall activity is weighted by the extent to which this activity is correlated with the time course of the stimulus boxcar. We used the software package BrainTools developed by Dr. Krish Singh (Smith *et al.*, 1998).

During the same recording sessions, we acquired a high-resolution, T1-weighted 3-D anatomical data set (T1-weighted MP-RAGE, magnetization-prepared, rapid acquisition gradient echo). Each subject's anatomical data were registered and normalized to the 3-D stereotactic atlas of Talairach and Tournoux (1988). The origin of this atlas is at the superior edge of the anterior commissure, and all regions of interest are reported in terms of millimeter deviation from this origin in the three image planes. The results of our ROI-analysis performed on each subject's data set were statistically analyzed using an ANOVA for repeated measures. Fourteen subjects participated (five female, nine male).

D. BOLD Responses to Optic Flow

Examples of BOLD responses in the condition with expanding optic flow are shown in Fig. 8. Clusters of activated pixels were found in striate and extrastriate cortex, in the ventral V3 area (V3b; also referred to as the kinetic occipital region KO) and in the V5/V5a complex in the occipitotemporal cortex. The center of V5/V5a activation in this subject is given in Talairach coordinates and corresponds well with those reported by other laboratories. An additional site of activation was located in the superior temporal sulcus (Talairach coordinates: $x = 43, y = 50, z = 16$ mm; not shown in Fig. 8). This location is in close agreement with that reported in association with biological motion paradigms (Puce *et al.*, 1998).

The results of the ROI analysis are summarized in Fig. 9. The findings are shown for four cortical areas under investigation. The differently shaded histograms show the results for the four different condi-

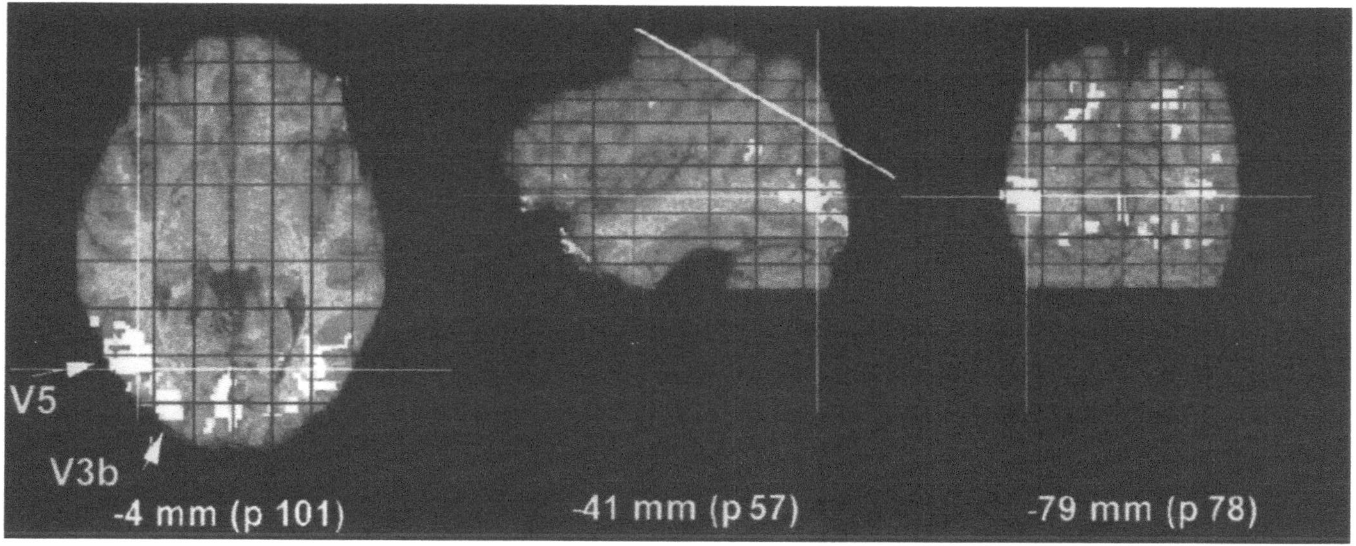

FIG. 8. Examples of BOLD responses in striate cortex (V1), extrastriate area KO (V3b) and the V5/V5a complex. The Talairach coordinates (and atlas page numbers) are given in the lower corner of each panel. Activation level is in white as in Fig. 4.

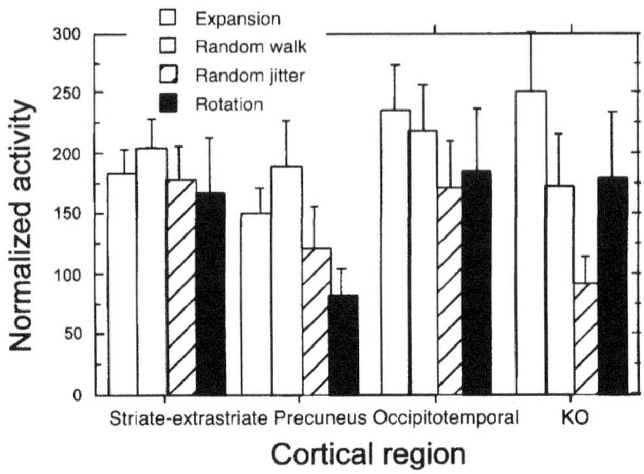

FIG. 9. Normalized activity (T2*-weighted signal multiplied by the z-transformed correlation coefficient) is shown for the four visual areas under investigation. The results are averaged over the 14 subjects for left and right hemispheres. Error bars show one standard error of the mean. The different columns show the results for the four types of optic flow studied.

tions of motion stimulation—expansion, random walk, random jitter, and rotation. Little response selectivity is found in the striate and immediate extrastriate regions. Areas in the precuneus, putatively corresponding to the dorsal parts of V3 and V3a, respond somewhat better to the random walk stimuli than to the other three conditions. Surprisingly, the V5/V5a complex showed little sensitivity to the flow patterns of the motion stimulation. In contrast, area KO in the ventral region of V3 (V3b) appears to show the greatest selectivity to optic flow. Area KO gave the overall best responses to expansion, followed by rotation and random walk. Random jitter caused little activity in KO.

Figure 10a shows the effect of the disparity manipulation, again for the four regions under investigation. Overall we found little or no differences in the BOLD responses in the striate-extrastriate regions, the precuneus, or the V5/V5a complex. There is some indication that KO might respond preferentially to the disparity gradients presented in the optic-flow fields. When the effect of disparity is analyzed for each of the four types of motion with respect to responses in KO, we found that disparity had a measureable effect in the rotational condition and some effect in expansion. No difference was found for the random conditions (Fig. 10b).

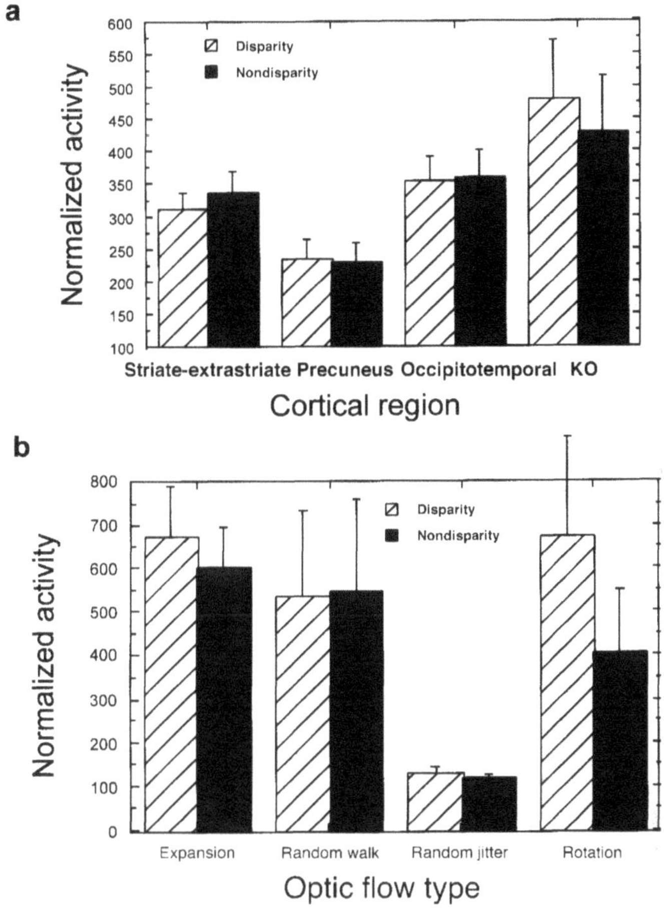

FIG. 10. Normalized activity (T2*-weighted signal multiplied by the z-transformed correlation coefficient). (a) The results for the four regions under study for the conditions with and without binocular disparity. (b) The findings for the kinetic occipital region (V3b) for the four optic flow conditions under study, otherwise as in panel (a).

III. Summary

In summary, we have reviewed electrophysiological and brain imaging studies of motion and optic-flow processing. Single-unit studies indicate that MST (V5a) is a site of optic-flow extraction and that this information can be used to guide pursuit eye movements and to estimate heading. The EEG and MEG studies point to a localized electrical dipole in occipitotemporal cortex evoked by visual motion. We have also discussed the evidence from functional imaging studies for response specificity of the rCBF and BOLD effects in posterior cortex to visual motion and optic flow. Focal attention modulates the amplitude of the BOLD signal evoked by visual motion stimulation. Retinotopic mapping techniques have been used to locate region borders within the visual cortex.

Our results indicate that striate (V1) and extrastriate areas (V2, V3/V3a) respond robustly to optic flow. However, with exception of a more pronounced response in V3/V3a to random walk, we found little evidence for response selectivity with respect to flow type and disparity in these early visual areas. In a similar fashion, the human V5/V5a complex responds well to optic flow, but these responses do not vary significantly with the type of flow field and do not seem to depend on disparity. In contrast, the kinetic occipital area (KO/V3b) responds well to optic-flow information, and it is the only area that produces more pronounced activation to the disparity in the flow fields. These initial results are promising because they suggest that the fMRI method can be sensitive to changes in stimulus parameters that define flow fields. More work will be required to explore the extent to which these responses reflect the neuronal processing of optic flow.

Eye position tracking is now possible during fMRI experiments. We have demonstrated that the eye movements affect the BOLD responses in motion-sensitive areas (Kimmig *et al.,* 1999). Further experiments in our laboratory are aimed at understanding the effects of eye movements on the neuronal coding of complex optic-flow fields (Schira *et al.,* 1999).

Acknowledgments

This work was supported by the Deutsche Forschungsgemeinschaft (GR988-15) and by the Hermann-und-Lilley Schilling Stiftung. The author thanks Dr. Roland Rutschmann and Dr. Hubert Kimmig for the good collaboration, Dr. K. Singh for BrainTools, and Bettina Gomer for her help in editing this manuscript.

References

Albright, T. D. (1984). Direction and orientation selectivity of neurons in visual area MT of the macaque. *J. Neurophysiol.* **52,** 1106–1130.

Albright, T. D., Desimone, R., and Gross, C. G. (1984). Columnar organization of directionally selective cells in visual area MT of the macaque. *J. Neurophysiol.* **51,** 16–31.

Andersen, R. A. (1995). Encoding of intention and spatial location in the posterior parietal cortex. *Cereb. Cortex* **5,** 457–469.

Andersen, R. A. (1997). Neural mechanisms of motion perception in primates. *Neuron* **18,** 865–872.

Anderson, S. J., Holliday, I. E., Singh, K. D., and Harding, G. F. A. (1996). Localization and functional analysis of human cortical area V5 using magneto-encephalography. *Proc. Roy. Soc. Lond.* **263,** 423–431.

Bach, M., and Ullrich, D. (1994). Motion adaptation governs the shape of motion-evoked cortical potentials. *Vision Res.,* **34**(12), 1541–1547.

Bandettini, P. A., Jesmanowicz, A., Wong, E. C., and Hyde, J. S. (1993). Processing strategies for time-course data sets in functional MRI of the human brain. *Magn. Reson. Med.* **30,** 161–73.

Barton, J. J. S., Simpson, T., Kiriakopoulos, E., Stewart, C., Crawley, A., Guthrie, B., Woods, M., and Mikulis, D. (1996). Functional MRI of lateral occipitotemporal cortex during pursuit and motion perception. *Ann. Neurol.* **40,** 387–398.

Beauchamp, M. S., Cox, R. W., and DeYoe, E. A. (1997). Graded effects of spatial and featural attention on Human area MT and associated motion processing areas. *J. Neurophysiol.* **78,** 516–520.

Bradley, D. C., and Andersen, R. A. (1998). Center-surround antagonism based on disparity in primate area MT. *J. Neurosci.* **18,** 7552–7565.

Britten, K., H., and van Wezel, R. J. A. (1998). Electrical microstimulation of cortical area MST biases heading perception in monkeys. *Nat. Neurosci.* **1,** 59–63.

Britten, K. H., Newsome, W., T., and Saunders, R., C. (1992). Effects of inferotemporal cortex lesions on form-from-motion discrimination in monkeys. *Exp. Brain Res.* **88,** 292–302.

Bruce, C. J., Goldberg, M. E., Stanton, G. B., and Bushnell, M. C. (1985). Primate frontal eye fields: 2. Physiological and anatomical correlates of electrically evoked eye movements. *J. Neurophysiol.* **54,** 714–734.

Cheng, K., Fujita, H., Kanno, I., Miura, S., and Tanaka, K. (1995). Human cortical regions activated by wide-field visual motion: An $H_2{}^{15}O$ PET study. *J. Neurophysiol.* **74,** 413–427.

Colby, C. L. (1998). Action-oriented spatial reference frames in cortex. *Neuron* **20,** 15–24.

Corbetta, M., Miezin, F. M., Dobmeyer, S., Shulman, G. L., and Petersen, S. E. (1991). Selective and divided attention during visual discriminations of shape, color and speed: Functional anatomy by positron emission tomography. *J. Neurosci.* **11,** 2383–2402.

Cornette, L., Dupont, P., Rosier, A., Sunaert, S., Van Hecke, P., Michiels, J., Mortelsmans, L., and Orban, G. A. (1998). Human brain regions involved in direction discrimination. *J. Neurophysiol.* **79,** 2749–2765.

Cox, R. W. (1996). AFNI: Software for analysis and visualization of functional magnetic neuroimages. *Comput. Biomed. Res.* **29,** 162–173.

Dale, A. M., Fischl, B., and Sereno, M. I. (1999). Cortical surface-based analysis. I. Segmentation and surface reconstruction. *Neuroimage* **9,** 179–194.

DeAngelis, G. C., and Newsome, W. T. (1999). Organization of disparity-selective neurons in macaque area MT. *J. Neurosci.* **19,** 1398–1415.

de Jong, B. M., Shipp, S., Skidmore, B., Frackowiak, R. S. J., and Zeki, S. (1994). The cerebral activity related to the visual perception of forward motion in depth. *Brain* **117,** 1039–1054.

Duffy, C. J., and Wurtz, R. H. (1997). Medial superior temporal area neurons respond to speed patterns in optic flow. *J. Neurosci.* **17,** 2839–2851.

Dupont, P., De Bruyn, B., Vandenberghe, R., Rosier, A. M., Michiels, J., Marchal, G., Mortelsmans, L., and Orban, G. (1997). The kinetic occipital region in human visual cortex. *Cereb. Cortex* **7,** 283–292.

Dupont, P., Orban, G. A., De Bruyn, B., Verbruggen, A., and Mortelsmans, L. (1994). Many areas in the human brain respond to visual motion. *J. Neurophysiol.* **72,** 1420–1424.

Engel, S. A., Glover, G. H., and Wandell, B. A. (1997). Retinotopic organization in human visual cortex and the spatial precision of Functional MRI. *Cereb. Cortex* **7,** 181–192.

Engel, S. A., Rumelhart, D. E., Wandell, B. A., Lee, A. T., Glover, G. H., Chichilnisky, E. J., and Shadlen, M. N. (1994). fMRI of human visual cortex. *Nature,* **369,** 525.

Felblinger, J., Müri, R. M., Ozdoba, C., Schroth, G., Hess, C. W., and Boesch, C. (1996). Recordings of eye movements for stimulus control during fMRI by means of electro-oculographic methods. *Magnet. Res. Med.,* **36,** 410–414.

Fischl, B., Sereno, M. I., and Dale, A. M. (1999). Cortical surface-based analysis. II: Inflation, flattening, and a surface-based coordinate system. *Neuroimage* **9,** 195–207.

Freitag, P., Greenlee, M. W., Lacina, T., Scheffler, K., and Radü, E. W. (1998). Effect of eye movements on the magnitude of fMRI responses in extrastriate cortex during visual motion perception. *Exp. Brain Res.* **119,** 409–414.

Friston, K. J., Holmes, A. P., Grasby, P. J., Williams, S. C. R., and Frackowiak, R. S. J. (1995). Analysis of fMRI time-series revisited. *Neuroimage* **2,** 45–53.

Gandhi, S. P., Heeger, D. J., and Boynton, G. M. (1999). Spatial attention affects brain activity in human primary visual cortex. *Proc. Natl. Acad. Sci. USA* **96,** 3314–3319.

Göpfert, E., R., M., Breuer, D., and Greenlee, M. W. (1999). Similarities and dissimilaries between pattern VEPs and motion VEPs. *Doc. Ophthalmol.,* in press.

Greenlee, M. W., and Smith, A. T. (1997). Detection and discrimination of first- and second-order motion in patients with unilateral brain damage. *J. Neurosci.* **17,** 804–818.

Greenlee, M. W., Rutschmann, R. M., Schrauf, M., and Smith, A. T. (1998). Cortical areas responsive to the direction, speed and disparity of optic flow fields: An fMRI study. *Neurosci. Abstr.* **24,** 530.

Kastner, S., De Weerd, P., Desimone, R., and Ungerleider, L. G. (1998). Mechanisms of directed attention in the human extrastriate cortex as revealed by functional MRI. *Science* **282,** 108–111.

Kimmig, H., Greenlee, M. W., Huethe, F., and Mergner, T. (1999). MR-Eyetracker: A new method for eye movement recording in functional magnetic resonance imaging (fMRI). *Exp. Brain Res.* **126,** 443–449.

Kraemer, F., and Hennig, J. (1999). Image registration algorithm for fMRI. In preparation.

Kubova, Z., Kuba, M., Spekreijse, H., and Blakemore, C. (1995). Contrast-dependence of the motion-onset and pattern-reversal evoked potentials. *Vision Res.* **35,** 197–205.

Lappe, M., Bremmer, F., Pekel, M., Thiele, A., and Hoffmann, K. (1996). Optic flow processing in monkey STS: a theoretical and experimental approach. *J. Neurosci.* **16,** 6265–6285.

Lappe, M., Pekel, M., and Hoffmann, K. P. (1998). Optokinetic eye movements elicited by radial optic flow in the macaque monkey. *J. Neurophysiol.* **79,** 1461–1480.

Lynch, J. C. (1987). Frontal eye field lesions disrupt visual pursuit. *Exp. Brain Res.* **68,** 437–441.

Müller, R., and Göpfert, E. (1988). The influence of grating contrast on the human cortical potential visually evoked by motion. *Acta Neurobiol. Exp.* **48,** 239–249.

Müller, R., Göpfert, E., Breuer, D., and Greenlee, M. W. (1999). Motion VEPs with simultaneous measurement of perceived velocity. *Doc. Ophthalmol.,* in press.

Müller, R., Göpfert, E., and Hartwig, M. (1986). The effect of movement adaptation on human cortical potentials evoked by pattern movement. *Acta Neurobiol. Exp.* **46,** 293–301.

Niemann, T., Lappe, M., Büscher, A., and Hoffmann, K.-P. (1999). Ocular responses to radial optic flow and single accelerated targets in humans. *Vision Res.* **39,** 1359–1371.

Newsome, W. T., and Paré, E. B. (1988). A selective impairment of motion perception following lesions of the middle temporal visual area (MT). *J. Neurosci.* **8,** 2201–2211.

Newsome, W. T., Wurtz, D. H., Dürsteler, M. R., and Mikami, A. (1985). Deficits in visual motion processing following ibotenic acid lesions of the middle temporal visual area of the macaque monkey. *J. Neurosci.* **5,** 825–840.

O'Craven, K. M., Rosen, B. R., Kwong, K. K., Triesman, A., and Savoy, R. L. (1997). Voluntary attention modulates fMRI activity in human MT-MST. *Neuron* **18,** 591–598.

Orban, G. A., Saunder, R. C., and Vandenbussche, E. (1995). Lesions of the superior temporal cortical motion areas impair speed discrimination in the macaque monkey. *Eur. J. Neurosci.* **7,** 2261–2276.

Orban, G. A., Dupont, P., De Bruyn, B., Vandenberghe, R., Rosier, A., and Mortelsmans, L. (1998). Human brain activity related to speed discrimination tasks. *Exp. Brain Res.* **122,** 9–22.

Page, W. K., and Duffy, C. J. (1999). MST neuronal responses to heading direction during pursuit eye movements. *J. Neurophysiol.* **81,** 596–610.

Pasternak, T., and Merigan, W. H. (1994). Motion perception following lesions of the superior temporal sulcus in the monkey. *Cereb. Cortex* **4,** 247–59.

Petit, L., Clark, V. P., Ingeholm, J., and Haxby, J. V. (1997). Dissociation of saccade-related and pursuit-related activation in human frontal eye fields as revealed by fMRI. *J. Neurophysiol.* **77,** 3386–3390.

Probst, T., Plendl, H., Paulus, W., Wist, E. R., and Scherg, M. (1993). Identification of the visual motion (area V5) in the human brain by dipole source analysis. *Exp. Brain Res.* **93,** 345–351.

Puce, A., Allison, T., Bentin, S., Gore, J., and McCarthy, G. (1998). Temporal cortex activation in humans viewing eye and mouth movements. *J. Neurosci.* **18,** 2188–2199.

Reppas, J. B., Niyogi, S., Dale, A. M., Sereno, M. I., and Tootell, R. B. H. (1997). Representation of motion boundaries in retinotopic human visual cortical areas. *Nature* **388,** 175–179.

Schiller, P. H., True, S. D., and Conway, J. L. (1979). Effects of frontal eye field and superior colliculus ablations on eye movements. *Science* **206,** 590–592.

Schira, M. M., Kimmig, H., and Greenlee, M. W. (1999). Cortical areas underlying motion perception during smooth pursuit. *Perception,* in press.

Sereno, M. I., Dale, A. M., Reppas, J. B., Kwong, K. K., Belliveau, J. W., Brady, T. J., Rosen, B. R., and Tootell, R. B. H. (1995). Borders of multiple visual areas in humans revealed by functional magnetic resonance imaging. *Science* **268,** 889–893.

Smith, A. T., Greenlee, M. W., Singh, K. D., Kraemer, F. M., and Hennig, J. (1998). The processing of first- and second-order motion in human visual cortex assessed by functional magnetic resonance imaging (fMRI). *J. Neurosci.* **18,** 3816–3830.

Talairach, J., and Tournoux, P. (1988). "Co-planar Stereotaxic Atlas of the Human Brain." Thieme Verlag, Stuttgart.

Tanaka, K., and Saito, H. A. (1989). Analysis of motion of the visual field by direction, expansion/contraction and rotation cells clustered in the dorsal part of MST of the macaque. *J. Neurophysiol.* **62,** 626–641.

Tanaka, K., Fukada, Y., and Saito, H. A. (1989). Underlying mechanisms of the response specificity of expansion/contraction and rotation cells in the dorsal part of the medial superior temporal area of the macaque monkey. *J. Neurophysiol.* **62,** 642–656.

Tanaka, K., Hikosaka, K., Saito, H., Yukie, M., Fukada, Y., and Iwai, E. (1986). Analysis of local and wide-field movements in the superior temporal visual areas of the macque monkey. *J. Neurosci.* **6,** 134–144.

Teo, P. C., Sapiro, G., and Wandell, B. A. (1997). Anatomically consistent segmentation of the human visual cortex for functional MRI visualization. Hewlett-Packard Labs, Technical Report, HPL-97-03, pp. 1–21.

Tootell, R. B., Hadjikhani, N., Hall, E. K., Marrett, S., Vanduffel, W., Vaughan, J. T., and Dale, A. M. (1998). The retinotopy of visual spatial attention. *Neuron* **21,** 1409–1422.

Tootell, R. B. H., Mendola, J. D., Hadjikhani, N. K., Ledden, P. J., Liu, A. K., Reppas, J. B., Sereno, M. I., and Dale, A. M. (1997). Functional analysis of V3A and related areas in human visual cortex. *J. Neurosci.* **17,** 7060–7078.

Tootell, R. B., Reppas, J. B., Dale, A. M., Look, R. B., Sereno, M. I., Malach, R., Brady, T. J., and Rosen, B. R. (1995a). Visual motion aftereffect in human cortical area MT revealed by functional magnetic resonance imaging. *Nature* **375,** 139–141.

Tootell, R. B., Reppas, J. B., Kwong, K. K., Malach, R., Born, R. T., Brady, T. J., Rosen, B. R., and Belliveau, J. W. (1995b). Functional analysis of human MT and related visual cortical areas using magnetic resonance imaging. *J. Neurosci.* **15,** 3215–3230.

Van Essen, D. C., and Zeki, S. M. (1978). The topographic organization of rhesus monkey prestriate cortex. *J. Physiol.* **277,** 193–226.

Van Essen, D. C., Maunsell, J. H. R., and Bixby, J. L. (1981). The middle temporal visual area in the macaque: Myeloarchitecture, connections, functional properties and topographic organization. *J. Comp. Neurol.* **199,** 293–326.

van Oostende, S., Sunaert, S., Van Hecke, P., Marchal, G., and Orban, G. A. (1997). The kinetic occipital (KO) region in man: An fMRI study. *Cereb. Cortex* **7,** 690–701.

Watson, J. D. G., Myers, R., Frackowiak, R. S. J., Hajnal, J. V., Woods, R. P., Mazziotta, J. C., Shipp, S., and Zeki, S. (1993). Area V5 of the human brain: Evidence from a combined study using positron emission tomography and magnetic resonance imaging. *Cereb. Cortex* **3,** 79–84.

Woods, R. P., Grafton, S. T., Holmes, C. J., Cherry, S. R., and Mazziotta, J. C. (1998a). Automated image registration: I. General methods and intrasubject, intramodality validation. *J. Comput. Assist. Tomogr.* **22,** 139–152.

Woods, R. P., Grafton, S. T., Watson, J. D., Sicotte, N. L., and Mazziotta, J. C. (1998b). Automated image registration: II. Intersubject validation of linear and nonlinear models. *J. Comput. Assist. Tomogr.* **22,** 153–165.

Zeki, S. M. (1971). Cortical projections from two prestriate areas in the monkey. *Brain Res.* **34,** 19–35.

Zeki, S. M. (1974). Functional organization of a visual area in the posterior bank of the superior temporal sulcus of the rhesus monkey. *J. Physiol. (London)* **236,** 549–573.

Zeki, S. (1978). Uniformity and diversity of structure and function in rhesus monkey prestriate visual cortex. *J. Physiol.* **277,** 273–290.

Zeki, S., Watson, J. D. G., Lueck, C. J., Friston, K. J., Kennard, C., and Frackowiak, R. S. J. (1991). A direct demonstration of functional specialization in human visual cortex. *J. Neurosci.* **11,** 641–649.

WHAT NEUROLOGICAL PATIENTS TELL US ABOUT THE USE OF OPTIC FLOW

Lucia M. Vaina* and Simon K. Rushton†

*Boston University, Brain and Vision Research Laboratory, Department of Biomedical Engineering and Neurology and Department of Neurology, Harvard Medical School, Brigham and Women Hospital & Massachusetts General Hospital, Boston, Massachusetts; †Cambridge Basic Research, Nissan Research & Development, Inc., Cambridge, Massachusetts and Department of Clinical Psychology, Astley Ainslie Hospital, Grange Loan, Edinburgh, Scotland, EH9 2HL, UK

I. Introduction

As we move through the environment, the pattern of visual motion on the retina provides rich information about our passage through the scene. It is widely accepted that this information, termed optic flow, is used to encode self-motion, orientation, visual navigation in three-dimensional space, perception of object movement, collision avoidance, visual stabilization, and control of posture and locomotion. The nature and properties of the mechanisms involved in the perception of optic flow have been extensively studied with physiological, psychophysical and computational methods and are the focus of other chapters in this volume. In this chapter, we describe research with neurological patients and what such research may tell us about the perception and control of locomotion.

II. Functional Architecture of Motion for Navigation

The posterior parietal cortex (PPC), a region encompassing several visually responsive areas (for a review, see Andersen, 1990, 1997; Andersen, *et al.*, 1996, 1997), is the primary neural substrate for the visual processing which could support guidance of movement using optic flow. Electrophysiological studies have demonstrated that the dorsal portion of

0074-7742/00 $30.00

the middle superior temporal area (MSTd) in the macaque PPC contains neurons highly responsive to specific kinds of optic flow patterns and to stereomotion (motion in depth). Microstimulation of MSTd neurons biases heading judgments (Britten and van Wezel, 1998) strengthening the view that this area is involved in mediating the perception of direction of heading. However, MSTd is not the only region of brain activity directly involved in heading and other tasks of visual navigation. Other areas in the parietal lobe (Maunsell and Essen 1983a, 1983b; Ungerleider and Desimone, 1986) of the macaque brain such as the ventral intraparietal cortex (VIP), the superior temporal polysensory area (STP), and area 7a are highly sensitive to optic flow stimuli and to speed and direction of motion (Duhamel *et al.*, 1997; Read and Siegel, 1997; Schaafsma *et al.*, 1997; Siegel and Read, 1997). It is thus possible that different forms of optic flow may activate neurons in different areas.

There is ample evidence for several motion streams, in addition to the classic V1-MT-MST, and thus it remains important to quantitatively explore response in these other areas to flow stimuli. For example, physiological studies report that both V2 and V3 contain a high percentage of direction selective cells, and they both have significant projections to directionally selective areas including MT, VIP and the floor of the superior temporal sulcus (area FST). Area V3 also contains neurons that respond to more global patterns of motion, (Gegenfurtner *et al.*, 1997), and FST neurons respond to complex object motion. Together with the classic V1-MT-MST visual motion processing stream, these results suggest a more loose hierarchy in the motion system of the macaque and a higher degree of parallelism than previously assumed. They also suggest a distribution of computational responsibilities among different areas.

Most studies to date have employed an abstraction of the flow field, a sparse field of small, fixed size, luminance-defined moving dots. The natural flow field, however, is characterized by a broader spatial, temporal, and chromatic profile and persisting, trackable objects with sharp edges and identifiable surfaces. The responses of visually responsive areas to the richer natural flow field remains to be characterized (Pekel *et al.*, 1996). Until that time, the data collated must be viewed as partial.

In humans, functional neuroimaging studies (PET and fMRI) have repeatedly reported that many regions in the human brain respond selectively to moving patterns (Dupont *et al.*, 1994; Tootell *et al.*, 1995a, 1996). Most studies have concentrated on the region in the ascending limb of the inferior temporal sulcus referred to as hMT+ because it functionally represents the human homologue of the macaque areas MT and MST (Tootell *et al.*, 1993a, 1993b, 1994, 1995a, 1995b; Cheng *et al.*, 1995; Dale *et al.*, 1995; Beauchamp and DeYoe 1996). The posterior part is bounded by area V3A, which is also quite sensitive to motion (Tootell *et al.*, 1997). An area anterior to V3A has been designated KO, to reflect sensitivity to the form of structured kinetic stimuli (Dupont *et al.*, 1997). A preliminary report of a PET study from Orban's group shows that hMT is activated in heading tasks, but only marginally (Peuskens *et al.*, 1999). The main areas of activation were seen in the human homologues of V2, V3 and V3A bilaterally, and in the posterior parietal regions. Additional motion responsive regions have been described in the occipital, occipital-parietal, occipital-temporal, and temporal and parietal regions. Taken together, these studies suggest that in humans, similar to the macaque, motion does not relate specifically to the classical distinction of the "dorsal" and "ventral" streams.

But the functional neuroimaging studies have told us little about the computations for which the visual areas in humans are responsible. Moreover, on their own, they seldom can relate to models of visual behavior. Studies of patients can help bridge this important gap.

III. Why Study Motion-Impaired Neurological Patients?

In a review of the work of Gibson and those who had followed his research program, Nakayama (1994) noted that the psychophysical work that had been carried out had not established more than the possibility that humans could use optic flow in natural locomotion. Since then, considerably more psychophysics, computational modeling, and neurophysiology has been done, but as Warren (1999) notes, the issue of what information and strategies are actually used has still to be properly addressed. A fundamental problem pointed out by Warren, is that many alternative models of the control of locomotion are functionally equivalent under normal circumstances (i.e., they all predict the same behavior). It requires inventive and often rather non-ecological manipulations of visual information to create circumstances under which models predict different behavior. We expand on this point in the next section. Warren and colleagues have recently begun to tackle this problem through the use of virtual reality technology (W. Warren, personal communication).

A complementary approach is the study of neurological patients. Such studies have possibly provided the earliest and certainly the most convincing evidence that, in humans, motion is a "special visual perception" (Riddoch 1917), and Goldstein and Gelb (1918) supported this position by a functional-anatomical architecture similar to that suggested by anatomical, physiological and behavioral studies in the macaque monkey.

More critically is that the same neurological studies reveal the role of motion in perception and action. For example, Goldstein and Gelb's (1923) patient, Schn, was unable to experience real or apparent movement in any condition; he was able to see correctly changes in position of objects and therefore could infer that they were moving. He also failed to perceive movement in depth. Patient Schn is very similar to the most extensively studied "motion blind" patient, L.M. (Zihl *et al.,* 1983, 1989, 1991; Shipp *et al.,* 1994). L.M. lives a very uncomfortable visual perceptual life because she fails to see movement of objects. For her, the movement of a car looks like a series of discrete "stills" advancing in time, and therefore she finds it disconcerting to cross the street in on-coming traffic. A car seems to her far away at one moment, and dangerously close the next, with no smooth progression in between. She is unable to perceive continuous motion, tea poured from a kettle appears to her as a static solid curved cylinder. She feels handicapped in social settings when conversing with other people because she cannot perceive and therefore read mobile facial expressions. As pointed out by Nakayama (1985) this suggests that motion processing could play a major role in reading facial expressions, something not hitherto suspected. However, and amazingly, L.M. can see and recognize without difficulty "biological motion" in the well-known Johansson's demonstration (Johansson 1973; McLeod *et al.,* 1996; Marcar *et al.,* 1997). She also can catch a ball thrown straight at her (P. McLeod, personal communication), although not when its trajectory is off to one side. What motion information, if any, is she using to perform these tasks?

Cases such as the "motion blind" patient, L.M., are very rare. Normally loss of motion sensitivity is more sparing and selective, and studies of such patients provide important insights into the functional organization of the human visual motion systems. For example, our patient, A.F. (Vaina et al., 1990a, 1990b), with bilateral lesions in the posterior parietal lobes, was severely impaired on several, but not all, low level motion tasks. Like L.M., he could promptly recognize the "biological motion" stimuli and could catch a ball thrown directly at him; he could even recognize 3-D structure from motion.

On the other hand, patient R.J. (Vaina *et al.,* 1999b), with a lesion in the anterior portion of the right temporal lobe, performed normally on both low level motion tasks and was able to recover both 2-D and 3-D structure from motion, but he failed to recognize the human figure depicted in Johansson's display. These reports tell us that there must be dissociations within the motion system. Another example is the demonstration that first- or second-order motion can be selectively impaired in patients (Vaina and Cowey, 1996; Vaina *et al.,* 1996a, 1998b, 1999a) and these studies provide compelling evidence for the hypothesis that first- and the second-order motion processes are mediated by different, parallel pathways. A novel model that accounts for these and other (Plant *et al.;* Greenlee and Smith, 1997) findings was put forward by Clifford and Vaina (1998).

Psychophysical studies of patients performance on an *a priori* defined set of motion tasks can provide an important benchmark for the biological validity of existing computational models. For example, the double dissociation of deficits on perception of motion discontinuity and motion coherence (Vaina *et al.,* 1994a, 1998a) rules out models based on Random Markov Fields and line processors (Koch *et al.,* 1989), popular in computer vision, as the only explanation of how the human brain processes motion discontinuity.

Furthermore, the performance of patients on specific perceptual tasks can reveal the necessary anatomical substrate for that task. For example, King *et al.* (1996), in a study of the residual visual sensitivity in patients deprived of the cortical pathway following either functional or anatomical hemispherectomy, found that they had neither electrodermal responses nor any conscious experience of the direction of motion in complex motion patterns (radial or circular). They interpreted their results as suggesting that the subcortical visual pathways which survive hemispherectomy and which can certainly subserve some visual processing are insufficient to process information about direction in complex motion.

Although the foregoing examples provide only a very incomplete list of neurological studies reporting particular dissociations of motion processes, they illustrate the importance of careful quantitative studies of neurological patients in elucidating the functional architecture of the human visual motion system. In particular, such studies often provide evidence for the existence of several motion subsystems and their computational characteristics. Even more importantly, neurological studies also tell us about the putative functional roles of these various subsystems, and, as Pollen (1999) noted they "open the way for improved models based on refutations or confirmations of the present set of proposals."

IV. The Radial Flow Field

During locomotion it is necessary for a person to control their direction of heading. Psycho-physical judgment tasks demonstrate that direction of heading can be recovered from the instantaneous velocity flow field (Warren *et al.,* 1991). Neural areas such as MST have been shown to be responsive to optic flow stimuli. Therefore, it has been assumed that MST computes the direction of heading which is then used to guide locomotion.

Although the preceding theory is compelling, it is necessary to question it. The simple combination of psychophysical data and a putative neural substrate does not prove that optic flow guides locomotion. There are a number of ways mobile observers could visually guide themselves around an environment. Consider walking to a target. First, observers could recover their instantaneous direction of heading and/or path from the velocity flow field and use this information for guiding locomotion. Second, they could use cruder velocity field-based iterative strategies, such as equalizing the flow to the left and right of the target (Srinivasan *et al.*, 1991; Duchon and Warren 1998) or canceling any rotational flow. Third, they could use the relative displacements of other objects in the scene (Cutting 1986; Cutting *et al.*, 1992) to determine their heading and use that for guiding locomotion. Fourth, they could use the egocentric direction of the target and a simple alignment strategy to reach their target (Rushton *et al.*, 1998) or target drift (Llewellyn, 1971).

The psychophysical and neurophysiological demonstrations of sensitivity to radial flow have been taken as powerful evidence that flow guides locomotion. However, this does not necessarily follow because, as argued earlier, there are a range of other tasks that optic flow might be used to perform. For example, radial flow may provide important information for the control of action and posture. Lee and colleagues (1975) showed that vision is the dominant source of information in the control of vertical posture with their "swinging room" experiments. In these experiments, an observer stands upright whilst a room is swung around them. It is found that the observer swings in synchrony with the room. This coordination of change of posture with the movement of the room may be through a process that nulls expansion and contraction in the global flow field. A recent elegant neurophysiological study of Duffy and Wurtz (1996) investigated the effects of optic flow on posture in the macaque monkeys. In particular, these authors wanted to learn whether presentation of the same optic flow patterns that elicit neuronal responses in area MST would also alter the animal's posture. Furthermore, they tested whether lesioning the areas MT and MST would alter posture. In the neurophysiological experiments, they observed in some trials postural sway that was dependent on the characteristics of the optic flow pattern. However, in many trials posture was not affected by the optic flow patterns, suggesting that other factors may also influence posture (e.g., vestibular and proprioceptive). Bilateral lesions in the MT and MST areas resulted in postural instability, and these authors suggested that the "disruption of the normal visual motion processing might have lead to degraded postural control signals."

For standing posture, early psychophysical studies (Ambland and Carblanc, 1980) reported peripheral dominance, but more recent experiments (Andersen and Dyre, 1989) suggest that postural sway can be elicited by central stimulation, using both translational and radial flow (in the central 15°). In a series of clever experiments, Stoffregen (1985, 1986) found that both radial and translational flow patterns induced postural sway when presented in central vision, whereas only translational flow was effective in the periphery. These results put forth an interesting hypothesis that, for postural control, the central retina is sensitive to both radial and translational flow, whereas the peripheral retina is sensitive only to translational flow. Bardy *et al.* (1996), investigating psychophysically postural responses to optic flow patterns at different eccentricities during walking, found that the structure of the optic flow is more important than the retinal locus of stimulation. In particular, they found that both central and peripheral vision can use both radial and translational flow patterns to control posture during locomotion. That radial pattern is of use in the periphery and is consistent with reports on accurate peripheral heading discrimination (Crowell and Banks, 1993), and avoidance of looming objects (Stoffregen and Riccio 1990). Investigating the neural substrate for control of posture from optic flow would be very challenging with imaging or using neurophysiological recording techniques. Patients provide far more accessible a route for investigation of this issue.

When approaching a wall at a constant velocity, radial flow indicates the time of collision with the wall (Lee, 1976). Local expansion and translation in the binocular array indicates that an object is approaching. Lee (1974, 1976), Regan and colleagues (Regan and Beverley, 1979; Regan and Vincent, 1995), and others have provided the behavioral data that support the role of such information in perception of time-to-contact, interceptive timing, and motion-in-depth. How might we get about investigating the neural basis of such performance? Transcranial-magnetic stimulation (TMS) provides one way in humans, but the type of tasks that could be performed is obviously constrained. Patients again offer a simple way forward. If a neurological patient loses the ability to perceive radial flow, is his/her control of locomotion impaired? If the patient has a deficit in perception of expansion, can he/she still perceive motion in depth? Does a loss of sensitivity to motion in depth co-occur with impairment on postural control? Examination of the patterns of deficits of actions can tell us about the functional independence or modularity and hierarchical organization of the perceptual control of actions. Thus insight is gained into questions as to whether there is a single flow center and what role flow has in the control of actions.

Most of the questions outlined in this section wait for case study data. In the next sections, we focus specifically on the perception of heading and the control of locomotion and discuss several relevant neurological case studies.

V. Impairment of Locomotion and Recovery of Locomotor Function

We (L. M. Vaina and G. Shepherd, in preparation) have recently studied a patient, M.Sh., an 81 years old woman who suffered a right occipital infarct in the region of the posterior cerebral artery involving the medial temporal and occipital lobes. The right cerebellum was spared. She had a left homonymous hemianopsia. Reading, colors, faces, and object recognition were all normal. Her constructions in 2-D and 3-D were very poor, and on formal neurological examination she demonstrated some left-side neglect from which she fully recovered in a few days. The patient was brought to our attention because, in spite of normal vestibular and cerebellar signs, her posture and gait were impaired. Her gait was unsteady and very slow, and when walking she noticeably veered to the right. On a series of psychophysical motion tests, she performed normally on direction discrimination in random dot kinematograms for stimuli presented in her normal field; however, her discrimination of radial flow and straight trajectory heading was severely impaired even when the stimuli were fully presented in her normal visual field. Her walking pattern is very similar to the reported behavior of patient W.V. studied by Rushton *et al.* (1998).

Patient W.V. is a 64-year-old male, who suffered right temporal-parietal damage associated with left homonymous hemianopsia and hemiparesis. Immediately after the lesion, the patient exhibited marked left visual neglect, first informally noted on the ward during normal activities and then clinically assessed. For example, on the cancellation task (Albert, 1973), he showed consistent left omissions, and right omissions with visual reversal. On the line-bisection task, he made rightward errors. Also on copying he omitted the left side of the figure. Further details of W.V.'s neurological and neuropsychological history, and some results of neglect testing are to be found in Shillcock, Kelly, and Monaghan (1998). Initially his mobility was restricted, but after extended physical therapy he regained his ability to walk unaided. It was when he returned for an out-patient appointment, proud to demonstrate his ability to walk, that his wife mentioned his curved trajectories. He was able to reach objects or places of interest; he just took a strange route to get to them. Rushton *et al.* (1998) investigated W.V.'s veering motion, but an explanation for this behavior was not immediately obvious. The veering of neglect patients has been reported previously. The reports were not particularly problematic for flow theories of heading because they described left-sided (assuming left neglect) collisions and being "magnetically" pulled toward right doorposts. Veering while walking through an open environment is more difficult to account for.

It has been reported that some patients with neglect misperceive the straight-ahead direction; this can be tested by giving the patient a laser pointer and having them point to a position on a wall that is directly ahead of them. Instead of pointing to the point directly ahead they may point to a position approximately 15° to one side (see Karnath, 1994). As noted in a previous section, one potential walking strategy is to ignore optic flow and instead simply place an object of interest directly ahead and walk forward. Under normal circumstances, this would lead to a straight trajectory toward an object of interest. However, if the observer misperceives the true straight-ahead direction, then it predicts a curving trajectory that eventually reaches the object. In contrast, if optic flow guides locomotion, then the misperception of the direction of straight ahead should not matter, all that is critical is the relative position of the target and the focus of expansion.

The misperception-of-egocentric-direction hypothesis was tested with normal subjects. The locomotion models were pitted against each other by displacing the perceived direction of objects in normal observers with prism glasses. This manipulation affects direction but not flow. It was found that observers took a veering trajectory when walking to a target. The trajectory was well described by the place-the-target-straight-ahead model and clearly very different to that expected from the flow model (Rushton *et al.* 1998).

To determine whether misperception of egocentric direction could account for W.V.'s behavior, it was arranged to assess his walking trajectories on his next out-patient visit. However, when he was tested, he proceeded to walk a perfectly straight trajectory to the target given him. He was then tested with prism glasses that had been brought along to "cure" his veering. With the prisms he continued to walk a straight course. This meant that we were unable to determine if the hypothesis regarding his earlier veering was correct. However, his later behavior provides further important information. Normals do not walk straight with prisms unless they undergo a period of adaptation. W.V.'s ability to walk straight with prisms indicates that it is possible to learn new strategies for locomotion around the environment.

VI. Heading Perception in the Presence of Objects

The real-world environment is cluttered with stationary and moving objects that may aid or hinder heading perception. In the presence of independently moving objects, it has been hypothesized that the visual motion system could (a) segment the scene using optic flow together with contrast, texture, and extraretinal signals to remove objects from the heading estimate (Adiv 1985; Hildreth 1992; Thompson *et al.*, 1993;

Hildreth and Royden 1998) or (b) simply pool across motion vectors to find the best-fit radial pattern. In the case of a motion-pooling mechanism, this would yield systematic heading errors. Warren and Saunders (1995) found that an object moving in depth does in fact bias perceived heading by a few degrees toward the object's center of motion, consistent with spatial pooling. In contrast, Royden and Hildreth (1996) observed that a laterally moving object produces a heading error in the opposite direction. Although these effects are small and only occur when the object obscures the heading point (global center of motion), they suggest that the scene may not be segmented prior to heading estimation. Our data from several patients (Vaina *et al.,* 1994b, 1996b) appear to support this hypothesis. For example, at the time when patient R.A. (Vaina *et al.,* 1996b) was impaired on perceiving segmentation from motion (Fig. 1b and d show his performance on motion discontinuity and on 2-D form from motion contrast) and failed to recover 3-D structure from motion, (Fig. 1h), his performance on straight-trajectory heading (Fig. 1j) was normal. His performance on spatial integration of motion signals (the motion coherence test) was also normal (Fig. 1h). If hypothesis (a) were correct, then it would be expected that damage to a mechanism early in the perception of heading process would lead to an impairment of heading perception. The lack of impairment suggests that to discriminate radial motion and straight-trajectory heading in an environment populated with objects, he might use hypothesis (b). In patients similar to R.A., it would be revealing to formally test heading in the presence of static or moving objects, by using methods similar to those used in psychophysics with normal subjects.

The functional deficit experienced by patient J.V. (Vaina *et al.,* 1999c) may reveal the role of objects in guiding locomotion. J.V. is a 63 years old, right-handed woman who had an embolic stroke, in the left occipital lobe, involving the inferior occipital gyrus. She has full visual fields by formal neuro-ophthalmological examination. On the first visit, 6 months after the stroke, she spontaneously reported that "I must be very careful how I walk and where I step, often I do not see a step or an object in front of me," "when I go shopping, I feel overwhelmed by people walking around me, and I seem to lose my way in a crowd where people move in different directions. But I do not have this problem when I am out there in the field."

Figure 2 shows her performance on direction (Fig. 2b) and speed discrimination (Fig. 2d) and on the discrimination of complex patterns of motion: translational, circular, and radial (Figures 2f, h, j). In all cases and for stimuli shown in either visual field, her performance was not significantly different from that of the normal population (the z scores between her thresholds on these tasks and the thresholds of the age-matched normal controls were less than 2).

These data illustrate her performance after 4 months of weekly visits to our laboratory for evaluation and training on tasks of visual motion perception. Initially she was impaired on speed discrimination, but she improved to normal level within a few weeks.

However, her performance on a straight-trajectory heading task and on 3-D structure from motion (Fig. 3b, d) remained significantly impaired, compared with the normal controls. Specifically, her performance did not change during the 9 months of weekly testing with these tasks. It is interesting, however, that the patient's self perception of her own abilities was that they improved.

She began driving, but noted that this was possible only in very familiar environments, where she was aware and could "predict" all the landmarks. She still did not feel comfortable driving in novel environments. First, it is remarkable that she could drive; second, it is surprising that she can even walk around. So, how could she control locomotion on foot given that she is not sensitive to the direction of flow? First, we noted earlier that the relative movement of objects in the scene provides information about direction of heading; second, the egocentric direction of a target object may be used. J.V. could potentially be relying on either of these sources of information, sensitivity to which may not be impaired. Thus her behavior and performance on the psychophysical tests provide evidence against the sole use of the velocity field.

A second problematic case for velocity-field-based models is the patient G.Z. (Vaina *et al.*, 1999c). She showed a clear-cut dissociation between normal judgment of heading when familiar landmarks were present in the display (a simulated room) and the complete failure to perceive heading in dynamic random dot stimuli. First, as we noted at the beginning of this chapter, natural scenes contain continuous surfaces and dynamic occlusions at depth edges, which computationally could simplify the determination of heading direction. This itself may account for G.Z.'s dissociation. However, the landmark display contained salient features and reference objects that can be tracked over time. The patient's account of how she made the decision about her heading revealed that she was tracking the visibility of the reference objects (objects of furniture). This suggests that she may be using one of the alternatives to the velocity field, the egocentric target direction, or relative motion of objects in the perception of the direction of locomotion. An interesting question arises, can we find the reciprocal case: patients that are sensitive to radial flow but topographically disoriented, (i.e., impaired on spatial relationships between objects in the environment or objects and themselves) but that can successfully locomote around the environment?

Two recent studies (Cammalleri 1996; Takahashi *et al.*, 1997) de-

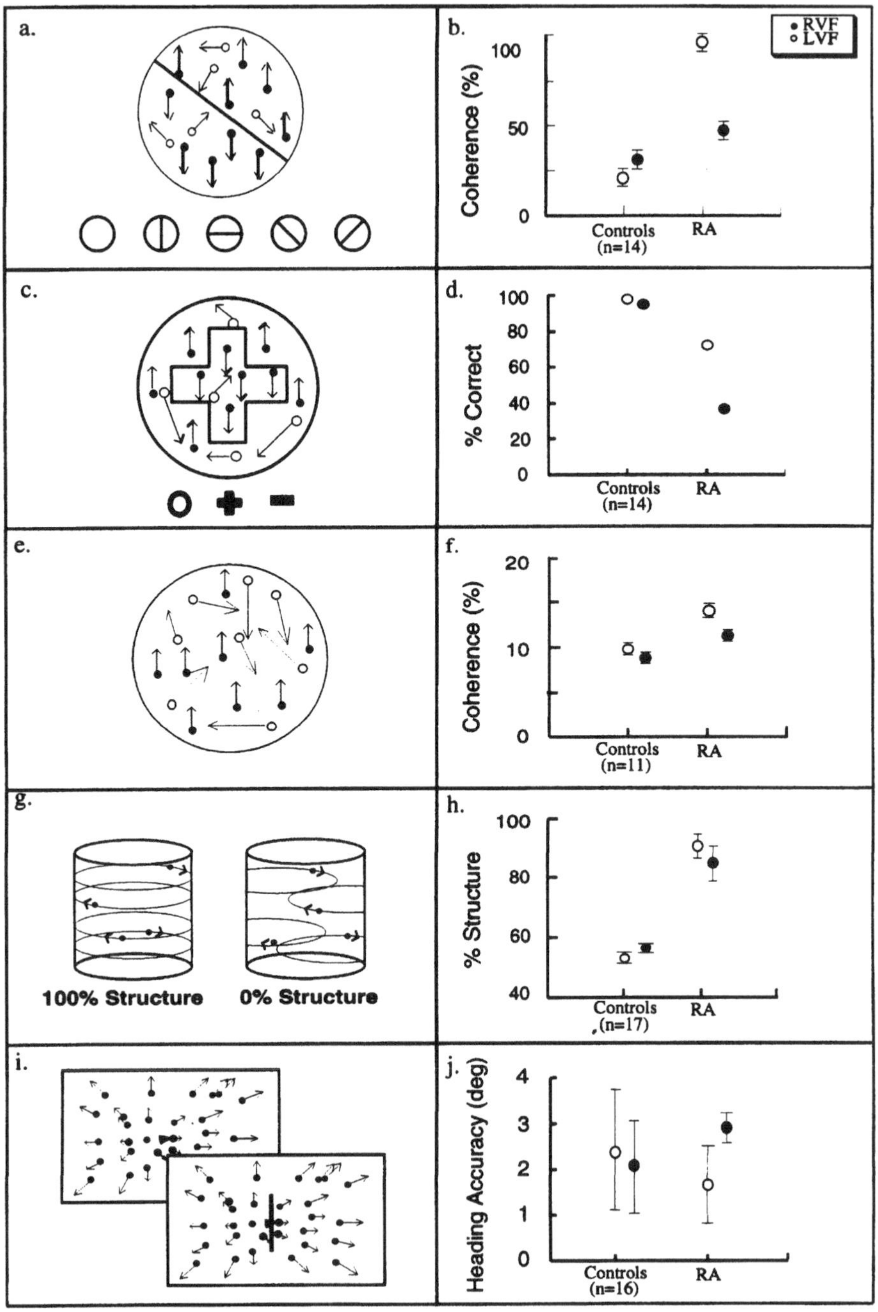
a.
b.
RVF
LVF
Coherence (%)
100
50
0
Controls
(n=14)
RA
c.
d.
% Correct
100
80
60
40
20
0
Controls
(n=14)
RA
e.
f.
Coherence (%)
20
10
0
Controls
(n=11)
RA
g.
100% Structure
0% Structure
h.
% Structure
100
80
60
40
Controls
(n=17)
RA
i.
j.
Heading Accuracy (deg)
4
3
2
1
0
Controls
(n=16)
RA

FIG. 1. Panels a, c, e, g and i on the left illustrate schematic drawings of the random dot motion stimuli. The corresponding panels b, d, f, h, and j show the data from patient R.A. and matched control subjects. The open circles indicate data from stimuli presented in the left visual field, for processing with the right hemisphere; and the filled circles indicate data from stimuli presented in the right visual field. (This convention is maintained in all the figures in this chapter.) (a) Discontinuity detection. In a single stimulus interval, the subject is asked to report whether the display is homogeneous or contains a motion discontinuity. In the discontinuous case, half of the signal dots move upward and the other half move downward. The figures on the bottom illustrate the types of displays, the first indicates that no discontinuity is present, and the other four show the orientation of the imaginary discontinuity border. The difficulty of the test is manipulated by varying the proportion of the signal and noise dots according to an adaptive staircase procedure. The signal and noise dots are defined according to the algorithm described by Newsome and Pare (1986) and Vaina *et al.* (1990). (b) The graph shows comparatively R.A.'s performance and that of 14 normal control subjects. The *y*-axis shows the percent coherence (the coherence threshold defined as the mean of the last six reversals in the adaptive staircase) necessary for reliably discriminating discontinuity in the display. (c) Discrimination of 2-D form from motion contrast. The display is similar to that described in (a), except that in the center of the display the oppositely moving signal dots either delineate an imaginary cross or an oblong when the shape is present. Constant stimuli were used in this task, and subjects were asked to determine whether the display was homogeneous, contained a central cross, or contained an oblong moving in opposite direction from the background. In 33% of the trials, the signal dots in the entire display moved homogeneously upward or downward. The proportion of noise dots was fixed to 0.05 which is well in the range of R.A.'s ability to discriminate direction of motion (as demonstrated in panel f). (d) The graph shows the performance (expressed as percent correct) of 13 normal control subjects and R.A. (e) Motion coherence. The display is similar to that portrayed in (a), except that no discontinuity is present. The subjects' task is to discriminate whether the motion is upward, downward, left, or right. The proportion of signal dots was manipulated by an adaptive staircase. This exact stimulus is described in detail in Vaina *et al.*, (1994b). (f) The graph shows the threshold (defined as the mean of the last six reversals in the adaptive staircase) expressed as percent of coherently moving dots required by normal observers and R.A. for direction discrimination. (g) 3-D structure from motion. In a spatially two-alternative forced-choice task, subjects are requested to determine which of the two random dot kinematograms (left or right) represents a "better" transparent rotating cylinder. The dots in one of the kinematograms, the "structured" display, are the orthographic projection of points on the surface of a hollow cylinder rotating with 30°/sec around its vertical axis. In the other kinematogram, the "unstructured" stimulus, the "noise" dots that corrupt the structure of the cylinder, move with the same angular velocity along a circular trajectory of the same radius as the "signal" dots, but their rotation axis is horizontally offset by a variable amount to the left or right of the "structured" stimulus axis of rotation. (h) The figure illustrates thresholds from R.A. and 6 matched normal control subjects, for the discrimination of the structured display. The *y*-axis represents percent structure necessary for reliable discrimination. (i) Straight-line heading (courtesy of C. Royden). The display consists of a random dot kinematogram which simulates what an observer would see if approaching two transparent planes of dots at a distance of 400 and 1000 cm. The simulated motion of the observer is pure translation. The simulated observer's speed is 200 cm/sec, and its direction varies to the left or right of the center of the display. After the motion sequence disappears, a vertical line appears at a given horizontal angle from the true heading. In a two-alternative choice task, subjects are asked to indicate whether their heading was to the left or to the right of the vertical line. (j) The graph displays the heading accuracy (in degrees of visual angle) of 16 normal observers and R.A.

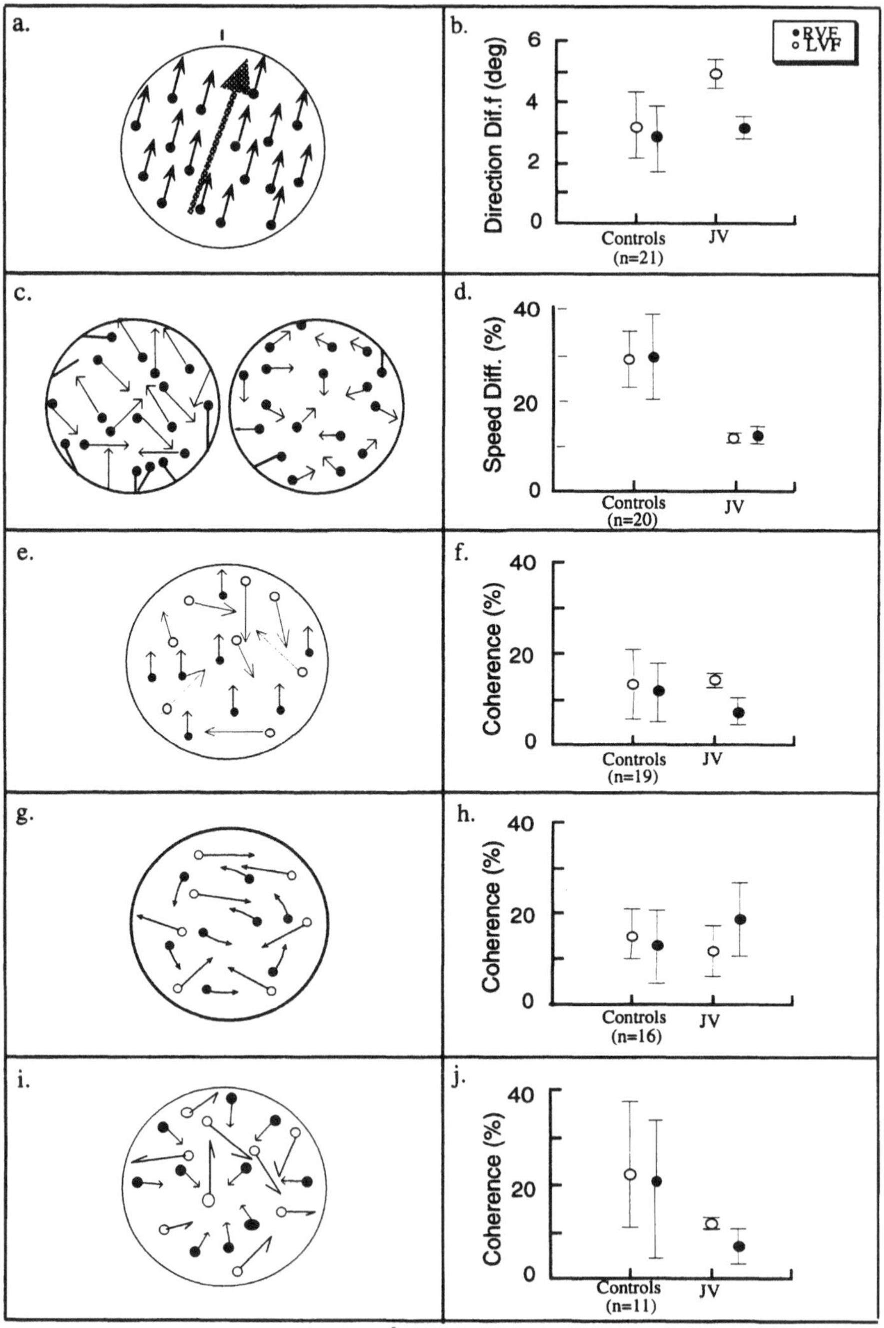
a.
b.
•RVF
○LVF
Direction Dif.f (deg)
0
2
4
6
Controls
(n=21)
JV
c.
d.
Speed Diff. (%)
0
20
40
Controls
(n=20)
JV
e.
f.
Coherence (%)
0
20
40
Controls
(n=19)
JV
g.
h.
Coherence (%)
0
20
40
Controls
(n=16)
JV
i.
j.
Coherence (%)
0
20
40
Controls
(n=11)
JV

scribed patients with lesions involving the posterior cingulate cortex, who do not have either of these two spatial deficits. They are both able to recognize landmarks and to relate to objects in the environment, but they have selectively lost their heading discrimination ability, as they cannot derive directional information from the landmarks they recognize. As pointed out by Aguirre *et al.* (1998), the posterior cingulate cortex in rodents contains head-direction cells, and they appear to fire to a combination of ego-motion, landmark, and vestibular cues.

We have described several patients that exhibit behavior that is problematic for the velocity field account of heading. None of the studies are conclusive, but they are suggestive and we have demonstrated or dis-

---

Fig. 2. Panels a, c, e, g, i display schematically the motion stimuli used. The graphs on the right (b, d, f, h, j) show the data from normal controls and subject J.V. for each of these tests. The data from the patient plotted here were obtained at 4 months after she began coming to the laboratory (10 months after her stroke). A more detailed description of the motion tests c an be found in Vaina *et al.* (1999). (a) Direction discrimination. The stimulus consists of a random dot kinematogram displayed in a circular aperture. In a two-alternative forced-choice task, subjects are asked to indicate whether the dots move to the left or to the right of an imaginary vertical line. The stimuli are displayed by an adaptive staircase procedure varying the difference (the angle) between the direction of the dots and the imaginary vertical. (b) The graph shows threshold on direction discrimination from 21 matched normal controls and J.V. (c) Speed discrimination. Random dot kinematograms are displayed in two apertures, one next to the other. Each dot trajectory changed randomly from frame to frame, but the speed was the same for all the dots within an aperture. The standard speed was presented in one aperture (randomly, top or bottom) and the test speed in the other. The standard speed was varied by an adaptive staircase procedure, and observers were asked to report in which of the two apertures the dots appeared to move faster. The test is described in detail in Vaina *et al.* (1990, 1998). (d) Data from the speed test from J.V. and 20 matched normal controls. The *y*-axis represents the subject's threshold expressed in the percent of speed difference between the standard and test speed. (e) Motion coherence. The test was replicated from Newsome and Pare (1986). A detailed description can be found in Vaina *et al.* (1990b). The display consists of a random dot kinematogram in which a variable proportion dots provide a directional signal (up, down, left, or right) and the reminder provide masking motion noise. The proportion of signal is varied by an adaptive staircase procedure. Subjects are requested to indicate the global direction of motion. (f) The graph represents the data from J.V. and 19 age-matched normal controls. The threshold shown on the *y*-axis, represents the percent coherence necessary for reliable direction discrimination of the stimulus. (g) Circular motion. The display is similar to the motion coherence, except that the direction of the signal dots is either clockwise or counterclockwise. (h) The graph shows thresholds, expressed as percent coherence, for J.V. and 16 matched normal controls. (i) Radial motion. The stimulus is similar to e and g, except that the direction of the signal dots is either expansion or contraction (shown in the figure). The proportion of signal dots (expanding or contracting) was varied by an adaptive staircase procedure. (j) The graph shows the thresholds, expressed as percent coherence, from J.V. and 11 matched normal controls.

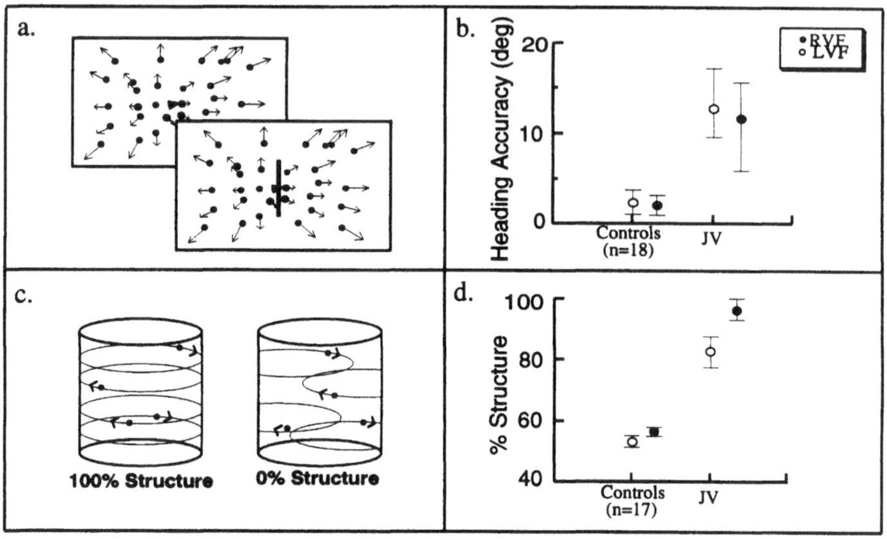

FIG. 3. (a) Straight-line heading. The stimulus and the task are described in Fig. 1i. (b) The graph displays the heading accuracy (in degrees of visual angle) of 18 normal observers and J.V. (c) The stimulus and the task are described in Fig. 1g. (d) The figure illustrates thresholds from J.V. and 17 matched normal control subjects, for the discrimination of 3-D structure from motion (indicate the structured display).

cussed how the hypotheses they provoke can be tested with normal subjects, further testing of patients with similar deficits or patients with the opposite pattern of behavior.

VII. Conclusion

In this chapter we highlighted two related issues: does optic flow guide locomotion (and if so which features of the flow field), and what else is flow used for (e.g., postural control, interceptive timing)? These points raise many questions. If optic flow is used to guide locomotion, postural control, interceptive timing, etc., then is there a single "flow system," or a number of specialized systems or routes? If object displacement is used to guide natural locomotion, then what neural areas are implicated in the necessary processing? Studying neurological patients allows us to get some early and important insights into these matters. Case studies naturally complement standard psychophysical experiments, manipulations or perturbations of behaviors in natural environments, and precise neurophysiology and computational modeling. However, lesions in human patients are not always where one would like them to be, and there are complicating factors which must be addressed. For example, functional damage is often not restricted to the lesioned area as retrograde synaptic degeneration (Sachdev *et al.*, 1990) or diaschisis may functionally affect brain areas that anatomically appear intact on structural imaging studies (Feeney and Baron, 1986). However, their primary value can be summed up as follows: if we listen to what patients tell and note their behavior, then they will teach us what we didn't expect, that we wouldn't predict, and can't explain! In such a well-developed incrementally progressing research area as optic flow, this is an invaluable contribution. Early and tentative conclusions derived from patient work suggest that it is important to review the validity of many assumptions and priorities associated with research on optic flow.

Acknowledgments

For the preparation of this manuscript, LMV was supported by a National Institute of Health grant EY-2RO1-0781, and SKR was supported in part by a grant from the UK MRC and by Nissan Research and Development, Inc. We thank Julie Harris, Alan Cowey, Mark Bradshaw, and Franco Giuliannini for valuable comments and Damien Henderson for drawing the figures and helping with the formatting of this manuscript.

References

Adiv, G. (1985). Determining three-dimensional motion and structure from optical flow generated by several moving objects. *IEEE Pattern Analysis and Machine Intelligence* **7,** 384–401.

Aguirre, G. K., Zarahn, E., and D'Esposito, M. (1998). Neural components of topographical representation. *Proc. Nat. Acad. Sci. U. S. A.* **95**(3), 839–46.

Albert, M. I. (1973). A simple test of visual neglect. *Neurology* **23,** 658–664.

Ambland, B., and Carblanc, A. (1980). Role of foveal and peripheral vision information in the maintenance of posture equilibrium in man. *Perceptual and Motor Skills* **51,** 903–916.

Andersen, G. J., and Dyre, B. P. (1989). Spatial orientation from optic flow in the central visual field. *Perception and Psychophysiology* **45,** 453–458.

Andersen, R. A. (1990). Visual and eye movement functions of the posterior parietal cortex. *Ann. Rev. Neurosci.* **12,** 377–403.

Andersen, R. A. (1997). Neural mechanisms of visual motion perception in primates." *Neuron* **18,** 865–872.

Andersen, R. A., Bradley, D. C., and Shenoy, K. V. (1996). "Neural Mechanisms for Heading and Structure-from-Motion Perception." Cold Spring Harbor Symposia on Quantitative Biology, Cold Spring Harbor Lab. Press.

Andersen, R. A., Snyder, L. H., Bradley, D. C., and Xing, J. (1997). Multimodal representation of space in the posterior parietal cortex and its use in planning movements. *Annu. Rev. Neurosci.* **20,** 303–330.

Bardy, B., Warren, W. J., and Kay, B. A. (1996). Motion parallax is used to control postural sway during walking. *Exp. Brain Res.* **111,** 271–282.

Bardy, B. G., Warren, W. H., and Kay, B. A. (in press). The role of central and peripheral vision in postural control during woring. *Perception and Psychophysics.*

Beauchamp, M., and DeYoe, E. (1996). fMRI of human parietal and occipital areas for processing visual motion and their graded modulation by spatial and featural attention. *Soc. Neurosci. Abstr.* **26,** 1198.

Britten, K. H., and van Wezel, R. J. A. (1998). Electrical microstimulation of cortical area MST biases heading perception in monkeys. *Nat. Neurosci.* **1,** 59–63.

Cammalleri, R. (1996). Transient topographical amensia and cingulate cortex damage:a case report. *Neuropsychologia* **34**(4), 321–326.

Cheng, K., Fujita, H., Kanno, I., Mura, S., and Tanaka, K. (1995). Human cortical regions activations by wide field visual motion: An H215O PET study. *J. Neurophys.* **74**(1), 413–426.

Clifford, C. W. G., and Vaina, L. M. (1998). A computational model of selective deficits in first- and second-order motion perception [ARVO Abstract]. *Invest. Ophthalmol. Vis. Sci.* **39**(4), S1077. Abstract no. 4982.

Crowell, J. A., and Banks, M. S. (1993). Perceiving heading with different retinal regions and types of optic flow. *Perception and Psychophysics* **53,** 325–337.

Cutting, J. E. (1986). "Perception with an Eye Towards Motion." MIT Press, Cambridge, MA.

Cutting, J. E., Springer, K., Braren, P. A., and Johnson, S. H. (1992). Wayfinding on foot from information in retinal, not optical, flow. *J Exp Psy: General* **121,** 41–72.

Dale, A., Ahlfors, S. P., *et al.* (1995). "Spatiotemporal Imaging of Coherent Motion Selective Areas in Human Cortex." Society for Neuroscience, San Diego.

Duchon, A. P., and Warren W. H. (1998). Interaction of two strategies for controlling locomotion [ARVO Abstract]. *Invest. Opthalmol. Vis. Sci.* **39**(4), S892. Abstract no. 4122.

Duffy, C. J., and Wurtz, R. H. (1996). Optic flow, posture and the dorsal visual pathway. *In* "Perception, Memory and Emotion: Frontier in Neuroscience." (T. Ono, B. McNaughton, S. Molotchnikoff, E. Rolls, and H. Nishijo, eds.), Elsevier Science, Oxford, New York, Tokyo.

Duhamel, J. R., Bremmer, F., BenHamed, S., and Graf, W. (1997). Spatial invariance of visual receptive fields in parietal cortex neurons. *Nature* **389,** 845–848.

Dupont, P., De Bruyn, B., *et al.* (1997). The kinetic occipital region in human visual cortex. *Cereb. Cortex* **7,** 283–292.

Dupont, P., Orban, G. A., *et al.* (1994). Many areas in the human brain respond to visual motion. *J. Neurophysiol.* **72**(3), 1420–1424.

Feeney, D. M., and Baron, J. C. (1986). Diaschisis. *Stroke* **17**(5), 817–830.

Gegenfurtner, K., Kiper, D., and Levitt, J. (1997). Functional properties of neurons in macaque area V3. *J. Neurophysiol.* 1906–1923.

Goldstein, K., and Gelb, A. (1918). Psychologische Analysen hirnpatologischer Falle auf Grund von Untersuchungen Hirnverletzter. I. Abhandlung. Zur Psychologie des optischen Wahrnehmungsund Erkennungsvorganges. *Zeitschrift fur die gesamte Neurologie und Psychiatrie* **41,** 1–142.

Greenlee, M. W., and Smith, A. T. (1997). Detection and discrimination of first- and second-order motion in patients with unilateral brain damage. J. Neurosci. **17,** 804–818.

Hildreth, E. C. (1992). Recovering heading for visually-guided navigation. *Vis. Res.* **32,** 1177–1192.

Hildreth, E. C., and Royden, C. S. (1998). Computing observer motion from optical flow. "High-Level Motion Processing: Computational, Neurobiological, and Psychophysical Perspectives" (T. Watanabe, ed). MIT Press, Cambridge, MA.

Johansson, G. (1973). Visual perception of biological motion and a model for its analysis. *Perception & Psychophysics* **14**(2), 201–211.

Karnath, H. (1994). Subjective body orientation in neglect and the interactive contribution of neck muscles and proprioception and vestibular stimulation. *Brain* **117,** 100–112.

King, S. M., Frey, S., Villemure, J. G., Ptito, A., and Azzopardi, P. (1996). Perception of motion-in-depth in patients with partial or complete cerebral hemispherectomy. *Behav. Brain Res.* **76**(1–2), 169–80.

Koch, C., Wang, H. T., and Mathur, B. (1989). Computing motion in the primate's visual system. *J. Exp. Biol.* **146,** 115–139.

Lee, D. N. (1974). Visual information during locomotion. *In* "Perception: Essays in Honor of James Gibson" (R. B. Macleod and H. Pick, eds). Cornell University Press, Ithaca, NY.

Lee, D. N., and Lishman, J. R. (1975). Visual proprioceptive control of stance. *J. Human Movement Stud.* **1,** 87–95.

Lee, D. N. (1976). A theory of visual control of braking based on information about time-to-collision. *Perception* **5,** 437–459.

Llewellyn, K. R. (1971). Visual guidance of locomotion. *J. Exp. Psychol.* **91,** 245–261.

Marcar, V. L., Zihl, J., and Cowey, A. (1997). Comparing the visual deficits of a motion blind patient with the visual deficits of monkeys with area MT removed. *Neuropsychologia* **35**(11), 1459–1465.

Maunsell, J. H., and Essen, D. C. V. (1983a). Functional properties of neurons in middle temporal visual area of the macaque monkey. I. Selectivity for stimulus direction, speed, and orientation. *J. Neurophysiol.* **49**(5), 1127–1147.

Maunsell, J. H. R., and Essen, D. C. V. (1983b). The connections of the middle temporal visual area (MT) and their relationship to a cortical heirarchy in the macaque monkey. *J. Neurosci.* **3**(12), 2563–2586.

McLeod, P., Dittrich, W., Driver, J., Perret, D., and Zihl, J. (1996). Preserved and impaired detection of structure from motion by a "motion blind" patient. *Visual Cognition* **3**(4), 363–391.

Nakayama, K. (1985). Biological image motion processing: A review. *Vis. Res.* **25,** 625–660.

Nakayama, K. (1994). James J. Gibson—An appreciation. *Psychol. Rev.* **101**(2), 329–335.

Newsome, W. T., and Pare, E. B. (1986). MT lesions impair discrimination of direction in a stochastic motion display. *J. Neurosci.* **8**(6), 2201–2211.

Pekel, M., Lappe, M., Bremmer, F., Thiele, A., and Hoffmann, K.-P. (1996). Neuronal responses in the motion pathway of the macaque monkey to natural optic flow stimuli. *NeuroReport* **7,** 884–888.

Peuskens, H., Rosie, A., Dupont, P., Sunaert, S., Mortelman, L., Van Hecke, P., and Orban, G. A. (1999). Heading judgments involve hMT/V5+. *Soc. Neurosci. Abst.* (submitted).

Plant, G. T., Laxer, K. D., *et al.* (1993). Impaired visual motion perception in the hemifield contralateral to unilateral posterior cerebral lesions in humans. *Brain* **116,** 1303–1335.

Pollen, D. A. (1999). On the neural correlates of visual perception. *Cereb. Cortex* **9**(1), 4–19.

Read, H. L., and Siegel, R. M. (1997). Modulation of responses to optic flow in area 7a by retinotopic and oculomotor cues in monkeys. *Cereb. Cortex* **7,** 647–661.

Regan, D., and Beverley, K. I. (1979). Visually guided locomotion: psychophysical evidence for a neural mechanism sensitive to flow patterns. *Science* **205,** 311–313.

Regan, D., and Vincent, A. (1995). Visual processing of looming and time to contact throughout the visual field. *Vis. Res.* **35**(13), 1845–1857.

Riddoch, G. (1917). Dissociation of visual perceptions due to occipital injuries with especial reference to appreciation of movement. *Brain* **40,** 15–57.

Royden, C., and Hildreth, E. (1996). Human heading judgments in the presence of moving objects. *Perception and Psychophysics* **58,** 836–856.

Rushton, S. K., Harris, J. M., Lloyd, M. R., and Wann, J. P. (1998). Guidance of locomotion on foot uses perceived target location rather than optic flow. *Curr. Biol.* **8**(21), 1191–1194.

Sachdev, M., Kumar, H., Jain, A., Goulatia, R., and Misra, N. (1990). Transsynaptic neuronal degeneration of optic nerves associated with bilateral occipital lesions. *Indian J. Ophthalmol.* **38**(4), 151–152.

Schaafsma, S. J., Duysens, J., and Gielen, C. C. (1997). Responses in ventral intraparietal area of awake macaque monkey to optic flow patterns corresponding to rotation of planes in depth can be explained by translation and expansion effects. *Visual Neurosci.* **14**(4), 633–646.

Shillcock, R., Kelly, M., and Monaghan, P. (1998). Processing of plaindromes in negelct dyslexia. *Neuroreport* (9), 3081–3083.

Shipp, S., de Jong, B. M., and Monaghan, P. (1994). The brain activity related to residual motion vision in a patient with bilateral lesions of V5. *Brain* **117,** 1023–1038.

Siegel, R, and Read, H. (1997). Analysis of optic flow in the monkey parietal area 7a. *Cereb. Cortex* (7), 327–346.

Srinivasan, M. V., Lehrer, M., Kirchner, W., and Zhang, S. W. (1991). Range perception through apparent image speed in freely-flying honeybees. *Vis. Neurosci.* **6,** 519–535.

Stoffregen, T. A. (1985). Flow structure versus retinal location in the optical control of stance. *J. Exp. Psychol.: Human Perception and Performance* **11,** 554–565.

Stoffregen, T. A. (1986). The role of optical velocity in the control of stance. *Perception and Psychophysics* **39,** 355–360.

Stoffregen, T. A., and Riccio, G. E. (1990). Response to optical looming in the retinal center and periphery. *Ecol. Psychol.* **2,** 251–274.

Takahashi, N., Kawamura, M., Shiota, J., Kasahata, N., and Hirayama, K. (1997). Pure topographic disorientation due to right retrosplenial lesions. *Neurology* **49,** (464–469).

Thompson, W. B., Lechleider, P., and Stuck, E. R. (1993). Detecting moving objects using the rigidity constraint. *IEEE Part A* **15,** 162–165.

Tootell, R., Dlae, A., Mendola, J., Reppas, J., and Sereno, M. (1996). FMRI analysis of human visual cortical area V3A. *NeuroImage* **3,** S358.

Tootell, R., Reppas, J., Dale, A., Look, R., Malach, R., Jiang, H.-J., Brady, T., and Rosen, B. (1995). Visual motion aftereffect in human cortical area MT/V5 revealed by functional magnetic resonance imaging. *Nature* **375,** 139–141.

Tootell, R., Reppas, J. B., Malach, R., and Al, E. (1994). Functional MRI of human V5/MT correlates with two different visual illusions. *Soc. Neurosci. Abstr.* **20,** 250–255.

Tootell, R. B., Mendola, J. D., Hadjikhani, N. K., Ledden, P. J., Liu, A. K., Reppas, J. B., Sereno, M. I., and Dale, A. M. (1997). Functional analysis of V3A and related areas in human visual cortex. *J. Neurosci.* **17,** 7060–7078.

Tootell, R. B. H., Kwong, K. K., Belliveau, J. W., Baker, J. R., Stern, C. E., Hockfield, S. J., Breiter, H. C., Born, R., Benson, R., Brady, T. J., and Rosen, B. R. (1993a). Mapping human visual cortex: Evidence from functional MRI and histology. *Invest. Opthalmol. Vis. Sci.* **34**(4), 813.

Tootell, R. B. H., Kwong, K. K., Belliveau, J. W., Baker, J. R., Stern, C. S., Savoy, R. L., Breiter, H., Born, R., Benson, R., Brady, T. J., and Rosen, B. R. (1993b). *Functional MRI (fMRI) Evidence for MT/V5 and Associated Visual Cortical Areas in Man.* Society for Neuroscience 23rd Annual Meeting, Washington, D.C., Soc. Neuro., Washington, DC.

Tootell, R. B. H., Reppas, J. B., Kwong, K. K., Malach, R., Born, R. T., Brady, T. J., Rosen, B. R., and Belliveau, J. W. (1995a). Functional analysis of human MT and related visual cortical areas using magnetic resonance imaging. *J. Neurosci.* **15**(4), 3215–3230.

Tootell, R. B. H., Reppas, J. B., Kwong, K. K., Malach, R., Born, R. T., Brady, T. J., Rosen,

B. R., and Belliveau, J. W. (1995b). Functional analysis of human MT and related visual cortical areas using magnetic resonance imaging. *J. Neurosci.* **15**(4), 3215–3230.

Ungerleider, L. G., and Desimone, R. (1986). Cortical projections of visual area MT in the macaque. *JCN* **248,** 147–163.

Vaina, L. M., and Cowey, A. (1996). Selective deficits to first or second order motion in stroke patients provides further evidence for separate mechanisms. **22,** 1718.

Vaina, L. M., Cowey, A., and Kennedy, D. (1999a). Perception of first- and second-order motion: Separate neurological mechanisms. *Human Brain Mapping* **7,** 67–77.

Vaina, L. M., Gross, C., and Zamani, A. (1999b). Impaired recognition of biological motion in patients with temporal lobe lesions. Submitted.

Vaina, L. M., Grzywacz, N., *et al.* (1990a). Selective deficits of measurement and integration of motion in patients with lesions involving the visual cortex. *Invest. Opthalmol. Vis. Sci.* **31**(4), 523.

Vaina, L. M., Grzywacz, N. M., and Kikinis, R. (1994a). Segregation of computation underlying perception of motion discontinuity and coherence. *NeuroReport* **5**(17), 2289–2294.

Vaina, L., Gryzwac, N., and Bienfang, D. (1994b). Selective deficits of motion integration and segregation mechanisms with unilateral extrastriate brain lesions. *Invest. Ophthalomol. Vis. Sci.* **35**(4), 1438.

Vaina, L., Grzywacz, N., *et al.* (1998a). Perception of motion discontinuity in patients with selective motion deficits. "High-Level Motion Processing Computational, Neurobiological, and Psychophsyical Perspectives." (T. Watanbe, ed). MIT Press, Cambridge, MA.

Vaina, L. M., Jakab, M., Beardsley, S., and Zamani, A. (1999c). Impaired complex motion perception in patients with bilateral posterior parietal lesions [ARVO Abstract]. *Invest. Ophthalmol. Vis. Sci.* **40**(4), S765. Abstract no. 4042.

Vaina, L. M., LeMay, M., *et al.* (1990b). Intact "biological motion" and "structure from motion" perception in a patient with impaired motion mechanisms: A case study. *Vis. Neurosci.* **5**(4), 353–369.

Vaina, L. M., Makris, N., and Cowey, A. (1996a). The neuroanatomical damage producing selective deficits of first or second order motion in stroke patients provides further evidence for separate mechanisms. *NeuroImage* **3**(3), 360.

Vaina, L. M., Makris, N., Kennedy, D., and Cowey, A. (1998b). The selective impairment of the perception of first-order motion by unilateral cortical brain damage. *Vis. Neurosci.* **15**(2), 333–348.

Vaina, L. M., Royden, C. S., *et al.* (1996b). Normal perception of heading in a patient with impaired structure-from-motion [ARVO Abstract]. *Invest. Ophthalmol. Vis. Sci.* **37**(3), S137. Abstract no. 2360.

Warren, W. (1999). Visually controlled locomotion, 40 years later. *Ecol. Psychol.* **10**(3–4), 177–219.

Warren, W., Blackwell, A., *et al.* (1991). On the sufficiency of the velocity field for perception of heading. *Biol. Cybern.* **65,** 311–320.

Warren, W. H. J., and Saunders, J. A. (1995). Perceiving heading in the presence of moving objects. *Perception* **24**(3), 315–331.

Zihl, J., Baker, Jr., C. L., and Hess, R. H. (1989). The "motion-blind" patient: Low-level spatial and temporal filters. *J. Neurosci.* **9**(5), 1628–1640.

Zihl, J., Cramon, D. V., Mai, N., and Schmid, C. (1991). Disturbance of movement vision after bilateral posterior brain damage. *Brain* **114,** 2235–2252.

Zihl, J., Van Cramon, D., and Mai, N. (1983). Selective disturbance of movement vision after bilateral brain damage. *Brain* **106**(2), 313–340.

INDEX

FSC
www.fsc.org
MIX
Papier | Fördert
gute Waldnutzung
FSC® C083411

Kolibri 360 GmbH
Ferdinand-Jühlke-Straße 7
99095 Erfurt, Deutschland
produktsicherheit@kolibri360.de